Synopsis of Gross Anatomy

John B. Christensen, Ph.D.

Chairman and Professor of Anatomy
St. George's University School of Medicine
Grenada, West Indies

Ira R. Telford, Ph.D.

Visiting Professor of Anatomy
Uniformed Services University of the Health Sciences
Bethesda, Maryland

Professor Emeritus of Anatomy
George Washington University School of Medicine and Health Sciences
Washington, D.C.

J. B. Lippincott Company   *Philadelphia*

London   *New York*   *Mexico City*   *St. Louis*   *São Paulo*   *Sydney*

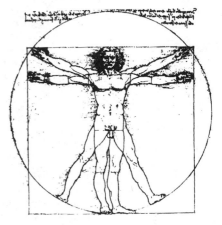

SYNOPSIS OF
Gross Anatomy

Fifth Edition
revised and enlarged

Sponsoring Editor: Sanford J. Robinson
Manuscript Editor: Brenda Lee Reed
Indexer: Ann Cassar
Design Coordinator: Anita R. Curry
                  Anne O'Donnell
Cover Design: Kevin Curry
Production Manager: Kathleen P. Dunn
Production Coordinator: George V. Gordon
Compositor: Bi-Comp, Inc.
Printer/Binder: R.R. Donnelley & Sons Company

Figures 2-16, 2-26, 3-10, 3-13, 3-15, 4-1, 4-2, 4-12, 4-15, 4-20, 5-3, 5-5, 5-10, 5-15, 6-5, 6-15, 6-16, 7-3, 7-8, 7-9, 7-16 to 7-18, 7-20, 7-24, 7-25, 7-28 to 7-35, 7-37 to 7-39, 7-41 to 7-43, 7-46, and 7-47 are borrowed from Langley LL: Dynamic Anatomy & Physiology, 5th ed, New York, McGraw-Hill Book Company, 1980.

5th Edition

Library of Congress Cataloging-in-Publication Data

Christensen, John B., 1927–
  Synopsis of gross anatomy.

  Includes index.
  1. Anatomy, Human.   I. Telford, Ira Rockwood,
1907–   .   II. Title. [DNLM: 1. Anatomy. QS 4 C549s]
AM23.2.C52   1987   611   87-14063
ISBN 0-397-50850-6

The authors and publisher have exerted every effort to ensure that drug selection and dosage set forth in this text are in accord with current recommendations and practice at the time of publication. However, in view of ongoing research, changes in government regulations, and the constant flow of information relating to drug therapy and drug reactions, the reader is urged to check the package insert for each drug for any change in indications and dosage and for added warnings and precautions. This is particularly important when the recommended agent is a new or infrequently employed drug.

Preface

SYNOPSIS OF GROSS ANATOMY has almost doubled in size since its initial publication. In the fifth edition, parts of the core text have been revised with a conscious effort being made to keep the basic descriptive text concise and synoptic. The increase in the number of pages consists primarily of elements that have been added to assist the student in the achievement of a better understanding of gross anatomy. Material not found in the previous edition includes specified learning objectives, lists of major anatomic and clinical points, review questions for self-testing, an expanded number of illustrations, and comprehensive arterial tables.

Learning objectives are interspersed throughout each chapter, and are related to the major subsections. (Thus, in the chapter on the upper extremity there is a set of objectives for the axilla.) The objectives are designed to enable the student to zero in on the information that should be mastered. A second new feature of this edition is the inclusion of lists of major anatomic and clinical points. These appear in each chapter, following the main text, and are also regionally oriented. Used in combination with the objectives, they will further enhance the student's grasp of key information. Review questions have been included at the end of each chapter; they correspond to the learning objectives, and the anatomic and clinical point lists. Thus, they can be used both for general review, and can be selected out and used, for immediate feedback, as a self-administered test during study. Forty five new illustrations have been added. A few replace art in the fourth edition; most are new and cover additional anatomic regions. At the suggestion of many students, a new series of tables has been developed for the arteries. This information has been treated in the same manner as the tables for other anatomic elements, with the information presented with the text on a regional basis rather than combined into one overwhelming, ten-page table.

We wish to acknowledge the numerous suggestions from students and colleagues. We are especially indebted to Dr. Robert L. Jordan, Dr. C. Vishnu Rao, and Dr. Lyle C. Deardon for their review of the text. As with all textbooks, the careful attention of editors Sanford Robinson and Brenda Reed of the JB Lippincott Company was an integral element in the production of this new edition.

J.B.C.
I.R.T.

Preface to the First Edition

This synoptic volume of regional anatomy presents the basic facts and concepts in the study of gross anatomy considered essential for students of medicine and associated sciences. It meets the need of the student for a concise, straightforward textbook, uncluttered by minutiae. This synopsis is intended not to replace selective reading in large conventional textbooks, but rather to give the student an initial appreciation of important body structures and relations.

Many original illustrations have been especially prepared to enhance this epitomized approach to the study of anatomy. The line drawings are keyed to the text and can be readily correlated with regional dissections.

For the student who finds gross anatomy difficult, this compact text may provide all that one can or need comprehend of the subject. However, for those who wish to pursue the subject more deeply, it will serve as a framework for the building of a broader and firmer foundation in anatomy.

Because of its regional approach, we suggest that this book could be used as (1) a study guide in conjunction with anatomic atlases and larger textbooks, (2) a companion text in gross dissection, or (3) a review of the fundamentals of gross anatomy.

In brief, we have endeavored to present, in the most succinct form, the essentials of human gross anatomy that we believe every medical student should know.

We gratefully acknowledge the kindness of our colleagues, Dr. Frank D. Allen, Associate Professor of Anatomy, George Washington University, and Dr. W. Montague Cobb, Chairman and Professor of Anatomy, Howard University, in reading and offering constructive suggestions for improvement of our original text.

We are deeply indebted to Mr. David S. Kern and his daughter Bonnie for the excellent rendering of most of the illustrations, to Mr. Michael S. Murtaugh for his splendid diagrammatic sketches and drawings, and to Dr. William A. Rush, Jr., Mrs. Margaret Dupree, and Miss Joan Ruback for their contributions to the artwork.

By the kind permission of various authors and their publishers, we have borrowed a few illustrations from the sources acknowledged in the individual legends.

J.B.C.

Washington, D.C.

I.R.T.

Contents

ONE

Introduction

Man has a natural curiosity about his body. It is first expressed when the infant becomes fascinated by his own hand movements. The study of gross anatomy is the continuation of this innate interest, but it goes far beyond bodily movements; it is the formal identification and study of the dissectable structures of the body and their interrelationships. To the degree that this inborn interest is cultivated and developed, anatomy will be a stimulating and rewarding study.

The purpose of gross anatomic dissection and study is to obtain direct exposure to the three-dimensional relationships of the body and to be able to visualize just how we are put together. This does not negate the use of oral descriptions or explanations in the form of lectures, written text material, illustrations, or various visual aids such as models and films. These, however, should be considered only as adjunct tools to the firsthand information obtained through dissection as the student moves toward the goal of gross anatomy study—to secure a working knowledge of, and appreciation for, the structure and form of the human body.

In gross anatomy a voluminous amount of factual data must be acquired during a short time. This introductory chapter will present certain basic concepts as a foundation to aid the student to assimilate the deluge of information that will follow. As regional dissection and study of the body proceed, these general concepts of the different systems should become recurrent themes. Understanding them will make the subsequent acquisition of information more meaningful and rewarding.

The brief accounts given here of concepts of the skeletal system and associated joints, muscular system, fasciae, body cavities, lymphatics, cardiovascular system, and the nervous system are in no sense complete, nor are they intended to be. Rather, they constitute an introductory, conceptual approach to the study of anatomy.

Anatomic Terminology

Anatomy introduces the student to the language of medicine and dentistry. It has been estimated that this language, which the physician must master, comprises about 10,000 terms, three-fourths of which are encountered in anatomy. The word roots or stems, prefixes, and suffixes are largely derived from Latin and Greek. Those who have studied these languages are well-equipped to understand and use anatomic terminology. If, however, the student's background in classical languages is deficient, this handicap can be overcome by learning certain fundamentals of vocabulary and linguistic principles. For example, the stem *myo* (Greek $\mu\hat{\upsilon}\sigma$) in the terms myocardium (*cardium* heart), myometrium (*metrium* uterus), myoglobin (*globin* protein), myoblast (*blast* immature cell), myocele (*cele* hernia), myoma (*oma* tumor), all refer to muscle. Thus, the astute student will pay special attention to new words and stems as they are encountered.

Terms of Reference

Terms of reference to the human body are standardized to refer to a rather arbitrary concept called the **anatomic position,** in which the body is erect, the face forward, the arms at the sides with the palms of the hands turned forward (Fig. 1-1). The terms listed in Table 1-1 are used to indicate the location of structures in the body with reference to the anatomic position irrespective of the position the body of a patient or the cadaver might assume. This does away with the necessity of using words such as over, under, below, and above, all of which can indicate two directions in the three-dimensional body and may thereby be confusing.

It is important to keep in mind that the above terms of reference locate a structure in its rela-

tionship to other structures of the body. For example, the descending aorta in the chest is located anterior to (in front of) the necks of the ribs and posterior to (behind) the heart; it is inferior to (below) the arch of the aorta and superior to (above) the diaphragm. It lies lateral to the vertebral column, but medial to the angle of the ribs.

Terms denoting planes of the body also refer to the anatomic position. Of the three basic planes of the body, the **sagittal plane** and **coronal** (frontal) **plane** are both vertical along the long axis of the body, while the **transverse** (horizontal or cross-sectional) **plane** is at right angles to the longitudinal axis (Fig. 1-2). These planes are used either in reference to the whole body, to a specific region of the body, or to a separate organ. In the latter case, if the structure has been removed from the body, the terms longitudinal and transverse may be substituted.

The sagittal plane separates the body into right and left segments. If sectioned in the median sagit-

Text continues on page 6

Table 1-1
Terms of Reference

Term	Synonym	Definition
Superior	Cranial, cephalic	Toward the head
Inferior	Caudal	Toward the feet
Anterior	Ventral, volar, palmar (latter two refer to hand)	Toward the front of the body
Posterior	Dorsal	Toward the back of the body
Medial		Toward the midline of the body
Lateral		Toward the side of the body
External	Superficial	Toward the surface of the body
Internal	Deep	Away from the surface of the body
Proximal		Toward the main mass of the body
Distal		Away from the main mass of the body
Central		Toward the center of the body
Peripheral		Away from the center of the body
Plantar		Sole of the foot
Palmar		Palm of the hand

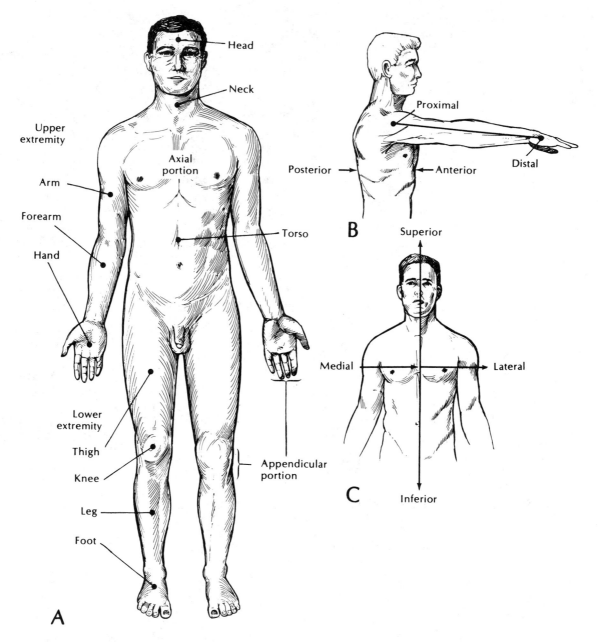

Figure 1-1.   *Anatomic position. (Langley LL, Telford IR, Christensen JB: Dynamic Anatomy and Physiology, 5th ed. New York, McGraw-Hill, 1980)*

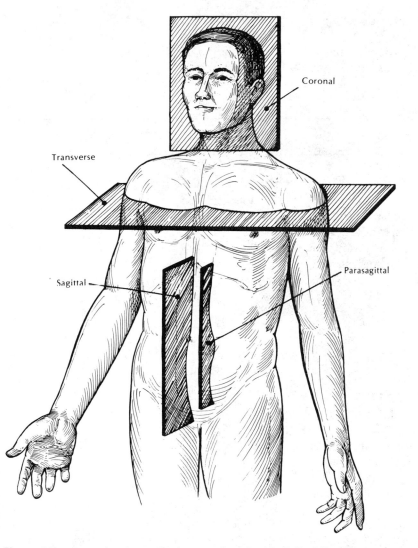

Coronal

Transverse

Sagittal

Parasagittal

Figure 1-2. *Anatomic planes of reference. (Langley LL, Telford IR, Christensen JB: Dynamic Anatomy and Physiology, 5th ed. New York, McGraw-Hill 1980)*

tal plane, the body would be divided into equal halves, except for unpaired organs; if sectioned in a parasagittal plane, unequal portions would be produced. A coronal section would separate the anterior (front) part of the body from the posterior (back) part. A transverse section would bisect a superior (upper) segment from an inferior (lower) segment.

Skeletal System

The skeletal system comprises approximately 206 bones and a number of cartilaginous components. The total number of bones is an approximation because a variable number of supernumerary or accessory bones may also be present. These additional elements most frequently occur as small (wormian) bones between the flat bones of the skull, as additional carpal or tarsal bones in the wrist and ankle, respectively, or as sesamoid bones within tendons.

Cartilage

Three types of cartilage occur in the body: hyaline, elastic, and fibrous. **Hyaline cartilage** is present on the articular surface of bones. In the adult it comprises the costal cartilages at the distal ends of ribs. **Elastic cartilage** has a greater resiliency due to embedded elastin fibers in its ground substance and is present in structures that undergo functional distortion, for example, the external ear and epiglottic cartilage. **Fibrocartilage** has an increased strength because there is a preponderance of collagen fibers in its ground substance. It is found in structures subjected to excessive stress, such as the weight-bearing intervertebral discs.

Bones

Although the exact number of bones may vary from individual to individual, the average complement, as tabulated below, may be subdivided into axial and appendicular portions of the skeleton.

Axial skeleton	
Skull	22
Ear ossicles	6
Hyoid bone	1
Vertebral column	26
Ribs and sternum	25
Subtotal	80
Appendicular skeleton	
Upper extremity	64
Lower extremity	62
Subtotal	126
Total	206

Functions

Bones provide the supportive framework of the body and protect vital organs. Their marrow cavities are the primary site of blood formation in the adult (especially the flat bones of the skull, scapulae, vertebrae, and ilia). They also afford storage of minerals, principally calcium and phosphorus. To the student in gross anatomy, bones are of interest for four central reasons.

1. They are sites of muscular attachments and thereby act as levers to provide movement.
2. Their morphologic characteristics and markings, including sexual differences.
3. The relationships of other structures to bones, principally vessels and nerves, which become clinically important in trauma.

4. The sequential appearance and fusion of epiphyses during the growth period, which are indicative of normal or of pathologic development.

Morphologic Characteristics

Bones are classified according to their shape, for example, long, short, flat, or irregular. In **long bones** the length exceeds the width, as in most bones of the extremities. Long bones consist of (1) a **shaft** (diaphysis), having an elongated marrow cavity with little internal trabeculation, (2) two ends or **extremities** known as **epiphyses,** which may or may not be separated from the diaphysis during the growth period by a plate of cartilage (the epiphyseal disc), and, (3) the **metaphysis,** the zone between the diaphysis and the epiphysis, which flares out from the shaft of the bone toward the epiphysis. During the growth period, the cartilaginous **epiphyseal disc** provides new cells for the increase in length of long bones. When growth ceases, this area ossifies and the epiphysis becomes continuous with the diaphysis. The metaphysis and the epiphysis have extensive internal trabeculation. The trabeculae are usually aligned in the direction of the stress and strain that has been placed on the bone.

Short bones are approximately equal in all three dimensions. Examples of short bones are the carpal bones and tarsal bones of the wrist and ankle. Internal trabeculation, similar to that seen in the epiphyses of long bones, is also present in short bones.

Flat bones include the scapulae, ribs, sternum, and the bones of the cranium. They are formed by two thin plates of compact bone with a minimal interval of trabecular bone between them that forms the interval of the marrow cavity. In the flat bones of the skull this area is referred to as the diploë.

Irregular bones, as their name implies, have a complicated configuration with numerous processes. Examples of irregular bones are the individual vertebrae of the vertebral column and the sphenoid and ethmoid bones of the skull.

Sesamoid bones are very small bones (with the exception of the relatively large patella) embedded within certain tendons. They usually occur in a tendon as it passes over an articulation. They can act as fulcrums to increase the mechanical advantage of the muscle action over the joint, and provide extra strength to limit trauma to the tendon from pressure or friction. The most constant sesamoid bones are the patella and those associated with the metacarpophalangeal joint of the thumb and metatarsophalangeal joint of the great toe.

Bony processes are discrete projections from the main body of the bone. The naming of processes indicates their morphologic characteristics. For example, crests or ridges are lineal elevations; sulci and grooves are lineal depressions; tubercles, tuberosities, or trochanters are circumscribed, roughened elevations of increasing size, and styloid processes or spines are spikelike projections. Foramina are holes that may pass either entirely through the bone or only through the cortex. They provide for the passage of nerves and vessels. Fissures are clefts between adjacent bones.

Experimental evidence indicates that the definitive form of a bone is dependent upon both genetic or intrinsic factors and physical or external factors. Embryologic transplant studies have shown that a given bone, for example, the femur, will develop its characteristic processes irrespective of chemical, hormonal, or extrinsic factors. Bone markings, however, are modified by mechanical factors. As a response to increased stress of the muscular attachments to bone, well-developed muscular individuals have more prominent processes than do less muscular individuals. Moreover, the orthopedic surgeon is able to compensate to a degree for developmental abnormalities by varying stress or tension on a given bone.

Sexual Characteristics

Sexual differences are reflected by the degree of massiveness of the skeleton, and by functional modifications. In the first instance, although variations occur in races and between individuals, the male skeleton, in general, is more massive, with more pronounced and larger processes, than is the female skeleton.

The most pronounced example of functional adaptation occurs in the female pelvis. The innominate bones are modified to provide minimal osseous impedence in the birth channel. This facilitates passage of the fetus through the pelvic cavity in parturition. Features of this adaptation include a greater relative and absolute width of the pelvic openings. This results in a circular inlet of the female pelvis, in contrast to an oval configuration in the male, as well as a divergence of the ischial spines and coccyx to increase the dimensions of the pelvic outlet.

Bones as Levers

The physician is also interested in the interplay of bones at articulations. The articular ends of a bone reflect the type of activity at the joint. Thus, the structure of bones associated with freely movable articulations are covered with hyaline (articular) cartilage, which has a smooth, glassy appearance. In contrast, bones forming nonmovable joints have rough surfaces at their junctional sites.

The disparity in area of smoothness of two bones forming a freely movable joint is also meaningful. For example, the extent of the articular surfaces at the distal end of the femur and at the proximal end of the tibia suggests that the knee is not a simple hinge joint, but rather that the surfaces must also slide across one another in movement.

Relationships of Bones to Other Structures

Bone markings frequently provide clues to the relationships of bones to other anatomic structures. If a muscle attaches to a bone over a relatively extensive area, the surface of the bone will appear smooth. Attachment of a muscle to a limited area, for example the attachment of the tendon of the deltoid muscle, results in a definitive

elevation, the deltoid **tuberosity.** Moreover, the size of the elevation will usually indicate the magnitude of stress that the muscle produces.

Study of the foramina of the skull should always parallel study of the nerves, arteries, and veins that traverse these foramina.

Sulci or grooves almost invariably reflect the relationship of soft structures to bone. For example, the intertubercular sulcus of the humerus lodges the tendon of the long head of the biceps. The depth of a sulcus that lodges a tendon is, moreover, related to the strength of the pull of the muscle upon that tendon. Grooves may also lodge nerves or vessels. There is a depression along the posterior aspect in the midshaft of the humerus that delineates the path of the radial nerve as it spirals around the bone. The relatively deep channels on the internal surface of the skull reflect the course of the meningeal blood vessels.

Organs also form depressions on bones. The bilateral, shallow concavities on the internal aspect of the occipital bone are due to the convexity of the cerebellar hemispheres that are situated in this area.

Centers of Ossification

Embryologically, bones may arise through either of two processes. **Intramembranous bone formation** is the direct transformation of mesenchyme into bone, as occurs in the flat bones of the skull. In **endochondral bone formation** a precursor cartilage model of the developing bone is laid down, with subsequent replacement of this cartilage model by bony elements. In long bones, such bone transformation from cartilage occurs initially at the midpoint of the diaphysis and is referred to as the **primary center of ossification. Secondary ossification centers** appear later in the epiphyses at the ends of the long bones. The cartilage between the primary and secondary centers decreases progressively in relative amount, but persists as the **epiphyseal disc,** a platelike zone, as long as growth in length is taking place.

Slipped Epiphysis

In a young person, injury to the extremities of long bones may cause an epiphyseal displacement. Such an injury may be serious since the epiphyseal plate (disc) is the growth center for the bone. If such injuries go untreated, the longitudinal growth of the affected bone may be retarded, or even arrested, resulting in permanent shortening of the limb.

Some basic concepts of bone growth are summarized below:

1. The age of a growing individual may be reliably estimated from a radiographic assessment of ossification centers.

2. Centers normally present at birth include the distal femur, proximal tibia, calcaneus, talus, cuboid, and the proximal end of the humerus.

3. Most epiphyses have fused (*i.e.,* the growth zone cartilage has become ossified) in the male by the age of 20. Both the appearance and fusion of epiphyses in the female precede those in the male by about two years.

4. A given long bone may have epiphyses at both ends or only at one end. The latter situation is seen in the metacarpals, metatarsals, and phalanges.

5. If a long bone has a single epiphysis, it usually occurs at the end of the bone that undergoes the greatest excursion in movement.

6. In long bones that have two epiphyses, the epiphysis that appears first is usually the last to fuse with the shaft and it contributes most to the growth in length of the bone.

7. The more rapidly growing ends of bones of the extremities are at the knee (for the femur, fibula, and tibia), at the shoulder (for the humerus), and at the wrist (for the radius and ulna).

8. Nutrient arteries entering the diaphyses reflect this disparity in growth rate at the two epiphyses by angling away from the more rapidly growing end (*mnemonic:* From the knee I flee, to the elbow I go).

Joints or Articulations

A **joint,** or **articulation,** is the contact or union between two or more bones or cartilages. In gross anatomy our primary interest in articulations has to do with the degree of motion occurring at joints. In studying joint movements we are concerned with (1) the type of joint (hinge, ball and socket, gliding) (2) the accessory structures associated with joints, which may act primarily to stabilize or, conversely, to allow maximal motion; (3) the muscles or their tendons that cross joints, and are the movers of the bones forming the joint; and (4) the blood and nerve supply of joints.

An articulation is classified by the degree of motion that it exhibits namely, immovable (synarthroidial), slightly movable (amphiarthroidial), and freely movable (diarthroidial) joints (Fig. 1-3). This classification also reflects, in the same order, the nature of the tissue interposed between the bony surfaces, (*i.e.,* fibrous, cartilaginous or synovial).

Fibrous Joints

Fibrous joints are characterized by bones that are united by a minimal amount of fibrous tissue (sutural and gomphoses) or by a sheet of fibrous tissue (syndesmosis). Three types of **sutural** joints are recognized: serrated, squamosal, and plane. In **serrated** sutures the margins of the bones interlock like the cogs of a wheel. The sagittal suture of the skull is an example of a serrated suture. In the articulation of the temporal bone with the parietal bone of the calvarium the margins of the bones overlap each other; this is an example of a **squamosal** suture. In the **plane** sutural joints smooth edges of the bones abut against each other as seen in the articulation of the vomer with the perpendicular plate of the ethmoid in the nasal septum.

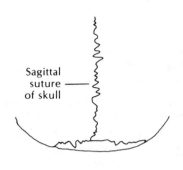

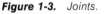

Figure 1-3. *Joints.*

The teeth, encased in the sockets of the alveolar processes of the maxilla and mandible, are examples of **gomphoses.** The interosseus membrane that unites the radius and ulnar of the forearm typifies a **syndesmosis.**

Cartilaginous Joints

Cartilaginous unions are differentiated as primary (synchondroses) or secondary (symphyses) cartilaginous joints. **Primary cartilaginous joints** are transitory and are frequently not considered to be true joints. They include the epiphyseal growth plates located between the epiphyses and diaphysis in growing bones (Fig. 1-4). Primary interest in these centers, apart from their being indicators of growth, is that they may be disrupted in traumatic injury. There may be a consequent separation of the epiphysis from the diaphysis, often resulting in arrested growth and deformity of the bone.

In **secondary cartilaginous joints** the bones are united by fibrocartilage. This type of union allows only a limited degree of movement. Symphyses are present between the adjacent vertebral bodies where fibrocartilage forms the intervertebral disc, and at the pubic symphysis where the intervening fibrocartilage plate permits a slight widening of the infrapubic angle during parturition under the influence of pregnancy hormones.

Synovial or Freely Movable Joints

The articular surfaces of bones that form a **synovial joint** are surrounded by an outer sleeve of connective tissue, the **joint capsule.** The joint capsule is lined by a **synovial membrane** that secretes a viscous, lubricating synovial fluid into the synovial cavity. The synovial fluid is interposed between articular surfaces, which, in all instances, are covered by hyaline (articular) cartilage. Such cartilage is not uniform in thickness, but is relatively thin in the central area of the articular surface and thicker toward the periphery. Examples of diarthrodial joints are the joints of the extremities.

There is a variety of subclassifications of synovial joints. Subtypes primarily reflect the shape of the articular surfaces, which is the prime determinant of the motion that can occur at a joint. For example, the surfaces may be either ovoid or saddle-shaped. In ovoid joints one surface is concave, the other convex. The convex surface has, invariably, a greater surface area than does the concave

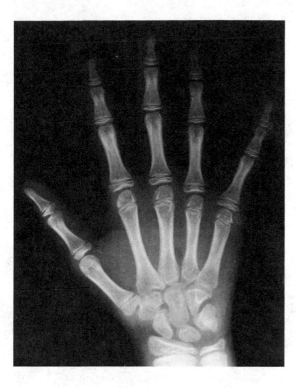

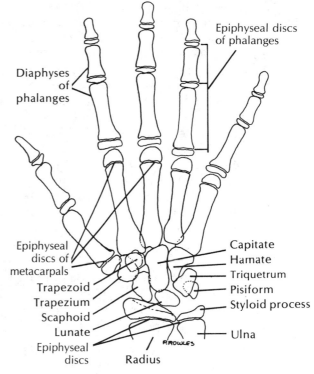

Diaphyses
of
phalanges

Epiphyseal discs
of phalanges

Epiphyseal
discs of
metacarpals

Trapezoid
Trapezium
Scaphoid
Lunate
Epiphyseal
discs

Radius

Capitate
Hamate
Triquetrum
Pisiform
Styloid process

Ulna

Figure 1-4. *Epiphyseal discs in the hand.*

surface. The opposed surfaces of saddle-shaped joints are reciprocally shaped to be congruous with a "male" or a "female" component. While few synovial joints provide movement in only one direction, most of them can be classified according to their major movement:

1. **Hinge** or **ginglymus joints** move essentially in a single axis at right angles to the long axes of the bones. The elbow is an example of this type of joint.

2. **Pivot** or **trochoid joints** allow movement in the vertical axis only. An example of this type of motion occurs in supination and pronation (turning over the hand) in which the head of the radius rotates (pivots) against the capitulum of the humerus.

3. **Condyloid joints** permit movement in two planes at right angles to each other. This is seen in the temporomandibular joint where the act of biting is combined with the grinding movement of the molars.

4. **Plane joints** provide a sliding movement in any direction over slightly curved surfaces. While the movement at these joints is multi-axial, their excursion is usually limited. Plane joints occur between the carpal and tarsal bones.

5. **Ball and socket joints** allow maximum freedom of movement in all directions. The hip and shoulder joints are examples.

The range of motion in joints is also determined by the degree of laxity of the joint capsule and the amount of slack present in the ligaments that bind adjacent bones together. Strain on a joint is taken up predominately by the ligaments. These reinforcements may be merely thickenings of the fibrous connective tissue of the joint capsule, or they may be bands of connective tissue separate from the capsule. The strength of a ligament is proportional to the strain it bears. The strongest ligament of the body, the iliofemoral ligament of the hip joint, is under constant strain in the erect position as it aids in the counteraction of the force of gravity.

In addition to the composition of a ligament the direction of the fibers also indicates its function. The direction of the fibers of the interosseous membrane in the forearm is oriented to counteract stress placed upon the radius and ulna. The ligamenta flava of the vertebral column, composed of predominately elastic tissue, aids the erector spinae muscles in counteracting the force of gravity.

As a limb is moved at a synovial joint in one direction, the capsule on the opposite side becomes taut. At the outset of a given action sufficient slack must exist to permit a normal range of movement. For example, in abduction at the shoulder, slack must be present on the medial aspect of the joint capsule to permit the movement.

The degree of tonicity or laxity of muscles that have tendons extending over a joint, as well as the amount of soft tissue adjacent to the joint, may physically impede its motion.

Intracapsular structures are present in many joints. These may be in the form of strengthening internal ligaments, as in the cruciate ligaments of the knee joints, or they may be associated with the functional muscles of the joint, as occurs in the long head of the biceps in the shoulder joint. They may exist as intra-articular discs, as in the temporomandibular joint, or they may modify the articular surfaces as in the meniscal cartilages of the knee. In some joints elaborations of the synovial lining increase its surface area and thereby increase its secretory capability (*e.g.,* around the knee joint).

Muscular System

Three types of muscle tissue are present in the body: **smooth muscle,** associated with blood vessels and organs; **cardiac muscle,** which forms the walls of the heart; and **skeletal muscle,** which forms the voluntary muscles of the body. Various types of muscle cells have specific histologic and functional characteristics, and differ in their innervation. In gross anatomy our interest is centered on skeletal muscle. Interest in smooth and cardiac muscle tissue is limited to the distribution of sympathetic and parasympathetic nerves that supply the organs that contain these types of muscle.

The muscular system comprises approximately 650 definitive skeletal or voluntary muscles, which afford us the only conscious control we have over our external environment. They are the motors of the body and produce their effect by virtue of their ability to contract. Muscles can decrease their length by as much as 50%. Inasmuch as both ends of a muscle are usually fixed to bones, movement of a bone results from contraction. Movement primarily occurs at the distal attachment of the muscle called the **insertion.** However, the attachment of insertion can sometimes be stabilized so as to produce movement at the proximal end of the muscle, the attachment of **origin.**

All voluntary muscles, however, are not under conscious control, for example, striated muscles of the middle ear and the musculature of the upper portion of the esophagus.

Basic Unit of Function and Structure

The functional and anatomic unit of the voluntary muscles, the **muscle fiber,** is an elongated, cylindrical, multinucleated cell covered with a tenuous membrane, the sarcolemma. With the light microscope, this cell (muscle fiber) can be seen to possess alternate light and dark cross-striated bands, and elongated, oval, peripherally located nuclei. These cells are the units that constitute the fleshy belly of a definitive voluntary muscle. Muscle fibers are separated from each other by a delicate connective tissue covering, the **endomysium.** Groups of muscle fibers are, in turn, bound together as a fasciculus or muscle bundle by a connective tissue investment, the **perimysium.** In addition to these connective tissue elements, the entire muscle is surrounded by the **epimysial fascial envelope,** a thickened connective tissue portion of deep fascia (Fig. 1-5).

Attachment of Muscles

Toward the end of a muscle, muscle fibers are replaced by collagen fibers that form **tendons.** They attach the muscle belly to structures to be moved, usually a bone. The manner of attachment of muscles varies considerably. Some muscles appear to rise directly from the surface of a bone, such as the intercostal muscles in which the tendinous attachment is very short. Muscle bellies may also share an attachment in common with adjacent muscles, for example, the flexor and extensor muscle masses at the medial and lateral aspects of the elbow joint. Some of the forearm muscles have long cylindrical tendons that extend from the bellies near the elbow to the digits. These tendons may be longer than their muscle bellies. Such thinning-out over the wrist and digits reduces tissue bulk opposite these joints, which otherwise would impede joint movement. Broad flat tendons are called **aponeuroses.** Examples of these are the aponeuroses of the anterolateral abdominal wall muscles, which extend from the inferior extent of the sternum to the level of the pubic symphysis as they pass to their insertion into the linea alba in the midline of the abdomen.

Intrinsic Architecture of Muscles

Muscles may be classified according to the arrangement of their fascicles to their tendons of attachment (Fig. 1-6). A **fusiform muscle** is a cigar-shaped muscle with the fleshy belly tapering at both ends as tendons. The fascicles of this type of muscle are essentially in parallel, so that its functional length approximates its actual length. Contraction of a muscle with this pattern results in decreasing its overall length by one-third to one-half. The movement of bones is maximal when such a muscle contracts. The power of strength that a contracting muscle is capable of generating is also a function of the arrangement of its fascicles. In some muscles the tendon extends into the fleshy belly and the fibers are oriented obliquely as they attach to their tendon. If the tendon is located along one side of the belly and all of the fascicles are oriented in the same oblique direction into the tendon, the muscle is described as being **unipennate.** In such a muscle the range of movement is decreased by the pattern of fascicles, but its power is increased; moreover, the direction of pull will be deviated toward the attachment of its fibers. In **bipennate muscles** the fascicles attach obliquely to both sides of a centrally placed tendon in a featherlike pattern. Such fascicles do not distort the directional pull of the tendon inasmuch as they exert equal pull from both sides of the tendon.

In the **multipennate muscles,** such as the deltoideus, the fascicles radiate out from a tendon in the central axis of a muscle. The arrangement of the fascicles of this muscle permits it to perform many different actions at the shoulder joint. The more anteriorly placed fascicles, working independently, can flex or internally rotate the humerus; the central fascicles abduct the humerus; while the anteriormost and posteriormost fascicles, working together, act to adduct the humerus. Fascicles may also be arranged in parallel to form

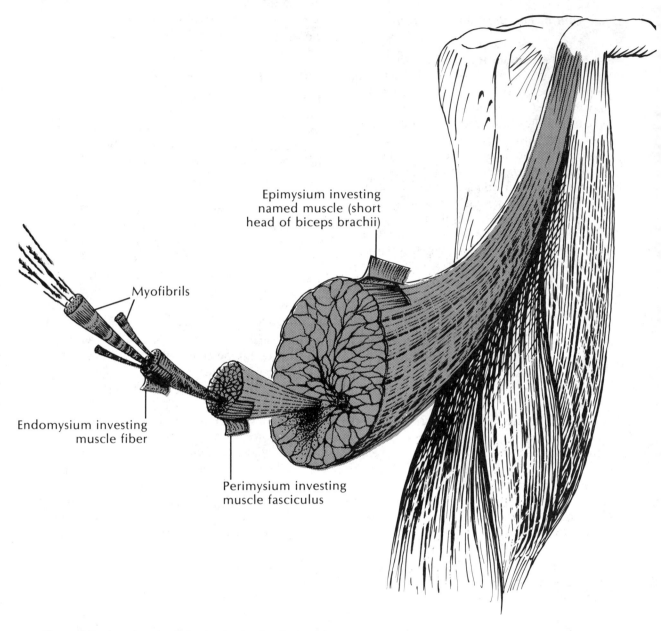

Epimysium investing
named muscle (short
head of biceps brachii)

Myofibrils

Endomysium investing
muscle fiber

Perimysium investing
muscle fasciculus

Figure 1-5.   *Investments of structural subunits of a muscle.*

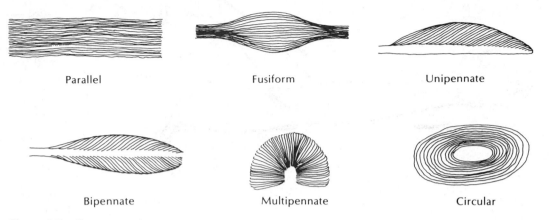

Parallel   Fusiform   Unipennate

Bipennate   Multipennate   Circular

Figure 1-6. *Gross muscular patterns.*

a flattened muscle, or circularly to form a **sphincteric** muscle.

Basic Muscle Actions

The action of a given muscle is designated according to the direction it moves a part of the body away from the anatomic position (Fig. 1-7). These movements are designated as flexion, extension, abduction, adduction, medial or lateral rotation, circumduction, supination, pronation, inversion, and eversion.

In the upper extremity, trunk, and at the hip joint, **flexion** is the action that opposes (brings together) anterior surfaces of segments of the body, thus reducing the angle at the joint. For example, flexion at the hand brings the anterior surface of the digits to the anterior surface of the palm, as in making a fist. At the elbow, flexion approximates the anterior surface of the forearm to the anterior surface of the arm.

Extension, in all instances, is the opposite action from flexion. In the upper extremity, extensor muscles open the clenched fist or straighten out the elbow to return the forearm to the anatomical position.

Due to the developmental rotation of the lower extremity, flexion at the knee opposes the posterior aspect of the leg to the posterior aspect of the thigh. At the ankle, to avoid confusion, the terms plantar flexon and dorsiflexion are substituted, respectively, for flexion and extension. **Plantar flexion** tends to bring the foot in line with the leg, whereas **dorsiflexion** approximates the dorsum of the foot on the tibia (shin bone).

Abduction is the action of moving the extremity laterally or away from the body, while **adduction** is the movement of a part toward the midline of the body. In movement of the digits in abduction and adduction a different line of reference is used. In the hand, the line of reference is the middle finger, in the foot the second toe. Thus, abduction spreads the fingers and toes; adduction brings them together.

Rotation is a pivoting movement. If this action is toward the body, it is internal or medial rotation; if it is away from the body, it is lateral or external rotation.

Circumduction is a combination movement in which the motion of an extremity describes a cone, the apex being at the shoulder or hip, the base being at the hand or foot.

Pronation and supination are actions limited to the hand and forearm. **Pronation** turns the palm downward; **supination** turns the palm upward.

Inversion and eversion are limited similarly to the foot. **Inversion** is turning the sole of the foot inward, as in standing on the lateral side of the

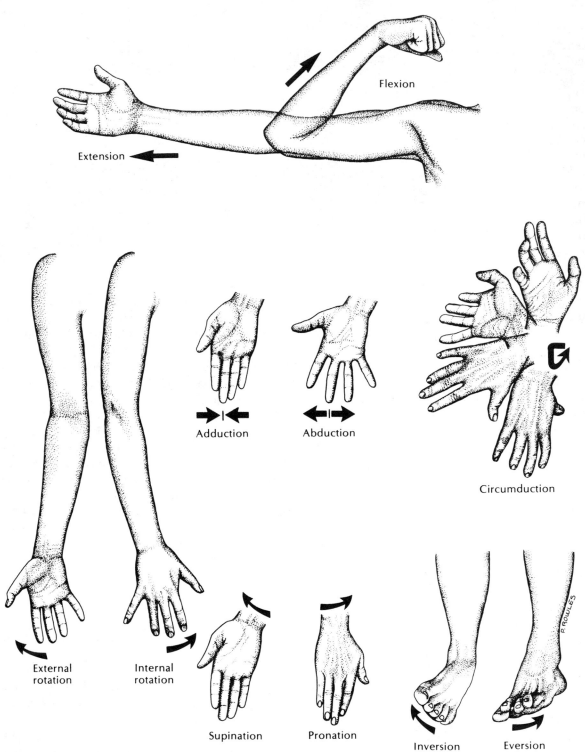

Figure 1-7. *Schematic of muscle action.*

foot. **Eversion** is turning the sole outward, as in standing on the medial side of the foot.

In a given action several muscles must act in concert if the movement is to be smoothly performed. Thus, in describing motion at a given joint, additional assignments beyond flexion, extension, abduction, and adduction are given to muscles. Although several muscles may contract to perform a given action, one muscle usually predominates and is called the prime mover. **Prime movers,** moreover, are subdivided as primary if they initiate an action, or secondary if they reinforce the action against resistance. Of the several actions a muscle may perform, the action in which it is most efficient is given as its prime function. All other muscles that contribute to the performance of a given action are referred to as **synergistic muscles.** Muscles that essentially oppose a given motion are called **antagonistic muscles.** For example, at the elbow joint, the biceps brachii and brachialis muscles act synergistically in flexing the elbow; the antagonistic triceps brachii muscle, an extensor, must relax concomitantly to balance the force of the contracting biceps and brachialis muscles for smooth movement to occur.

Stabilizing or **fixating muscles** are also necessary to perform many body movements. This is especially important in muscles that cross two joints. For example, in the action of the biceps brachii, flexion of the elbow occurs after the shoulder has been fixed, which gives the biceps a rigid base from which it can exert its pull.

In different movements a given muscle may function in all four roles given above, that is, it may perform as a prime mover, an antagonist, a synergist, or as a stabilizing muscle.

Muscular Compartments

In the extremities, muscles with similar actions are usually contained within a fascial compartment and are usually innervated by a common nerve. For example, in the thigh there are three well-defined compartments: an anterior compartment for the extensors of the leg, a posterior compartment for the flexors of the leg, and a medial compartment for the adductors of the thigh. The primary innervation of these muscle compartments are the femoral, sciatic, and obturator nerves, respectively.

Muscle Nomenclature

The names of muscles are orientation aids to the beginning student in anatomy if one considers the implication of the terms in the name. Most muscles are named from anatomic characteristics that indicate shape, size, location, action, attachment, or fiber direction. The rhomboid muscles attached to the scapula are so named from their shape; they are differentiated from each other on the basis of their size, as the rhomboideus major and rhomboideus minor. The muscles on the dorsum of the scapula are named according to their location, with respect to the spine of the scapula, as the supraspinatus and the infraspinatus. The name of the levator scapulae muscle indicates its action, the elevation of the scapula. The sternocleidomastoideus is named from its origin on the sternum and the clavicle (cleido) and its insertion on the mastoid process. The external abdominal oblique muscle of the abdominal wall is named because of its superficial location and the oblique direction of its muscle fibers.

Fasciae

Fasciae occur mostly as membranous sheets that are present between the tela subcutanea (superficial fascia) beneath the skin, and the subserous fascia adjacent to the lining membranes of the serous body cavities. Fascia is also present as loose, connective-tissue packing material containing variable amounts of fat that fill in the spaces between adjacent organs. Elsewhere fascia forms discrete membranous sheets or tubular enclosures, which are grossly dissectable. These sheets may separate organs or may enclose groups of organs. In the latter instance, fasciae function to hold structures in their proper locations.

Fasciae have three major subdivisions: the superficial (subcutaneous) fascia, the deep fascia, and the subserous fascia. Typically, the superficial fascia consists of loose areolar connective tissue, whereas deep fasciae form membranous sheets of varying density, which often enclose muscles or surround organs. The continuity of all fasciae can be demonstrated; however, by fusing and splitting they form cleavage planes or compartments throughout the body.

Superficial Fascia

The **superficial fascia** (subcutaneous tissue or tela subcutanea) is immediately deep to the skin. It invests the entire body and varies in thickness in different areas. Deep to the skin on the back of the hand, this layer is quite sparse, while over the lower abdominal wall it is markedly increased in thickness as a heavy layer of fat, the **panniculus adiposus.** Here two distinct layers may be identified, the superficial **fatty** layer (**Camper's fascia**), and a deeper, **membranous** layer (**Scarpa's fascia**). In obesity, fat in the superficial layer increases throughout the body.

Obesity
In obesity, although fat accumulates in the body cavity, for example, around and between organs, in mesenteries, where it is most clinically significant, it is most apparent in the superficial fascia. In the "fat man" at the circus, fat in the superficial fascia is responsible for his sixty-inch waist.

Superficial arteries, veins, lymphatics, and nerves course in the superficial fascia. Hair folli-

cles, and sebaceous, sweat, and mammary glands, and the muscles of facial expression are all embedded in this layer.

Deep Fasciae

The **deep fasciae** of the body are present as a series of laminae. The most superficial layer is the **external investing layer** of deep fascia; the deepest layer is the **internal investing layer.** These two layers form, respectively, a continuous external covering of the body and a continuous lining of the body cavities. Intervening or intermediate laminae between the external and internal investing layers vary in number depending on the number of muscles that constitute the body wall at a given area.

While deep fascia is continuous, individual laminae are encountered throughout the study of gross anatomy. This apparent paradox of the continuity of deep fascia with designation of individual fascial laminae may create difficulty in gaining a correct concept of deep fascia. An example of this arrangement is the make-up of the abdominal wall in which there is a lineal fusion of the deep fascia along the lateral border of the quadratus lumborum muscle (Fig. 1-8). Anterior to this line of fusion (which extends between the twelfth rib and the iliac crest), six laminae split off to envelop each of the three muscles that form the lateral abdominal wall. Anteriorly, a similar vertical line of fusion occurs at the lateral extent of the rectus abdominis muscle as the semilunar line. In the interval between these linear fusions, **fascial envelopes** surround each of the lateral abdominal wall muscles. The **fascial clefts** created between contiguous layers of the superimposed muscles permit each muscle to contract independently.

Essentially bloodless planes exist between the fascial clefts. The arteries, nerves, and veins supplying a given muscle usually enter at a definitive site. Thus, considerable surgical manipulation can be undertaken without damage to the primary blood and nerve supply of a muscle.

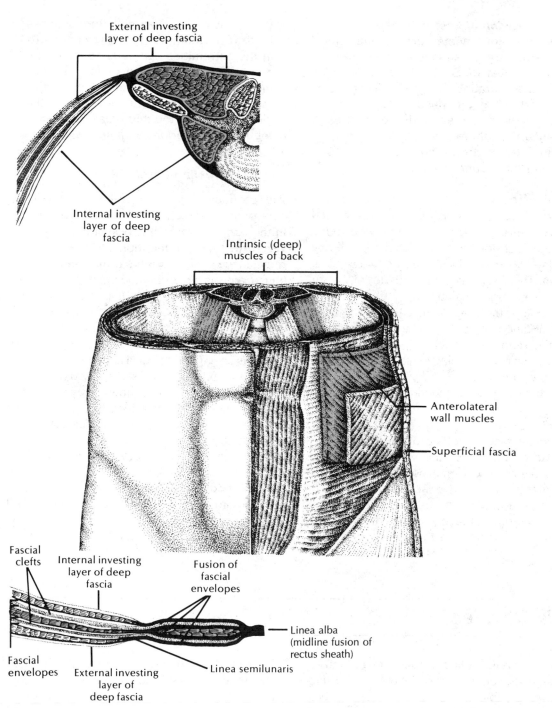

External investing
layer of deep fascia

Internal investing
layer of deep
fascia

Intrinsic (deep)
muscles of back

Anterolateral
wall muscles

Superficial fascia

Fascial
clefts

Internal investing
layer of deep
fascia

Fusion of
fascial
envelopes

Linea alba
(midline fusion of
rectus sheath)

Fascial
envelopes

External investing
layer of
deep fascia

Linea semilunaris

Figure 1-8. *Typical body wall section.*

Lines of Fusion of Fascial Sheets

Lines of fusion sheets, for example, the linea alba, are essentially avascular and are often sites for surgical incisions because the operating field is relatively free of blood. Surgeons favor fascial junctional areas for the anchoring of sutures because the inherent tensile strength of fascia and the strong, fibrous union that results from wound healing.

In other regions of the body, splitting and fusion of contiguous fascial layers form **fascial compartments.** In the upper and lower extremities deep extensions from the external investing fascial layer extend to bones to form intermuscular septa. These septa separate muscles of the extremities into functional groups.

Over joints, the external investing layer of fascia is modified. At the knee this modification forms the retinaculi, which help to support this joint. At the wrist and ankle joints, the deep fascia forms thickened transverse bands that bind the long flexor and extensor tendons in place.

Fascial Clefts

Clinically, fascial clefts provide routes for spread of infection. Fusion of laminae of deep fascia also forms pockets or compartments where the accumulation of pus, tissue fluid, or blood may occur. An understanding of fascial compartmentalization is important in the diagnosis and treatment of the accumulation of these fluids. Shortening of fascial components in the extremities can cause contracture deformities of the hand or foot. Fascial membranes may be used in corrective surgery. For example, in hernial repair the iliotibial tract in the thigh may be used to rebuild the anterior abdominal wall.

Internal Investing Fascia

This continuous fascial layer forms the deepest layer of the body wall as it lines all body cavities. In the cervical region it is described as the **prevertebral layer;** in the thorax as the **endothoracic;** in the abdomen as the **endoabdominal** or transversalis layer; and in the pelvis as the **endopelvic** or supra-anal layer. It is the layer in the thoracic, abdominal, and pelvic cavities to which the parietal pleura or peritoneum is attached.

The **subserous fascia,** between the internal investing layer of deep fascia and the serous membrane, is composed of loose areolar tissue. It varies in thickness and is continuous, by way of visceral ligaments (mesenteries), with the external surfaces of viscera in the thorax, abdomen, and pelvic cavities. Clinically, the composition of this layer permits the surgeon to separate, usually by blunt dissection, the parietal serous membrane from the body wall. Thus, this layer is present in the abdominal cavity between the peritoneum and endoabdominal fascia and in the thoracic cavity between the pleura and endothoracic fascia.

Body Cavities

Except for certain hollow organs and air sinuses in the skull, the entire area of the body internal to the skin is occupied by tissues. Thus, terms such as tissue spaces, fascial spaces, serous cavities, and body cavities refer not to voids, but rather to areas filled with fluids, tissues, or organs. Extracellular or tissue spaces exist between cells, between contiguous layers of tissues, and between adjacent structures, such as organs, muscles, tendons, and fascial membranes. In abnormal conditions, such an area may fill with excess tissue fluid, which results in edema or swelling, often accompanied by severe pain.

Facilitating mechanisms, which prevent friction between adjacent structures that move against one another, are modifications of the above so-called spaces. The simplest are the fascial spaces or, more correctly, the **fascial clefts.** These clefts, present between fascial membranes, are filled by loose connective tissue and tissue fluid. Where adjacent fascial membranes surround muscles, the resultant cleft between the contiguous muscle fasciae permits a limited amount of independent movement of muscles.

In addition to this functional role, fascial clefts are clinically significant in that they permit the surgeon to identify, isolate, manipulate, or transect a specific muscle. They also provide a bloodless plane for surgical incisions since major arteries and nerves to a muscle cross the fascial cleft at definite sites. Most fascial clefts also form channels or pathways that may facilitate the spread of infection, blood, tissue fluid, or air; conversely, they may fuse to form pockets to limit the spread of these substances. For example, urine extravasated from a ruptured urethra may extend, by way of a fascial cleft, into the abdominal wall, but is limited from extending laterally into the thigh by the fusion of superficial fascia to deep fascia along the ischiopubic rami.

Bursae

Bursae are more elaborate structures that facilitate motion. They are modifications of tissue spaces that coalesce to form a closed sac, lined with a secretory serous (synovial) membrane. These sacs are usually relatively small and are interposed between structures where there is need for a greater degree of motion than is permitted by fascial clefts. Bursae are located mostly around joints and are inconstant in number and size.

Subcutaneous bursae are found between the skin and underlying deep fascia. An example of this type is present between the skin and the olecranon process at the elbow. Bursae also occur where a muscle slides over a bone, as in the case of the large **submuscular bursa** between the gluteus maximus muscle and the underlying ischial tuberosity. Numerous bursae are located around the knee and often communicate with the synovial cavity of the knee joint.

Bursitis
Bursitis is inflammation of a bursa. Subcutaneous bursae over bony protuberances are exposed to frequent pressure and trauma. When the trauma is repetitive and prolonged, the bursa becomes swollen, inflamed, and painful. Occupational or sport activities predispose certain bursae to injury. Colorful, descriptive terms have been given to these conditions, for example, housemaid's knee (prepatellar bursitis), weaver's bottom (ischial tuberosity bursitis), tailor's ankle (bursitis over the lateral malleolus), and tennis elbow (olecranon bursitis).

Inflammatory processes (bursitis) may increase the secretion of fluid by the serous membrane into the closed bursal sac. The resultant increase of fluid exerts pressure, causes swelling and may be extremely painful.

Tendon Sheaths

Tendon sheaths are more complex structures that are modifications of the simple closed bursal sac. Tendon sheaths are elongated sacs, lined with a synovial (serous) secreting membrane into which tendons have invaginated. They occur mainly at the ankle, wrist, and along the plantar and palmar surfaces of the digits. They are associated with the long tendons of muscles attaching to bones of the hands or feet that have their muscle bellies located in the forearm or leg.

Inflammation of the synovial sheath may develop into adhesions between the layers of the synovial sheath. If this occurs, movement of the bones to which the tendon is attached becomes severely limited (see tendosynovitis, page 84).

Synovial Cavities

Synovial cavities are present in all freely movable joints of the body. The articulating surfaces of the bones of these joints are surrounded by a cufflike sleeve of connective tissue called the joint capsule, which extends well beyond the articulating surfaces to attach to the periosteum of the bones forming the joint. The internal surface of the fibrous joint capsule is lined by a synovial membrane that secretes fluid into the joint cavity. Along the line of attachment of the connective tissue capsule, the synovial membrane reflects onto the surface of the bone to cover the entire portion of the bone within the joint capsule, except the articular surfaces. The latter are covered by hyaline cartilage. In those cavities having internal ligaments, such as the knee, hip, and shoulder, the synovial membrane reflects onto the ligaments so as to exclude them from the synovial cavity. The fluid secreted by the synovial membrane is somewhat viscous in consistency and serves as a lubricant between the articular surfaces.

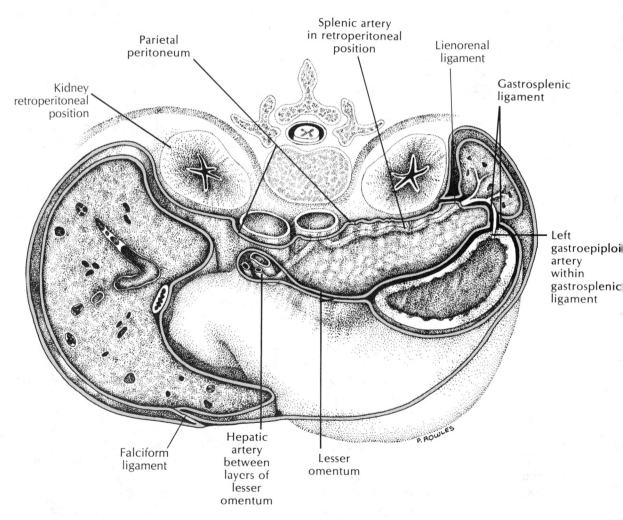

Figure 1-9. *Visceral ligaments (mesenteries).*

Serous Cavities

Serous cavities of the body facilitate movement between organs. They are present in the major body cavities. In the thorax, the pericardial cavity surrounds the heart, and a separate pleural cavity surrounds each lung. The peritoneal cavity is located for the most part in the abdominal cavity, but extends into the pelvic cavity. Serous cavities (pericardial, pleural, and peritoneal) are lined by serous membranes and are closed cavities, except for the peritoneal cavity in the female. In the female, the peritoneal cavity communicates by way of the uterine tube with the uterine cavity, which in turn opens to the exterior through the vagina.

As organs develop in the embryo they invaginate into the serous cavities, carrying the serous membrane with them. This membrane becomes intimately adherent (fused) to the external surface of the organ and is designated as the **visceral portion,** for example, visceral pleura, visceral pericardium, or visceral peritoneum. That portion of the serous membrane remaining adherent to the wall of the body cavity is called the **parietal layer.** As organs invaginate into serous cavities they essentially obliterate the lumen of the cavity. Normally, there is a limited space between opposing visceral and parietal layers of a serous cavity. This interval is filled with a minimal amount of fluid secreted by the serous membrane.

At several sites within the abdominal cavity two or more layers of serous membrane become fused to suspend an organ from the body wall. This fusion of layers forms the **visceral ligaments** or **mesenteries** (Fig. 1-9). Visceral ligaments of the peritoneal cavity are of special importance. They provide the only pathway for vessels and nerves to reach organs, since no structures penetrate the serous membrane lining the cavity.

In the abdomen, organs that are not suspended from the body wall by a visceral ligament are described as being retroperitoneal in position. Examples of retroperitoneal organs are the kidneys, the duodenum, and the pancreas.

Peritonitis
Peritonitis is inflammation of the peritoneum. It may be general or localized. It is characterized by an accumulation of a large amount of peritoneal fluid (ascites) containing fibrin and many leukocytes (pus). In the supine patient, the infected fluid tends to collect at two sites, the pelvic cavity, and in the right posterior subdiaphragmatic space. Tapping or draining of excess fluid from the abdomen is accomplished by inserting, under local anesthesia, a trocar and cannula or a needle and plastic tubing through the anterior abdominal wall, and aspirating the peritoneal fluid (paracentesis). When a patient with peritonitis is examined, stretching of the infected parietal peritoneum is very painful. The pain is especially severe when digital pressure over the inflamed area is suddenly released, causing the abdominal wall to rebound suddenly. Clinically, this is called rebound tenderness, which localizes the site of inflammation of the parietal peritoneum and often occurs over the infected organ.

Cardiovascular System

The cardiovascular system transports blood to and from capillary beds for the exchange of respiratory gases, nutrients, and metabolites. This exchange occurs in the fluid that passes through the walls of capillaries to bathe the surrounding cells. Most of the fluid that enters tissue spaces is retrieved by the capillaries to return to the heart through the veins. After nourishing the cells, a portion of the tissue fluid drains into the lymphatic vessels as lymph, which follows a different route in returning ultimately to the venous system.

The circulatory system comprises the heart, arteries, capillaries, veins, and lymphatics. However, in dissection capillary and lymphatic beds, as well as the smaller arteries and veins, are not visible. Therefore, in gross anatomy the functional level of the cardiovascular system is not investigated.

Heart

The clinician often speaks of right and left hearts. What is meant by this seemingly ambiguous expression is that the right chambers of the heart receive venous blood, low in oxygen content, which is pumped toward the lungs to be oxygenated. In contrast, the left cardiac chambers receive oxygenated blood from the lungs and pump it by way of the aorta to the rest of the body. Thus, the expression right and left hearts refers to a functional concept as well as to anatomic divisions.

The heart is a muscular pump whose prime function is to propel the blood into capillary beds in all parts of the body. The **right atrium** receives deoxygenated blood by way of the superior and inferior venae cavae and veins draining the heart. From the right atrium, the blood passes through the right **atrioventricular orifice,** which is guarded by the **tricuspid valve,** into the **right ventricle.** Contraction of the right ventricle forces the blood past the **semilunar valve** that guards the **orifice of the pulmonary trunk,** and into the lungs. Oxygenated blood returns to the heart through the pulmonary veins, which empty into the **left atrium.** Upon contraction of the left atrium, the blood passes through the left atrioventricular orifice, which is guarded by the **bicuspid (mitral) valve,** into the **left ventricle.** Contraction of this chamber forces the blood past the cusps of the **semilunar valve** of the **aortic orifice** for distribution, by way of the aorta, throughout the body.

The functional role of each chamber is reflected in the thickness of its muscular wall. The atrial walls are very thin, having only to propel blood from the atria into the ventricles. The relatively massive wall of the left vetricle is approximately three times as thick as the wall of the right ventricle. This is consistent with its much greater work load; this chamber must develop sufficient pressure to drive the blood to all parts of the body, as opposed to the short distance the right ventricular blood must travel to reach the lungs.

Arteries

Arteries carry blood away from the ventricles of the heart to the beds of the capillary vessels. At each bifurcation, the units of this distributing system decrease in size, from a diameter of over 30 mm at the pulmonary trunk and aorta to arterioles that may be as small as 0.5 mm. In the pulmonary circuit, the arteries carry blood low in oxygen content, whereas in the systemic circulation, they carry highly oxygenated blood.

Arteries are classified as **elastic** or **large arteries, muscular** or **medium-sized arteries,** and **arterioles.** The walls of arteries are much thicker than the walls of veins of corresponding diameter, thus they are less apt to collapse as blood pressure falls during diastole. As branching arteries decrease in diameter, the elastic tissue component of their walls diminishes and is replaced by smooth muscle. At the arteriolar level, all elastic tissue has been essentially replaced.

Aneurysm
An aneurysm is a thin, weakened section of the wall of an artery or a vein that bulges outward forming a balloonlike sac, or causing a permanent dilation of the blood vessel. The aneurysm dilates to a larger and larger size until the vessel wall becomes so thin it may burst and cause massive hemorrhage with shock, severe pain, stroke, or death, depending on which vessel is involved. Aneurysms commonly involve the circle of Willis,

which can cause stroke and mental impairment, or the thoracic and abdominal aorta. Although atherosclerosis is the most common cause, syphilis, congenital vessel defects, and trauma may also produce life-threatening aneurysms.

The structure of the walls of an artery suggests its function. For example, an elastic artery (aorta, pulmonary trunk) may be stretched lengthwise and distended in diameter to accommodate the large volume of blood that accompanies each ventricular contraction. During ventricular relaxation the elasticity of the walls of large arteries helps to maintain blood pressure and thereby effect a continuous flow through the smaller arteries.

In gross anatomy, most of the study of the cardiovascular system is spent in considering the pattern of arterial distribution. For a particular vessel we study its point of origin, its relationship to other structures along its course, its possible anastomoses with other vessels, and its area of distribution.

Structures of the body usually retain the blood supply that they acquired during development so that the pattern of arterial distribution is relatively constant; however, minor variations do occur. The most frequent deviations from the regular pattern include:

1. If a reduction in the size of a given artery occurs, an adjacent artery supplying the same general area will usually be increased in size. An example of this may be the uneven size of the two vertebral arteries contributing to the blood supply of the brain.
2. Two vessels that normally arise as direct branches from a larger artery may arise from a common trunk. This occurs frequently with the anterior and posterior humeral circumflex arteries in the upper part of the arm.
3. As an artery bifurcates, some secondary vessels may arise from the parent trunk. An example of this would be the common interosseous artery, normally a branch of the ulnar artery (formed as the brachial artery bifur-

cates at the elbow), arising directly from the brachial artery.

In general, the arterial supply to the upper and lower limbs follows analogous patterns.

Capillaries

While capillaries are much too small (7–10 μ) to be visualized at the dissection table, their role in circulation should be mentioned because exchanges between the blood and tissue fluid occur across their one-cell-thick walls.

Capillaries form a complex, anastomotic network between arterioles and venules. Many of their channels parallel one another and not all channels in a network function simultaneously. More metabolically active tissues (glands and muscles) have more extensive capillary beds than do less active tissues (tendons and ligaments). Anastomoses occur between arterioles and venules so that in different physiologic conditions the blood supply to various organs may be changed rapidly. Thus, in vigorous exercise more blood courses through muscles, while during digestion the blood supply to the gastrointestinal tract is increased (Fig. 1-10).

Avascular structures (cartilage, cornea, epidermis) have no capillary beds. Modifications in the vascular communication between arterioles and venules, in the form of sinusoids, occurs in some organs (spleen, liver, bone marrow, pituitary). **Sinusoids** are wide, irregular channels, and are partially lined by phagocytic cells.

Veins

Venous channels originate opposite the arteriolar side of capillary networks as small-caliber venules. They coalesce to form veins of increasing diameter that ultimately empty into the atria of the heart. Veins are usually more superficial, have thinner walls, are more numerous, and are of larger caliber than their companion arteries. Blood flow through veins is slower and under much less pressure than through arteries of comparable size.

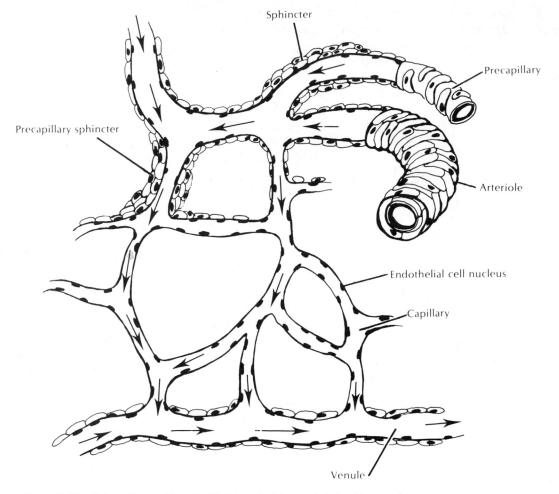

Figure 1-10. *Schematic capillary bed fed by arterioles and drained by venules.*

The pattern of venous tributaries is much more variable than that of corresponding arteries. Veins of the pulmonary circulation drain into the left atrium, while subsystems of venous drainage for the systemic circulation draining the body wall, abdominal viscera, and vertebral column empty into the right atrium. These subsystems include the venous drainage from the gastrointestinal tract that forms the **hepatic portal system,** which drains through the portal vein into the liver. The

azygos venous system drains blood from the wall of the thoracic cavity and provides some drainage of the abdominal wall through the ascending lumbar veins. Extensive venous plexuses surround the vertebral column and the spinal cord as the **vertebral plexus of veins.** Anastomoses between these subsystems occur freely. The anastomoses in the rectal, umbilical, and lower esophageal regions and the abdominal mesenteries are of special clinical importance in obstruction of the inferior vena cava flow or in pathologic conditions of the liver, in which venous flow through this organ is impeded (portosystemic anastomosis).

As in the arterial distributions, the venous return from the upper limb is analogous to that of the lower limb. Veins that accompany major deep arteries in the upper and lower extremity usually occur in pairs called **venae comitantes** (*i.e.,* each artery has two companion veins named for the artery they accompany). Veins from the hand and foot coalesce to form two major superficial vessels, the **cephalic** and **basilic** in the upper extremity; veins from the foot form the superficial **greater** and **lesser saphenous** in the lower extremity (these veins provide easy access to the circulatory system for intravenous infusion or drawing of blood).

Many veins are provided with **valves** to insure a unidirectional flow of blood. These are especially important in the limbs as venous blood returns to the heart against the force of gravity. Valves are usually situated in major vessels just distal to their larger tributaries. This arrangement limits retrograde blood flow. Incompetence of valves, as seen especially in the inferior extremity, results in an engorgement and dilatation of veins. If this becomes chronic, varicosities of the veins may occur, which are often removed surgically. Venous return from the extremities is facilitated by muscular contraction that compresses the deep veins and forces the blood toward the heart.

Varicose Veins of the Lower Extremity

One of the most common disorders of the vascular system is the dilation, elongation, and tortuosity of the superficial veins of the lower limb, called varicose veins. The principal cause of this affliction is an increased venous pressure with dilation of the veins, largely owing to valvular incompetence and gravity from the upright position. The increased venous pressure and varicosities cause edema of the skin and subcutaneous tissue with decreased blood flow and poor healing. This venous stasis and edema predispose the skin to develop ulcers following minor skin abrasions or injury. If the deep veins are patent and can provide the venous return from the foot and leg, then surgical removal or stripping of the superficial varicose veins decreases the venous pressure in the skin and subcutaneous tissue, decreases the edema, and allows the stasis ulcers to heal.

Portal Systems

Portal systems occur in the circulation wherever an extra set of capillaries is interposed in the circuit between blood leaving and blood returning to the heart. A prime example is the hepatic portal circulation. Arterial blood to the gastrointestinal system reaches the capillary beds in the stomach and intestine and is drained by tributaries that ultimately form the portal vein. The portal vein transports blood to the liver, where it passes through a second set of capillaries (sinusoids). The hepatic veins, which empty into the inferior vena cava, are ultimately formed from this second capillary plexus.

Portosystemic Venous Anastomosis

Tributaries of the portal vein communicate with systemic veins in several locations, particularly the lower end of the esophagus, rectum, and umbilicus. Ordinarily these channels are collapsed with little blood flow, because blood drains from the abdominal viscera through the portal vein to the liver. If venous flow through the liver is impeded by a blood clot or scarring in the liver (cirrhosis), the portal vein pressure rises markedly. This portal hypertension dilates the veins of the portal system, including some or all of these anastomotic channels. This can cause varicosities of the umbilicus (caput medusae), esophagus (esophageal varices), and rectum (hemorrhoids).

Anastomoses

An **anastomosis** is a union between the distal ends of blood vessels that permits free communication between the involved vessels. Anastomoses may occur between arteries, between veins, or between arterioles and venules. In the latter case, the blood bypasses the capillary network since it is transported directly from arterioles to venules. **Arteriovenous anastomoses** are widely distributed, but are found especially in the intestine, kidney, and skin.

Anastomoses may occur between relatively large vessels, for example, in the circle of Willis (located between branches of the internal carotid artery and branches of the basilar artery at the base of the brain), in the arterial arches in the hands and feet, and in the arterial arcades of the intestine. The latter form end-to-end anastomoses, providing, in essence, a dual blood supply to all segments of the gut and its accessory organs.

Most anastomoses occur between small vessels or develop from capillary networks or arterioles that supply the same region. Such potential alternate pathways are clinically significant in that they provide for the development of collateral circulation.

Collateral Circulation

Collateral circulation is the mechanism whereby blood may flow to an organ or region after its normal course has been blocked. This is necessary when occlusion of a vessel from a blood clot, foreign body, tumor, or ligation occurs. If the segment distal to the occlusion is to remain viable it must receive a new blood supply. When collateral circulation develops, the blood bypasses the obstruction by an anastomosis, and may even flow in both directions in the anastomotic artery to supply all tissue distal to the occlusion.

Lymphatics

The lymphatic portion of the vascular system consists of a series of lymph channels. It differs from the cardiovascular system proper in that it does not have a pumping apparatus, does not form a complete circuit (lymph travels only in one direction), and is a system of vessels with lymph nodes interposed along their course.

Lymph coursing through the vessels is a clear, slightly straw-colored fluid. It is formed by filtration from blood capillaries as tissue fluid and is similar in composition to blood plasma. The lymph contains lymphocytes (formed largely in the lymph nodes) enzymes, antibodies, and lipids. In intestinal lymphatics, lipids form a milky white, fatty emulsion, the chyle.

Lymphatic vessels begin as an extensive, diffuse network of blind-ending **lymphatic capillaries** that drain tissue fluid from nearly all tissues and organs. These capillaries are lined with endothelium, have variable diameters, and are usually slightly larger than blood capillaries. Lymph capillaries are absent in the central and peripheral nervous systems, and in avascular tissues, such as the cornea, hyaline cartilage, and the epidermis of the skin.

Lymph capillaries coalesce to form large collecting vessels that contain valves giving them a beaded appearance. The smaller vessels form extensive, diffuse anastomosing superficial plexuses in the skin and on the surface of organs, and deep plexuses within organs. While lymphatic vessels largely accompany the venous drainage, the superficial and deep lymphatic plexuses of a given organ usually follow different pathways. In cancer, a metastasis in the pelvic region may either pass by lymphatics superficially to the groin or deeply into the pelvic cavity, depending upon the site of the initial lesion.

Coalescence of larger lymphatic vessels forms lymphatic trunks. The largest, the **thoracic duct,** begins at the **cisterna chyli.** The latter is a large, irregular lymph sac located at the level of the second lumbar vertebra within the abdominal cavity. From this site the thoracic duct ascends through

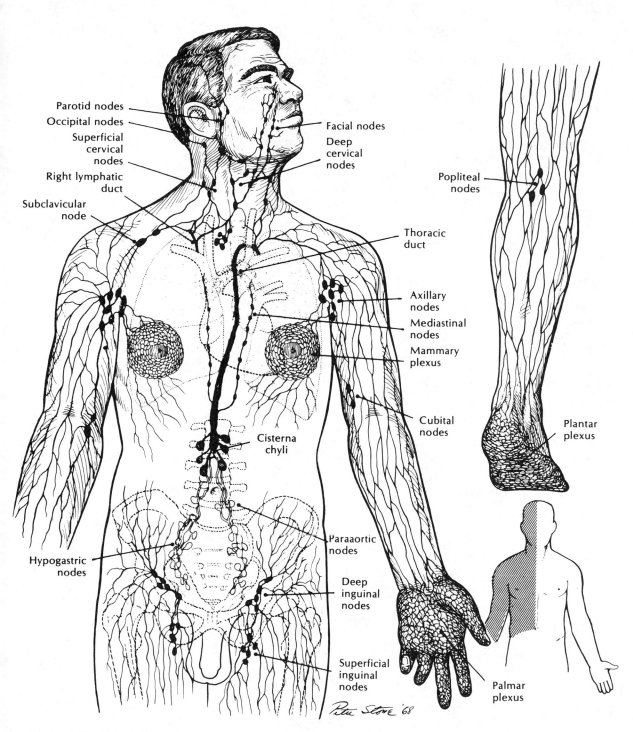

Figure 1-11. Principal lymph vessels and nodes. In the figure at the lower right, cross-hatching shows the area drained by the right lymphatic duct. (Langley LL, Telford IR, Christensen JB: Dynamic Anatomy and Physiology, 5th ed. New York, McGraw-Hill, 1980)

The labels in the figure are:

- Parotid nodes
- Occipital nodes
- Superficial cervical nodes
- Right lymphatic duct
- Subclavicular node
- Facial nodes
- Deep cervical nodes
- Thoracic duct
- Axillary nodes
- Mediastinal nodes
- Mammary plexus
- Cubital nodes
- Cisterna chyli
- Hypogastric nodes
- Paraaortic nodes
- Deep inguinal nodes
- Superficial inguinal nodes
- Palmar plexus
- Popliteal nodes
- Plantar plexus

the thoracic cavity to terminate by emptying into the left subclavian vein near its junction with the left internal jugular vein. Tributaries emptying into the thoracic duct drain the lower half and the left upper quadrant (left side of the head, left upper extremity, and left half of the thorax) of the body. The **right lymphatic duct,** a much shorter vessel, drains the upper right quadrant of the body. It is formed in the vicinity of the right internal jugular vein and the right subclavian vein as they join to form the right brachiocephalic vein. Lymph vessels draining the right side of the head, right upper extremity, and right side of the thorax coalesce to form the right lymphatic duct.

Lymph nodes are flattened, bean-shaped structures varying in size from 0.5 mm to 2 cm to 3 cm in length. They are interposed along the course of the lymph vessels and act as filtering stations for the lymph. In addition to screening out foreign particulate matter and detoxifying pathogenic bacteria, they produce lymphocytes and antibodies.

Normally, lymph nodes are only palpable in the axillary (armpit) and inguinal (groin) regions. However, in response to inflammation, they enlarge considerably and can be palpated easily in other parts of the body, such as under the mandible, at the elbow and knee, and along the lateral aspect of the neck.

Large lymph nodes are usually found in aggregates or chains extending along principal veins and are named from their anatomic location. In Figure 1-11 the major groups of nodes are shown.

Metastasis

Knowledge of the location of the lymph nodes and the direction of lymph flow is important in the diagnosis and prognosis of spread of carcinoma (metastasis). Cancer cells usually spread by way of the lymphatic system and produce aggregates of tumor cells where they lodge. Such secondary tumor sites are predictable by the direction of lymph flow from the organ primarily involved.

Nervous System

The highly specialized nervous system, together with the endocrine system, provides the remarkable and essential coordination necessary for the well-being of a complex living organism, such as a human being. As the major integrative system of the body, the nervous system functions to provide an awareness, through stimuli, of both the internal and external environment of the body, to make possible voluntary and reflex activities between the various structural elements of the organism, and to balance the organism's response to environmental changes.

The complexity of the brain and spinal cord is such that it is usually taught as a separate course, neuroanatomy. Gross anatomy courses are concerned with the distributing nerves which constitute the peripheral nervous system. However, to gain a basic understanding, some aspects of the central nervous system must be appreciated.

Neuron

The functional and anatomic unit of the nervous system is the **neuron** or **nerve cell.** This highly specialized cell is composed of a cell body and one or more nerve cell processes. A change in electric potential along the membrane of a process constitutes a nerve impulse.

Functionally, neurons are classified as **afferent** or **sensory** if they transmit impulses from com-

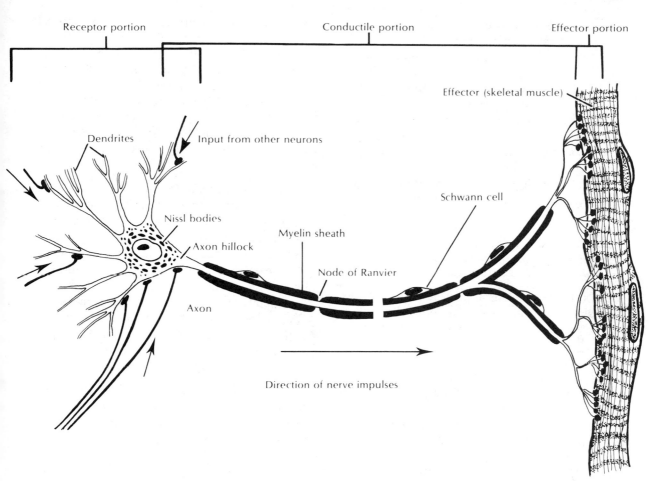

Figure 1-12. *Typical large motor neuron showing receptor, conductile, and effector regions. (Copenhover WM, Kelly DE, Wood RL: Bailey's Textbook of Histology, 17th ed. Baltimore, Williams & Wilkins, 1978)*

ponent parts of the body toward centers of integration in the brain and spinal cord, and **efferent** or **motor** if they transmit impulses from the central nervous system toward the distal parts of the body (Fig. 1-12).

The peripheral termination of each sensory nerve cell process is structurally modified as a **specialized receptor ending.** Structurally different endings are responsive to different modalities of sensations, such as pain, temperature and pressure. Receptor endings, located in the skin,

and special sense organs (eye, ear, nose, and tongue) provide contact with the external environment. Receptors in muscles, joints, ligaments, and viscera (organs) receive stimuli to provide information on the internal environment of the body.

At **synaptic junctions** (the sites where the process of one neuron comes in close proximity to the process or cell body of another) impulses are transmitted along neuronal processes to form conduction pathways. It is the complexity of the conduction pathways and the interplay between the various integrative centers of the central nervous system that necessitate special emphasis on this system of the body.

For descriptive purposes, the nervous system may be divided into a central portion, composed of the brain and the spinal cord, and the peripheral portion formed by the nerves of the body. The autonomic portion of the nervous system supplies smooth muscle, cardiac muscle and gland cells and overlaps, for gross anatomical study, both central and peripheral portions.

Central Nervous System

Cell bodies of neurons in the central nervous system are located in discrete layers or sites collectively referred to as the **gray matter.** In the cerebral and cerebellar hemispheres the cells are situated, for the most part, at the surface of these structures in the cerebral or cerebellar cortex. In contrast to this arrangement, deeply placed clusters of gray matter are called **nuclei.** Functionally, several nuclei may form an **integrative center,** such as the respiratory center.

All gray matter of the spinal cord is located centrally. In a cross section of the spinal cord this accumulation of nerve cell bodies forms a gross configuration of the letter H. Two legs of the H extend dorsally as the **dorsal horns** (or along the length of the cord as the **dorsal columns**). Cell bodies of the dorsal horns are associated with sensory impulses. The other two legs of the H extend ventrally as the **ventral horns** or **ventral columns,** which are composed of motor neuron cell bodies.

Processes of neurons in the central nervous system constitute the **white matter.** Generally, they are segregated into sensory and motor components and are gathered together into bundles as **spinal cord** or **brain tracts,** which subserve a specific function. For example, most fibers responsible for transmitting pain and temperature sensations, although present in most peripheral nerves, clump together in the spinal cord as the lateral spinothalamic tract, to transmit this modality as they ascend to higher centers in the brain. Fibers transmitting motor impulses from the brain to the muscles, such as the corticospinal tract, follow entirely separate pathways from the sensory fibers. Additional fiber tracts in the central nervous system connect integrative centers with each other.

Rhizotomy and Cordotomy

For relief of intractable pain, cutting of the sensory nerve roots, rhizotomy, or sectioning of the pain pathways in the spinal cord, cordotomy, may be indicated. In the latter procedure the anterolateral columns of the spinal cord are divided surgically. Immediate relief is achieved below the level of the section on the side opposite the incision. Bilateral cordotomy is necessary if the pain is present on both sides of the body.

The supportive tissue of the central nervous system is made up of specialized connective tissue cells called **glia.** They are not involved in the transmission of nerve impulses.

Peripheral Nervous System

The peripheral nervous system consists of the distributing nerves of the body and small clusters of nerve cell bodies located in **ganglia.** A distinction is made in the peripheral nervous system between the nerves supplying body wall structures and the extremities, the **somatic portion,** and the nerves supplying the viscera of the body, the **autonomic portion.**

Nerves are formed by cell processes of neurons. Nerve cell bodies present in ganglia of the peripheral nervous system are limited in number. All cell bodies in a definitive ganglion are either motor or sensory. The sensory ganglia are located in two sites: either interposed on the dorsal root of a spinal nerve (to be described later), or on those cranial nerves that transmit sensory impulses. Cell bodies of motor neurons in the peripheral nervous system are associated with ganglia of the autonomic portion.

Distributing nerves are also classified as to their origin from the central nervous system. Thirty-one pairs of nerves originating from the spinal cord are called **spinal nerves.** Twelve additional pairs of nerves originate from the brain stem and are called **cranial nerves.** Nerves derived from the spinal cord contain processes that transmit both motor and sensory impulses. Some nerves arising from the brain stem are similar to the spinal nerves in this respect, while others transmit only sensory or only motor impulses.

Supportive connective tissue investments surrounding the distributing peripheral nerves are the **endoneurium,** which is a connective tissue ensheathment around an individual nerve fiber; the **perineurium,** which surrounds a bundle of nerve fibers or a nerve fascicle; and the **epineu-** **rium,** which is a connective tissue investment of the entire nerve.

Typical Spinal Nerve

To avoid confusion in studying the peripheral nervous system, it is important to appreciate the distinction between the roots of origin of a spinal nerve and the rami or distributing branches of a spinal nerve. A typical spinal nerve is formed by fibers that have their cell bodies limited to a definitive block of the spinal cord termed a **spinal cord segment.** Rootlets arise from the dorsal aspect of each successive segment of the spinal cord and coalesce to form the **dorsal roots.** Similarly, rootlets arising from the ventral aspect of the spinal cord segment form the **ventral root** of a spinal nerve (Fig. 1-13).

Dorsal roots transmit only sensory fibers. The cell bodies of these fibers are located in the ganglia interposed on the dorsal roots called the **dorsal** or **spinal root ganglia.** The ventral root of a spinal nerve transmits only motor fibers. The cell

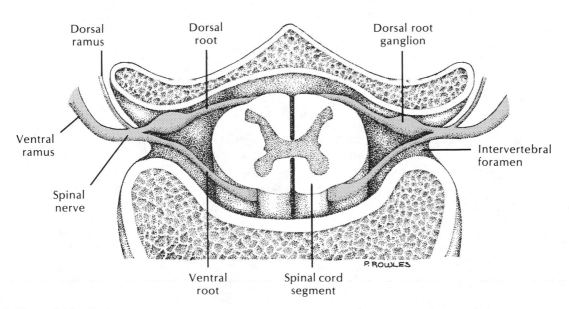

Figure 1-13. *Typical spinal nerve.*

bodies of these fibers are located in the gray matter of the spinal cord, either in the ventral horn if the fibers are destined for skeletal muscle, or in the lateral horn if the fibers (autonomic) supply smooth muscle, cardiac muscle, or glands.

As the dorsal and ventral roots traverse the intervertebral foramen to leave the vertebral canal, they unite to form the definitive **spinal nerve.** A tubular sheath of the meninges (dura, arachnoid, and pia) extends a short distance along the spinal nerve as it exits from the vertebral canal. Distally, this sheath fuses with the epineurium of the spinal nerve.

Shortly after the spinal nerve is formed it divides into two branches, a **dorsal** and a **ventral ramus.** The posterior area of the body is supplied by dorsal rami, and the anterior area and the extremities by ventral rami. Except for the second cervical nerve, the ventral rami are much larger branches than dorsal rami and the area they supply is much more extensive than that supplied by the dorsal rami. The latter supply the intrinsic or deep muscles of the back, joints of the vertebral column, and the skin over the back of the trunk, neck, and head. The ventral rami of the second through eleventh thoracic spinal nerves become intercostal nerves (T_{12} is called the subcostal nerve). These nerves supply the intercostal muscles, the anterolateral muscles of the abdominal wall, and the skin on the anterior aspect of the trunk.

Innervation to the upper and lower limbs is derived from ventral rami of the spinal nerves, which form plexuses. A **somatic nerve plexus** is formed as several ventral rami branch and coalesce to create a nerve network adjacent to their emergence from the vertebral canal. These networks, formed by the splitting and subsequent uniting of nerve components, vary in their complexity because the process may be repeated several times. The pattern of formation is sufficiently constant in different individuals so that each component of a given plexus can be consistently identified. The formation of a plexus results in a mingling of fibers from different spinal cord segments being distributed by a definitive nerve formed from the plexus. For example, the radial nerve, a branch of the brachial plexus, contains fibers from nerve cell bodies located in the fifth, sixth, seventh, and eighth cervical segments and the first thoracic segment of the spinal cord.

Referred Pain

Referred pain is the localization of visceral pain on a body surface often far removed from the organ involved. Nerves to the painful surface area arise from the same spinal segments as the nerve fibers supplying the viscus in question. For example, cardiac pain is often felt over the anterior chest wall and radiates down the inside of the left axilla and upper extremity. Gallbladder disease may elicit severe pain in the upper right abdominal region and over the right infrascapular region. One explanation for referred pain is that pain fibers from the body wall and those from the viscera share a common neuronal pool in the spinal cord. Bombardment of visceral pain impulses into this common area excites the somatic pain fibers whose response is misinterpreted by the brain as cutaneous pain in the peripheral dermatome supplied by this particular peripheral nerve or nerves.

Skin over the entire body is supplied **segmentally,** that is, each spinal nerve innervates a single constant segment of the skin. With the exception of the first cervical nerve, all spinal nerves supply branches to the skin. The skin segment supplied by a given spinal nerve is called a **dermatome.** In the neck and trunk the dermatomes form consecutive bands of skin. In the trunk there is an overlap of adjacent dermatome nerve supply so that to denervate a given dermatome three consecutive nerves must be transected or injected. Most of the skin of the face and scalp is supplied by the trigeminal (fifth cranial) nerve in which dermatomes may be assigned to each of its three divisions.

Shingles

Shingles, or herpes zoster, is a viral infection of the dorsal (sensory) root ganglion. Small vesicles (blisters) and discoloration of the skin occur over the dermatomal pattern of the skin, supplied by nerves from the involved ganglion. These lesions develop as circular bands of blisters around the neck and trunk and somewhat vertical bands along the extremities.

The **cervical plexus** is derived from ventral rami of the first through fourth cervical nerves and supplies the skin and muscles of the neck, and the thoracic diaphragm. The latter illustrates the basic concept that migrating muscles (those undergoing developmental shifting) carry their nerve supply with them; thus, in this case, the diaphragm migrated from the cervical region where it had its embryonic origin.

The **brachial plexus,** derived from the ventral rami of the fifth through eighth cervical nerves and the first thoracic nerve, supplies the skin and muscles of the upper extremity, including (with the exception of the trapezius) those muscles of the upper extremity that originate from the chest wall and the back.

The **lumbosacral plexus,** derived from ventral rami of all lumbar and sacral nerves, supplies muscles of the posterior abdominal wall and pelvis, and muscles and skin of the inferior extremity.

As distributing branches of plexuses supply their respective structures, the following generalizations may be made:

1. The nerve supply to a muscle usually joins the vascular supply to form a neurovascular pedicle and enters the deep aspect of the muscle at the proximal end of its fleshy belly.

2. If a nerve pierces a muscle, it usually sends branches to supply the muscle.

3. Nerves supplying a muscle send cutaneous branches to the skin overlying the muscle.

4. Nerves supplying muscles extending over joints also supply the joint.

5. Nerves from plexuses follow a common channel with vessels to enter the extremity. In the upper extremity this is the cervicoaxillary canal. In the lower extremity the femoral canal leads into the anterior compartment of the thigh, the obturator canal into the medial compartment of the thigh, and the greater sciatic foramen transmits nerves into the posterior compartment of the thigh.

Throughout most of their course, nerves (and vessels) usually traverse fascial clefts between muscles. This relationship is most evident in the thorax and abdomen. Here the intercostal and thoracoabdominal nerves course in the cleft or plane between the second and third layers of muscles of the body wall, the internal intercostal and innermost intercostal muscles in the thorax, or the internal abdominal oblique and the transversus abdominis in the abdominal wall.

Cranial Nerves

Most of the twelve pairs of **cranial nerves** arise directly from the brain stem. They traverse foramina in the floor of the cranial cavity as they course to their area of distribution. They transmit fibers whose cell bodies are located in discrete clumps within the brain called **nuclei.** Cranial nerves may differ from spinal nerves in that some transmit only sensory fibers while others transmit only motor fibers. Still others are similar to spinal nerves in that they transmit both motor and sensory fibers.

All cranial nerves transmitting general sensory fibers have ganglia interposed on the nerve, similar to the dorsal root ganglia of spinal nerves. Except for the spinal accessory and vagus nerves, the cranial nerves are limited in their distribution to structures of the head and neck.

Autonomic Nervous System

The autonomic portion of the peripheral nervous system supplies motor innervation to **smooth muscle, cardiac muscle,** and **glands.** There is a basic anatomic difference between the impulse pathway of somatic motor fibers to skeletal muscles, and the autonomic pathway to smooth muscle, cardiac muscle, and glands. From the central nervous system motor impulses to skeletal muscle fibers are transmitted along neurons that have their cell bodies in the ventral horn of the spinal cord or in motor nuclei of cranial nerves. Thus, a motor impulse to a skeletal muscle utilizes only one neuron to pass from the central nervous system to motor end plates in muscle fibers.

In contrast to the above, impulses destined for involuntary muscles or glands require two neurons in passing from the central nervous system to the effector organs. The first neuron in this pathway, termed the **preganglionic neuron,** has its cell body in the central nervous system. The second, termed the **postganglionic neuron,** has its cell body in a ganglion outside the central nervous system. Thus, impulses passing along preganglionic fibers activate postganglionic neurons at ganglia. Postganglionic fibers in turn activate the smooth or cardiac muscle fibers and glandular cells.

The **autonomic nervous system** is further subdivided into **sympathetic** (Fig. 1-14) and **parasympathetic portions.** Each of these subdivisions is distinctive in the location of the preganglionic cell bodies, the location of the ganglia, and in their function.

Sympathetic (Thoracolumbar) Division

This portion of the autonomic nervous system is also called the **thoracolumbar portion** because its preganglionic nerve cell bodies are located in the lateral horn (intermediolateral cell column) of the thoracic and upper two lumbar segments of the spinal cord. Processes from these nerve cell bodies leave the central nervous system by way of the ventral roots of the twelve thoracic and upper two lumbar spinal nerves. The preganglionic fibers contained in the fourteen pairs of nerves arising from these spinal cord segments form short filaments called **white rami communicantes.** These extend from the spinal nerve to a ganglion

situated adjacent to the body of a respective thoracic or lumbar vertebra. Such a ganglion is part of a chain of ganglia that extends along the entire length of the vertebral column from the base of the skull to the tip of the coccyx.

Once the preganglionic fiber passes into the chain ganglion by a white ramus communicantes, it may do one of three things. It may synapse on a postganglionic nerve cell body in the ganglion at that level. The process of the postganglionic nerve cell body can then return by a **gray ramus communicans** to the spinal nerve, to be distributed as a component of the spinal nerve. The second alternative is to ascend or descend by communications between adjacent ganglia to a higher or lower level (thereby creating the chain). This is necessary because the preganglionic outflow from the spinal cord is limited to the thoracolumbar level. Gray rami associated with these higher or lower ganglia will then transmit the postganglionic fiber to the spinal nerve at their levels. Thus, all spinal nerves receive gray rami communicantes and, therefore, transmit postganglionic sympathetic fibers to their areas of distribution.

The third alternative pathway for preganglionic fibers is to traverse the sympathetic chain ganglia without synapsing and to pass as a component of a **splanchnic nerve** to **preaortic ganglia.** These ganglia, containing postganglionic nerve cell bodies, are situated adjacent to major arteries arising from the aorta. The fibers of such a postganglionic nerve cell body contribute to the formation of autonomic plexuses that surround the major branches of the aorta. Extending along the vessel, they innervate the viscera supplied by the arteries they accompany. The major sympathetic plexuses include the carotid plexus, the celiac plexus, the superior mesenteric plexus, the inferior mesenteric plexus, and the hypogastric plexus.

Parasympathetic (Craniosacral) Division

This portion of the autonomic nervous system is also referred to as the **craniosacral portion** since its preganglionic nerve cell bodies are lo-

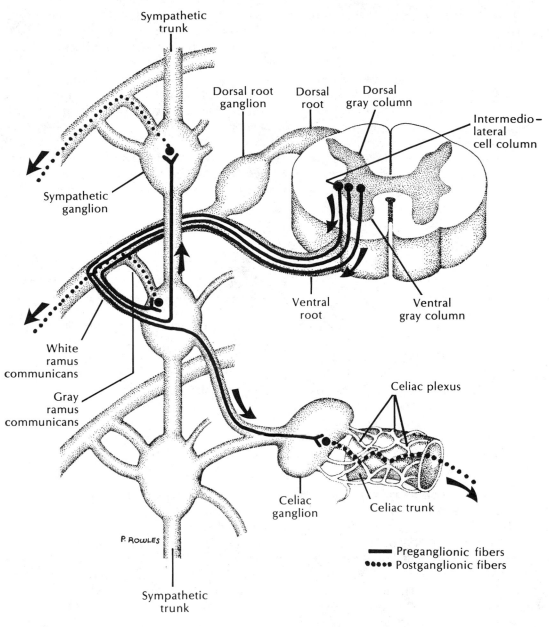

Sympathetic
trunk

Dorsal root
ganglion

Dorsal
root

Dorsal
gray column

Intermedio-
lateral
cell column

Sympathetic
ganglion

Ventral
root

Ventral
gray column

White
ramus
communicans

Gray
ramus
communicans

Celiac plexus

Celiac
ganglion

Celiac trunk

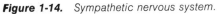

Sympathetic
trunk

—— Preganglionic fibers
•••• Postganglionic fibers

Figure 1-14. *Sympathetic nervous system.*

cated in the brain stem or in the second, third, and fourth sacral segments of the spinal cord.

Processes from the preganglionic cell bodies, located in the sacral region of the spinal cord, leave the cord by way of the ventral root of their respective spinal nerves. From the sacral spinal nerves they form the **pelvic splanchnic nerves** or **nervi erigentes.** Preganglionic fibers carried by the nervi erigentes may unite with sympathetic plexuses along their course toward the postganglionic nerve cell body. The latter cells are widely dispersed on the surface, or in the substance of the organ they innervate. The sacral portion of the parasympathetic system supplies pelvic organs, the descending portion of the colon and the genitalia.

In the brain stem, parasympathetic preganglionic nerve cell bodies are located in the nuclei associated with certain cranial nerves, namely, the oculomotor (III), facial (VII), glossopharyngeal (IX), and vagus nerves (X). Fibers from these cell bodies travel through their respective cranial nerves to parasympathetic ganglia located in the head, except for the vagus nerve.

The Skin

The **skin** or **integument** is a sturdy, elastic, movable envelope that covers the entire body, blending with the sensitive mucous membranes of the mouth, nose, eyes, and anal and urogenital openings. It consists of two distinct layers, the **epidermis,** an outer epithelial investment of closely packed stratified squamous epithelium, and the **dermis,** a deeper layer of dense irregular collagenous fibers, interlaced with blood vessels, nerves, and lymphatics.

Langer's Lines (Lines of Cleavage)
Langer's lines (lines of cleavage) are faint linear clefts in the skin indicative of the direction of the underlying collagen fibers. If a sharp rounded object penetrates the skin, it leaves an elongated slit instead of a circular wound. This is because the collagenous fibers of the dermis of the skin are arranged in parallel rows, which are separated along their length during the injury rather than disrupted. The perceptive surgeon will realize that an incision running parallel to the collagen fibers will heal with only a fine scar. However, if the incision is across the rows, the collagen is disrupted, and the would tends to gape open and to heal in a broad, thick scar. The direction of these rows of collagen—"Langer's lines"—are quite consistent in most individuals, being longitudinal in the limbs and circular in the neck and trunk. These lines are especially evident on the palmer surfaces of the fingers where they run parallel to the long axis of the digit.

Skin is more, however, than a mere protective covering against injury, loss of body fluids, and temperature changes. It is also an extensive sense organ equipped with exquisite nerve endings that inform us of our external environment, for example, pain, temperature, touch, and pressure. The many functions of the skin are summarized by Whitnall,

". . . even with our ingenious modern machinery we cannot create a tough but highly elastic fabric that will withstand heat and cold, wet and drought, acid and alkali, microbic invasion, and the wear and tear of three score years and ten, yet effect its own repairs throughout, and even present a seasonable protection of pigment against sun's rays. It is indeed the finest fighting tissue."

The skin has four appendages—hair, nails, sebaceous (oil) glands, and sweat glands. **Hairs** are dead, keratinized cells that protrude from follicles. The follicles are oblique invaginations of epidermis into the dermis and grow from a bulbous, dilated end, the dermal papilla. **Nails** are hardened, flat thickenings of the dead cells from the outermost layer of the epidermis. Growth occurs at the proximal end of the nail, the nail bed.

Sebaceous glands develop as downgrowths from hair follicles into the dermis, and, therefore, they are nearly always associated with hair. Their oily secretion, sebum, makes the skin essentially waterproof. **Sweat glands** are single, tubular glands present over the entire body, except the lips and glans penis. Their excretory ducts open onto the surface of the skin through small pores. Sweat is an effective coolant of the skin. A sedentary person excretes about 500 ml/day, a hard-working manual laborer, about a liter each hour.

TWO

Superior Extremity

The superior extremity consists of the shoulder, arm, forearm, and hand. The latter is specially adapted for prehension. The muscles of the shoulder and arm act to place the grasping hand in almost any desired position. The rich nerve supply of the fingertips makes the hand a sensitive tactile organ. The muscles of the hand permit complex activity of the digits, which are moved primarily by muscles in the forearm. The superior extremity articulates with the trunk at the small **sternoclavicular joint.** The sternoclavicular joint is a bicameral joint with an intra-articular disc. The superior extremity is firmly anchored to the chest by several muscles that cover the thorax. Thus, the superficial muscles of the trunk (back and pectoral regions) acting upon this member must be considered in any description of the upper extremity.

Superficial Back and Scapular Region

Objectives

At the completion of the study of the superficial back, shoulder, and pectoral regions the student should be able to

▶ *Palpate the surface landmarks of the back and shoulder*

▶ *Give the location, function, nerve supply, and blood supply of the superficial muscles of the back and muscles of the shoulder*

▶ *Define the "rotator cuff"*

▶ *List the component parts of the breast; give its blood supply, innervation, lymphatic drainage, and relationship to superficial and deep fasciae*

▶ *Locate the muscles of the pectoral region and give their origin, insertion, action, innervation, and blood supply*

▶ *Locate the axilla; relate its boundaries to anatomical structures and list its contents*

Surface Anatomy

The most superior structure palpable in the midline of the upper portion of the back is the spinous process of the seventh cervical vertebra, or **vertebra prominens.** Above this level, the spinous processes of the cervical vertebrae lie deep to the **ligamentum nuchae,** while inferiorly all vertebral spinous processes are palpable.

The **spine** of the scapula is subcutaneous through most of its extent, although its medial triangular portion is covered by the trapezius muscle. Laterally it is continuous with the **acromion,** which forms the point of the shoulder. The **medial (vertebral) border** and the **inferior** and **superior angles** of the scapula can be felt deep to the superficial musculature of the back. In a muscular individual, there is a diamond shaped depression between the two scapulae formed by the lack of muscle fibers in the aponeurosis of the trapezius muscle in this area.

Inferolaterally the **crests of the ilia** project as bony ridges below the waist and are palpable posteriorly to the **posteriosuperior iliac spines.** Further inferiorly, in the midline of the lower portion of the back, the posterior surface of the **sacrum** is subcutaneous, and at its inferior extent the **coccyx** can be felt in the cleft between the buttock. On each side the **lumbar triangle (of Petit),** a small area low in the back, is bounded by the crest of the ilium, the anterior border of the latissimus dorsi, and the posterior extent of the external abdominal oblique muscles.

Cutaneous Innervation

The cutaneous innervation to the shoulder and upper pectoral region is derived from **supraclavicular nerves** from the cervical plexus. Inferior to the first intercostal space, cutaneous branches of the intercostal or thoracoabdominal nerves, branches of the ventral rami of spinal nerves, supply the anterolateral body wall. The **intercostal nerves** are ventral rami of spinal nerves. They supply segmental bands of skin over the ribs and intercostal spaces, with branches of a single spinal nerve sending overlapping twigs to adjacent skin areas. From these nerves **lateral cutaneous branches** penetrate the skin near the midaxillary line and divide into anterior and posterior branches. The intercostal nerve then continues anteriorly in the intercostal space and terminates as **perforating branches** emerging just lateral to the sternum to divide into **medial** and **lateral cutaneous twigs.**

Skin over the back of the neck is supplied by cutaneous branches of **dorsal rami** of cervical spinal nerves. Below the shoulder region the dorsal rami of thoracic spinal nerves divide into medial and lateral branches. Above the level of the sixth thoracic vertebra the **medial branch** supplies the skin over the back, while the **lateral branch** is primarily muscular. Below this level the distribution of these branches is reversed.

Superficial Back

Muscles

The muscles of the back are arranged in layers (Fig. 2-1; Table 2-1). Those of the first and second layers, although related topographically to the back, afford attachment of the upper limb to the vertebral column. Except for the trapezius muscle, they are innervated by ventral rami of spinal nerves. These are the functional muscles of the upper extremity. The first and most superficial layer is composed of the trapezius and the latissimus dorsi muscles.

The **trapezius muscle,** with its companion of the opposite side, forms a large trapezoid over the upper portion of the back. Its anterior border in the cervical region gives the sloping contour to the neck and bulges in the action of shrugging the shoulders. Inferiorly, in the midline it overlaps the superior extent of the origin of the latissimus dorsi. With the scapula drawn forward, the lateral border of the trapezius, the superior border of the latissimus dorsi, and the medial border of the scapula bound the **triangle of auscultation,** an area often used in listening to respiratory sounds with the stethoscope. In this position the underlying ribs become essentially subcutaneous.

The **latissimus dorsi muscle** gives the lateral taper to the chest. With the teres major, the latissimus dorsi forms the posterior wall of the axilla, or armpit. Inferiorly, its lateral fibers interdigitate with those of the external abdominal oblique. Its inferior border spirals or turns under in passing to its insertion on the humerus. The latissimus dorsi, acting with the pectoralis major muscle, returns the flexed arm to the anatomic position, as in the action of rowing a boat or climbing a rope.

Deep to the trapezius, a sheet of three rela-

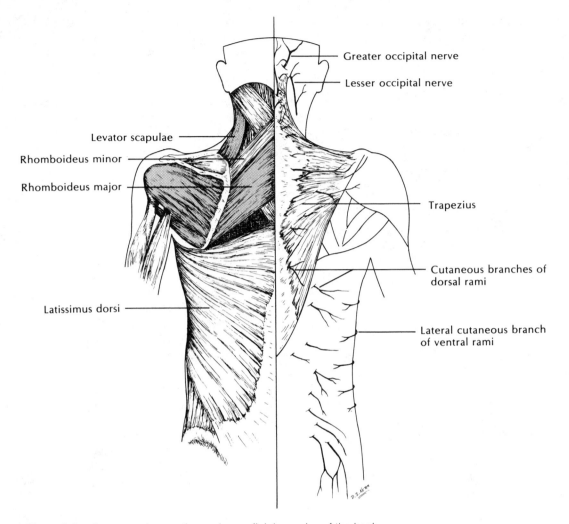

Figure 2-1. *Cutaneous innervation and superficial muscles of the back.*

tively thin straplike muscles, the **levator scapulae** and the **rhomboidei minor** and **major,** insert sequentially into the posterior lip of the medial border of the scapula. The levator scapulae and the rhomboideus minor are usually fused at their insertion and are therefore somewhat difficult to differentiate. These muscles elevate the scapula and draw it toward the midline. The rhomboideus major draws the inferior angle of the scapula superiorly to depress the lateral angle, which assists in adduction of the arm.

A fourth muscle, the **serratus anterior,** inserts onto the anterior lip of the medial border of the scapula. From its origin on the upper eight or nine ribs this muscle follows the contour of the thoracic cage as it passes to its insertion. It acts to

Table 2-1
Muscles of the Superficial Back

Muscle	Origin	Insertion	Action	Nerve
Trapezius	External occipital protuberance, superior nuchal line, ligamentum nuchae, seventh cervical and all thoracic spinous processes	Anterior border of spine of scapula, acromion, and lateral third of posterior border of clavicle	Adducts and rotates scapula; upper part elevates scapula; lower part depresses scapula	Spinal accessory and sensory twigs from third and fourth cervical nerves
Latissimus dorsi	Spinous processes of all vertebrae below sixth thoracic; lumbodorsal fascia, crest of ilium, and lower three or four ribs	Floor of intertubercular groove of humerus	Adducts rotates medially, and draws arm posteriorly	Thoracodorsal
Levator scapulae	Transverse processes of first through fourth cervical vertebrae	Posterior lip of medial border of scapula	Elevates scapula and inclines head	Twigs from cervical plexus and dorsal scapular
Rhomboideus major	Spinous processes of second through fifth thoracic vertebrae	Posterior lip of lower half of medial border of scapula	Adducts and laterally rotates scapula	Dorsal scapular
Rhomboideus minor	Spinous processes of seventh cervical and first thoracic vertebrae	Root of spine of scapula	Adducts and laterally rotates scapula	Dorsal scapular
Serratus anterior	Digitations from lateral surfaces of upper eight ribs	Anterior lip of medial border of scapula	Holds scapula to chest wall; draws scapula anteriorly; and rotates inferior angle laterally	Long thoracic

hold the scapula onto the rib cage and acts in rotation of the scapula. Loss of its action results in a flaring out of the medial border of the scapula (winged scapula).

Fasciae

The superficial fascia over the back has no special characteristic features. The deep fascia, in addition to forming muscular envelopes, specializes in the lower back as the thickened thoracolumbar fascia, which will be described with the deep muscles of the back and the abdominal musculature.

Arteries and Nerves

The arterial supply (Table 2-2) to the trapezius, levator scapulae, and rhomboidei muscles is derived from the transverse cervical artery, a branch of the thyrocervical trunk from the first part of the subclavian artery (Fig. 2-2). The **transverse cervical artery** crosses the posterior triangle of the neck to reach the anterior border of the trapezius. It divides into a **superficial branch,** ramifying on the deep surface of the trapezius, and a **deep branch** (dorsal scapular), passing parallel to the medial border of the scapula, deep to and supplying the levator scapulae and rhomboidei.

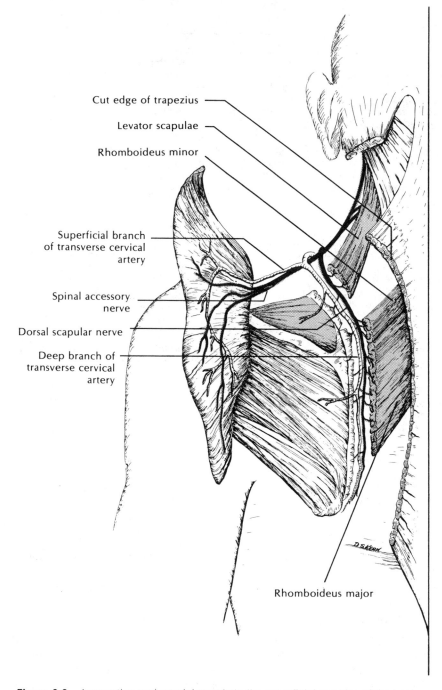

Cut edge of trapezius

Levator scapulae

Rhomboideus minor

Superficial branch
of transverse cervical
artery

Spinal accessory
nerve

Dorsal scapular nerve

Deep branch of
transverse cervical
artery

Rhomboideus major

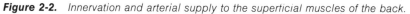

Figure 2-2. *Innervation and arterial supply to the superficial muscles of the back.*

Table 2-2
Arterial Supply to the Superficial Back and Shoulder Region

Artery	Origin	Course	Distribution	Anastomoses
Suprascapular	Thyrocervical trunk	Crosses posterior triangle of neck to scapular notch; passes superficial to transverse scapular ligament	Supraspinatus and infraspinatus	Scapular circumflex, dorsal scapular, superficial branch of transverse cervical
Transverse cervical	Thyrocervical trunk	Crosses posterior triangle of neck to anterior border of trapezius	Trapezius	Suprascapular, scapular circumflex, dorsal scapular
Dorsal scapular	Thyrocervical trunk or third part of subclavian	Crosses posterior triangle of neck to vertebral border of scapula	Levator scapulae, rhomboideus major and minor	Superficial cervical, scapular circumflex
Subscapular	Axillary artery	Descends to divide into scapular circumflex and thoracodorsal	See below	No direct anastomoses
Scapular circumflex	Subscapular	Curves around axillary border to dorsum of scapula	Subscapularis and infraspinatus	Dorsal scapular, suprascapular
Thoracodorsal	Subscapular	Descends vertically to reach latissimus dorsi	Latissmus dorsi	Lateral thoracic, perforating branches of intercostals

Innervation of the trapezius is from the **spinal accessory nerve** together with twigs from the third and fourth cervical nerves; the latter are probably sensory in function. The spinal accessory emerges from the deep surface of the sternocleidomastoideus and crosses the posterior triangle of the neck, passing deep to the trapezius to ramify on the deep surface of the muscle. The levator scapulae and rhomboidei are supplied by the **dorsal scapular nerve,** a branch of the brachial plexus. After piercing the scalenus medius muscle, this nerve passes deep to, and supplies, the levator scapulae and rhomboidei, passing parallel to the medial border of the scapula along with the deep branch of the transverse cervical artery.

The latissimus dorsi receives its blood supply from the **thoracodorsal artery,** a terminal branch of the subscapular from the third part of the axillary. This vessel passes along the lateral border of the scapula to its distribution in the muscle. The nerve supply of the latissimus dorsi is from the **thoracodorsal nerve,** a branch of the posterior cord of the brachial plexus. It passes anterior to the subscapularis and teres major muscles to descend along the lateral border of the scapula to its termination in the muscle.

The serratus anterior is supplied by the **lateral thoracic artery,** a branch of the second part of the axillary. This vessel arises near, and descends along, the lower border of the pectoralis minor muscle to ramify on the superficial surface of the serratus anterior muscle. Innervation of the latter is by the **long thoracic nerve,** a branch of the brachial plexus, which descends posterior to the brachial plexus to follow the lateral thoracic artery into the muscle.

Scapular Region

Muscles

Three muscles, the **supraspinatus, infraspinatus,** and **subscapularis,** originate from, and cover, the three shallow fossae of the scapula (Table 2-3). The tendons of these muscles, as well as the tendon of the teres minor, pass over the glenohumeral articulation deep to the deltoideus muscle to fuse with and strengthen the joint capsule as they insert into the greater and lesser tubercles of the humerus. Their insertions form the **musculotendinous (rotator) cuff** of the shoulder joint. The supraspinatus initiates, then assists the deltoideus in abduction of the arm. The infraspinatus and subscapularis rotate the arm laterally and medially, respectively.

Rotator Cuff

The principal stability of the shoulder joint is not provided by the configuration of its articular surfaces or its ligaments, but rather from several stout tendons that cross the joint. These tendons insert on the greater and lesser tubercles of the humerus and fuse to form a rotator (musculotendinous) cuff. The cuff greatly strengthens the joint, except in its inferior aspect. Dislocations (subluxations) frequently occur inferiorly.

The superficial **deltoideus muscle** forms the lateral mass of the shoulder. From its extensive origin on the clavicle, the acromion, and spine of the scapula, it acts primarily as an abductor, but segments working independently also function in

Table 2-3
Muscles of the Shoulder

Muscle	Origin	Insertion	Action	Nerve
Supraspinatus	Medial two-thirds of supraspinous fossa	Superior portion of greater tubercle of humerus	Initiates abduction of arm and augments deltoid function of abduction	Suprascapular
Infraspinatus	Medial three-fourths of infraspinous fossa	Midportion of greater tubercle of humerus	Main lateral rotator of arm	Suprascapular
Subscapularis	Medial two-thirds of subscapular fossa	Lesser tubercle of humerus	Principal medial rotator of arm; aids in flexion, extension, and adduction of arm	Upper and lower subscapular
Deltoideus	Lateral third of anterior border of clavicle, acromion, and posterior border of spine of scapula	Deltoid tuberosity of humerus	Main abductor of arm; aids in flexion, extension, adduction, and medial and lateral rotation of arm	Axillary (circumflex)
Teres major	Posterior surface of inferior angle and lower portion of lateral border of scapula	Medial lip of intertubercular groove of humerus	Adducts and rotates arm medially	Lower subscapular
Teres minor	Upper portion of lateral border of scapula	Inferior portion of greater tubercle of humerus	Rotates arm laterally and acts as weak adductor of arm	Axillary (circumflex)

adduction, extension, flexion, and internal and external rotation of the arm.

The teres major, teres minor, and long head of the triceps have a lineal origin along the lateral border of the scapula, whereas the omohyoideus arises adjacent to the scapular notch. Note that with respect to muscular attachments to the scapula, all the muscles attaching to the fossae, including the omohyoid attachment at the scapular notch, are attachments of origin. Muscles attaching to the medial border are all insertions, whereas muscles attaching to the lateral border are all origins, including the supraglenoid tubercular attachment of the long head of the biceps. The trapezius and deltoideus attach parallel to each other, to both the spine of the scapula and the clavicle. They form a U-shaped or V-shaped inner insertion for the trapezius and outer origin for the deltoideus.

Arteries and Nerves

The supraspinatus and the infraspinatus muscles are supplied by branches of the same artery and nerve (Fig. 2-3; see Table 2-2). The artery, the **suprascapular,** crosses the posterior triangle of the neck to arrive at the scapular notch. Here it passes over the transverse scapular ligament, which bridges the scapular notch, then terminates as **supraspinatus** and **infraspinatus branches** to these respective muscles. The **suprascapular nerve,** a branch of the brachial plexus, also crosses the posterior triangle of the neck to the scapular notch, but passes deep to the transverse

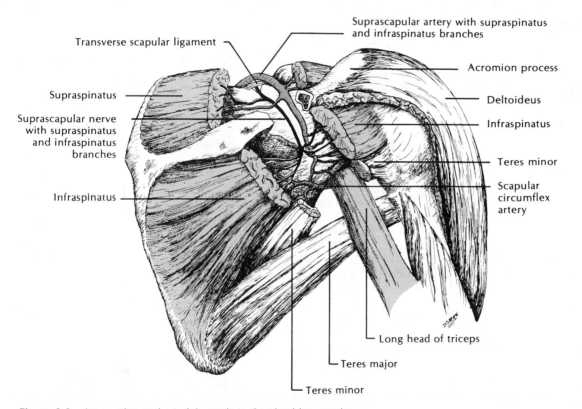

Figure 2-3. *Innervation and arterial supply to the shoulder muscles.*

scapular ligament, after which its distribution follows the arterial supply.

The subscapularis, teres major, and teres minor muscles receive their major blood supply from a single vessel, the **scapular circumflex,** a terminal branch of the subscapular artery from the third part of the axillary. The **lower subscapular nerve,** a branch of the posterior cord of the brachial plexus, innervates the teres major and sends twigs to the subscapularis, which receives additional innervation from the **upper subscapular nerve** of the brachial plexus. The teres minor is supplied by a branch of the **axillary nerve** from the posterior cord of the brachial plexus.

Blood is distributed to the deltoideus muscle via the **anterior** and **posterior humeral circumflex arteries** and small deltoid branches of the thoracoacromial. The humeral circumflex arteries, from the third part of the axillary, pass anterior and posterior, respectively, around the neck of the humerus. The larger posterior humeral circumflex passes through the **quadrangular space** with the **axillary nerve** which innervates the deltoideus.

Scapular Anastomoses
If the distal segment of the subclavian artery or the proximal segment of the axillary artery needs to be ligated, scapular anastomoses provide a route for a collateral blood supply of the upper extremity. Arteries involved in these anastomatic collateral channels are branches of the thyrocervical trunk; the subscapular artery; and other smaller arteries, such as the intercostal, pectoral, lateral thoracic, and thoracodorsal.

Pectoral Region

Surface Anatomy

On the anterior surface of the thorax the clavicle demarcates the chest from the neck. It is subcutaneous along its entire length from the manubrium of the sternum to its articulation with the acromion process of the scapula.

Fracture of the Clavicle
The clavicle serves as a strut between the point of the shoulder (acromion process) and the sternum that allows the arm to swing away from the body. Because of its position, the clavicle transmits forces from the upper extremity to the trunk. If such forces are excessive, as in falling on one's outstretched arm, the clavicle may be fractured—in fact it is the most frequently broken bone in the body.

A depression in the midline at the base of the neck, the **suprasternal notch,** is bounded inferiorly by the superior border of the manubrium and laterally by the tendons of the sternal heads of the sternocleidomastoid muscles. In the midline, a palpable ridge demarcates the **sternal angle (of Louis)** and corresponds to the junction of the manubrium with the body of the sternum. Laterally, this junction affords articulation for the costal cartilage of the second rib and, therefore, can be utilized for accurate superficial determination of rib number or intercostal space. The **nipple** of the breast in the male usually lies over the fourth intercostal space, about a hand's breadth from the midline, but the level is variable in the female. The inferior and lateral margins of the thoracic cage are easily palpable. In the midline the **xiphoid process** can be felt at the inferior extent of the body of the sternum.

Angle of Louis

The angle of Louis (sternal angle) is an important topographical landmark of the anterior thorax. It is a transverse ridge at the junction of the manubrium and body of the sternum that marks the level of articulation of the costal cartilage of the second rib with the sternum. When counting ribs in a physical examination the physician should begin with the second rib at the angle of Louis, since the first rib is behind the clavicle and is difficult to palpate.

Muscles

The **pectoralis major,** a large fan-shaped muscle, covers the anterior chest wall from an extensive origin on the clavicle, sternum, and ribs (Table 2-4, Fig. 2-4). Near its insertion into the humerus it forms the anterior wall of the axilla. Its superior border meets the deltoideus to form the **deltopectoral triangle,** a small triangular depression bounded by the anterior border of the deltoid, the superior border of the pectoralis major, and the midportion of the clavicle. It contains fat, the deltopectoral lymph nodes, the cephalic vein, and the deltoid branch of the thoracoacromial artery.

The **pectoralis minor,** lying immediately deep to the pectoralis major, is a much less extensive muscle and acts in forced respiration by elevating the chest wall. As it passes toward its insertion, it lies superfical to the axillary artery dividing it, for descriptive purposes, into three parts. The small **subclavius muscle** lies deep to the clavicle. In fractures of the clavicle this muscle may afford protection for the deeper lying subclavian vessels and the brachial plexus.

Fasciae

Over the pectoral region, the **superficial fascia** contains abundant fat, especially in the female, where it surrounds the mammary gland and gives the gross configuration to the breast.

Deep to the external investing layer of deep fascia in the pectoral region an additional lamina of deep fascia specializes as the **clavipectoral fascia.** This specialization is attached to the clavicle, encloses the subclavius muscle, fuses to span the gap between the clavicle and the pectoralis minor, then separates to enclose the latter. Lateral to the pectoralis minor, the fascia thickens and fuses to the external investing layer of fascia to form the **suspensory ligament of the axilla,** which passes to the floor of the axilla where it

Table 2-4
Muscles of the Pectoral Region

Muscle	Origin	Insertion	Action	Nerve
Pectoralis major	Clavicular head from medial half of clavicle; sternal head from sternum and costal cartilages; abdominal head from aponeurosis of external abdominal oblique	Lateral lip of intertubercular groove of humerus	Flexes, adducts, and medially rotates arm	Lateral and medial pectorals
Pectoralis minor	Anterior aspect of third, fourth, and fifth ribs	Coracoid process of scapula	Draws scapula inferiorly and elevates ribs	Medial (and lateral) pectoral
Subclavius	Junction of first rib and costal cartilage	Inferior surface of clavicle	Draws clavicle inferiorly and anteriorly	Nerve to subclavius

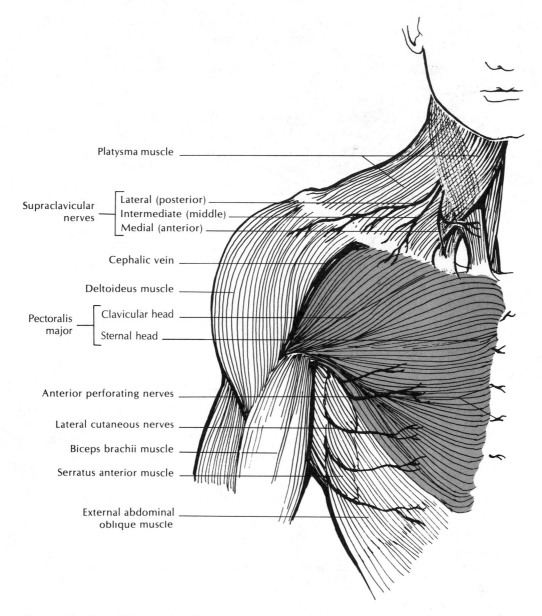

Figure 2-4. *Superficial muscles of anterior chest wall and arm.*

blends with the axillary fascia. Between the subclavius and the pectoralis minor the clavipectoral fascia is pierced by branches of the thoracoacromial artery, the medial and lateral pectoral nerves, the cephalic vein, and lymphatic vessels.

Arteries and Nerves

Blood is supplied to both pectoralis major and minor muscles by **pectoral branches** from the thoracoacromial trunk, the **lateral thoracic,** and **perforating branches** of the anterior intercostal arteries. Additional supply to the pectoralis major is derived from the **perforating branches** of the internal thoracic artery. The **medial** and **lateral pectoral nerves,** branches of the medial and lateral cords of the brachial plexus, respectively, pierce the clavipectoral fascia to innervate the pectoral muscles. The medial pectoral nerve passes through the pectoralis minor to the pectoralis major to supply both muscles. The lateral pectoral nerve courses medial to the pectoralis

Table 2-5
Nerve Distribution to Scapular and Pectoral Regions

Nerve	Origin	Course	Distribution
Supraclaviculars	Third loop of cervical plexus	From posterior border sternocleidomastoid penetrates platysma to run in superficial fascia	Skin over shoulder, clavicle and first intercostal space
Intercostal nerves	Ventral rami of spinal nerves	Traverse intercostal spaces; lateral and anterior cutaneous branches penetrate muscles to reach skin	Skin and muscles of respective intercostal spaces
Spinal accessory	Spinal root from upper five segments of spinal cord; cranial root from brain stem	From jugular forearm bisects posterior triangle to reach anterior border of trapezuis	Trapezius and sternocleidomastoid
Dorsal scapular	Ventral ramus of C_5 (upper root of brachial plexus)	Penetrates scalenus medius to run along deep aspect of vertebral border of scapula	Levator scapulae and rhomboids
Suprascapular	Upper trunk of brachial plexus	Crosses posterior triangle to reach scapular notch	Supraspinatus and infraspinatus
Long thoracic	Roots of brachial plexus (C_5, C_6, C_7)	Descends along lateral thoracic wall	Serratus anterior
Subscapulars	Posterior cord of brachial plexus	From axilla to anterior aspect of subscapularis	Subscapularis; lower subscapular also supplies teres major
Thoracodorsal	Posterior cord of brachial plexus	Runs along anterior border of latissimus dorsi	Latissimus dorsi
Nerve to subclavius	Upper trunk of brachial plexus	Runs on deep surface of subclavius	Subclavius
Axillary	Posterior cord of brachial plexus	Traverses quadrangular space to reach deep surface of deltoid	Deltoid and twigs to teres minor
Lateral pectoral	Lateral cord of brachial plexus	Penetrates clavipectoral fascia to reach deep surface of pectoral muscles	Pectoralis major and minor
Medial pectoral	Medial cord of brachial plexus	Penetrates clavipectoral fascia to reach pectoral muscles	Pectoralis major and minor

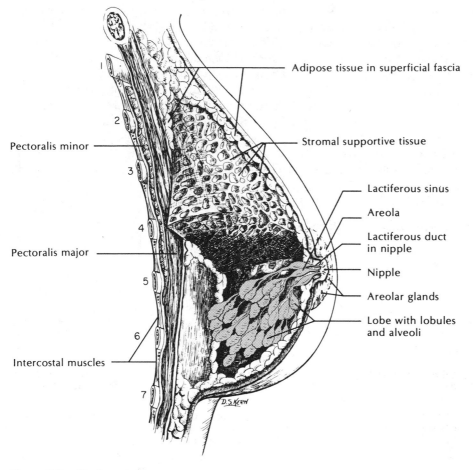

Adipose tissue in superficial fascia

Stromal supportive tissue

Lactiferous sinus

Areola

Lactiferous duct in nipple

Nipple

Areolar glands

Lobe with lobules and alveoli

Pectoralis minor

Pectoralis major

Intercostal muscles

D.S. KERN

Figure 2-5. *The breast.*

minor, usually supplying only the pectoralis major, although there is a communication between the two nerves so, in a technical sense, both can be said to supply both muscles.

The blood supply to the subclavius is the **clavicular branch** of the thoracoacromial artery. The **nerve to the subclavius** is a branch of the upper trunk of the brachial plexus as the latter passes deep to the clavicle (Table 2-5; Table 2-6).

Breast

The female **breast,** a modified sweat gland, is located in the superficial fascia of the pectoral region, where it rests upon the deep fascia covering the pectoralis major muscle (Fig. 2-5). It normally extends between the second and sixth ribs and from the lateral border of the sternum to the axilla.

Table 2-6
Arterial Supply to Pectoral and Region and Axilla

Artery	Origin	Course	Distribution	Anastomoses
Axillary	Continuation of subclavian at lateral border of first rib	Traverses axilla deep to pectoralis minor	Gives six branches in axilla (see below)	No direct anastomoses
Superior thoracic	Axillary	Superficial fascia over first intercostal space	Skin and superficial fascia over first intercostal space	Perforating branches of intercostals
Thoracoacromial trunk	Axillary	Pierces clavipectoral fascia	Divides into four branches to shoulder and pectoral region	No direct anastomoses
Acromial	Thoracoacromial trunk	Ascends toward acromion process	Trapezius, supraspinatus, deltoideus	Deltoid, clavicular
Clavicular	Thoracoacromial trunk	Ascends to subclavius muscle	Subclavius, clavicular head of pectoralis major	Acromial, pectoral
Deltoid	Thoracoacromial trunk	Courses in deltopectoral triangle	Deltoideus, clavicular head of pectoralis major	Clavicular, acromial
Pectoral	Thoracoacromial trunk	Passes between pectoralis major and minor	Pectoral muscles, serratus anterior, breast	Lateral thoracic, thoracodorsal, perforating branches intercostals
Lateral thoracic	Axillary	Descends on superficial aspect of serratus anterior	Serratus anterior, pectoral muscles, latissimus dorsi, and breast	Pectoral, thoracodorsal, perforating branches intercostals
Subscapular	Axillary	Descends along axillary border of scapula	Divides into scapular circumflex and thoracodorsal	No direct anastomoses
Scapular circumflex	Subscapular	Curves around axillary border of scapula	Subscapularis, infraspinatus	Suprascapular, dorsal scapular
Thoracodorsal	Subscapular	Descends to reach latissimus dorsi	Latissimus dorsi, serratus anterior, external abdominal oblique	Lateral thoracic, pectoral
Anterior and posterior humeral circumflexes	Axillary	Curves around respective aspects of surgical neck of humerus	Deltoideus	With each other and deltoid branch of profunda brachii

Axillary Tail (of Spence)

The axillary tail is a rather constant extension of the breast tissue through an opening in the axillary fascia into the anteromedial part of the axilla. Thus, the axillary tail is not enclosed by the superficial fascia like the rest of the mammary gland. Breast tissue extends along the anterior axillary line beyond where the edge of the breast is normally seen.

The breast consists of glandular tissue, stroma, and fat. The **mammary gland** has no distinct capsule and is composed of fifteen to twenty **lobes** radiating from the **nipple.** Each lobe has a single **lactiferous duct** that converges toward the **areola.** It dilates near its termination to form a secretory reservoir, the **lactiferous sinus,** then constricts to form a duct that opens individually on the surface of the nipple. The stroma of the mammary gland consists of fibrous connective tissue that loosely envelops the entire gland, extending into the gland to enclose the parenchyma. Condensations of the collagenous fibers form the **suspensory ligaments** (**of Cooper**), which extend from the skin through the mammary gland to the underlying deep fascia.

Physical Signs of Cancer of the Breast

Cardinal physical signs of cancer of the breast can be recognized by the patient. Any palpable mass in the breast may be malignant. When cancer cells invade the gland and enlarge, they often attach to the suspensory ligaments of Cooper (retinaculum cutis) and produce shortening of the ligaments causing depression or dimpling of the overlying skin. If the neoplasm attaches to and shortens the lactiferous ducts, the nipple may become retracted or inverted. When advanced cancer of the breast invades the deep fascia of the pectoralis major muscle, contraction of this muscle causes a sudden upper movement of the whole breast.

The arterial supply to the breast is from **pectoral branches of the thoracoacromial,** the **lateral thoracic,** perforating branches of the **internal thoracic,** and the **anterior intercostal** arteries.

Lymphatic drainage from **subareolar** and **circumareolar plexuses** is continuous with the general cutaneous drainage of the thoracic region and may drain toward the neck or the abdomen. **Perilobular** and **interlobular lymphatic plexuses** in the breast proper drain the deeper tissue and communicate with the subareolar plexus. The deeper drainage of the breast is regional. The lateral half of the mammary gland drains to the axillary and pectoral nodes. Five sets of lymph nodes are present in the axilla including the lateral, medial, anterior, and posterior nodes related to the walls of the axilla; and the central or apical nodes at the apex. Lymph drainage from the medial side passes to nodes along the internal thoracic artery or may cross the midline to the opposite breast. Inferiorly the lymph may flow toward the abdomen and drain into the nodes in the upper portion of the abdomen.

Cancer of the Breast

Cancer of the breast is the most common metastatic neoplasm in women. It spreads by direct invasion into adjacent tissues, but principally, the cancer cells travel through lymphatics and blood vessels to many parts of the body. The most common pathway is along the lymphatics leading to the axillary lymph nodes. To remove the breast carcinoma with a wide margin of tissue for the direct invasion spread and with the axillary lymph nodes for lymphatic metastases, radical mastectomy may be performed. It includes removal of the entire breast with the underlying fascia, pectoralis major and minor muscles, and all of the lymph nodes of the axilla.

Arm and Forearm

Objectives

*After completion of the study of the arm
and the forearm the student should be
able to*

▶ *Palpate surface landmarks of the arm,
forearm and hand*

▶ *Draw and label the brachial plexus, give
the general distribution of its branches,
and describe a dermatome*

▶ *Describe the muscular compartments of
the arm; list the muscles in each
compartment and give the innervation of
each muscle*

▶ *List the branches of the axillary and
brachial arteries that supply structures in
the arm*

▶ *List the flexor and extensor muscles in
the forearm and give their general
location and action*

▶ *Define the extensor and flexor
retinaculae; describe their role and
relationships*

▶ *Give the distribution of the radial,
median, and ulnar nerves in the forearm*

▶ *Describe the course and give the
branches of the radial and ulnar arteries
in the forearm*

Surface Anatomy

The **biceps brachii muscle** makes the anterior, and the **triceps** the posterior, bulge on the arm. Medial and lateral intermuscular septa are extensions of the external investing fascia of the arm that separate the flexor (anterior) and extensor (posterior) compartments. These septa pass deeply to attach to the humerus and are responsible for the **superficial grooves** on either side of the arm. At the elbow, the **medial** and **lateral** epicondyles are easily palpable. The **olecranon process** of the ulna forms the posterior prominence of the elbow. Pressure on the medial side of the olecranon process elicits a tingling sensation ("funny bone"), which demonstrates the superficial position of the ulnar nerve passing along the ulnar groove on the medial epicondyle.

The **antecubital fossa** forms a triangular depression anterior to the elbow joint. The **tendon of the biceps** can be palpated within its boundaries. The **lacertus fibrosus** (bicipital aponeurosis), a strong band of fibrous tissue, passes inferomedially from the tendon to the deep fascia of the forearm. Forming a connection between the basilic and cephalic veins, the **median cubital vein** can be seen crossing the fossa superficially. Just distal to the elbow, the increased width of the forearm results from a massing of the bellies of the **flexor muscles** of the wrist and fingers **medially,** and the **extensor group laterally.**

The posterior border of the **ulna** is subcutaneous along its entire length, as are its prominent head and styloid process at the wrist. The radius is palpable in its distal half and its **styloid process** can be felt at the lateral side of the wrist. Extension of the thumb results in a prominent ridge on the dorsum of the wrist formed by the **tendon of the extensor pollicis longus** muscle. With the thumb extended, this tendon, with that of the **abductor pollicis longus** and the **extensor pollicis brevis,** forms a lateral depression, the "anatomic snuff box." The radial artery courses along the floor of the "snuff box." When the wrist is flexed against resistance its palmar aspect reveals, in the midline, the **tendon of the palmaris longus,** and about a centimeter laterally, the **tendon of the flexor carpi radialis.**

The palmar aspect of the hand reveals transverse creases at the metacarpophalangeal joints. Note that the webbing of the fingers is distal to these articulations. The **thenar eminence** (ball

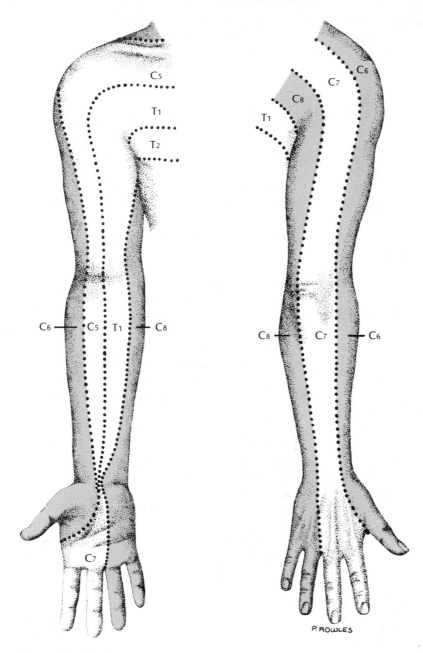

Figure 2-6. *Spinal nerve dermatomes of upper extremity.*

of the thumb) is formed by the small muscles to the thumb, the **hypothenar eminence** (heel of the hand) by the small muscles to the little finger. The interval between the thumb and the index finger contains the adductor muscle of the thumb anteriorly and the first dorsal interosseous muscle posteriorly. The skin of the palm is thickened by subcutaneous fat, and is firmly bound to deeper structures. Over the dorsum of the hand the skin is more delicate and freely movable. Upon maximal extension of the wrist and fingers, the tendons of the **extensor digitorum muscle** form prominent ridges on the dorsum of the hand.

Cutaneous Innervation

The cutaneous innervation to the arm, forearm, and hand is supplied by branches of the brachial plexus with some contribution from the cervical plexus (Figs. 2-6, 2-7, 2-8). The branches from the cervical plexus are the **supraclavicular nerves,** which supply skin over the upper portion of the deltoideus muscle.

Four cutaneous nerves are distributed to the skin of the arm. The **lateral brachial cutaneous,** a branch of the axillary nerve, supplies skin over the lower half of the deltoideus and the long head of the triceps. From the radial nerve the **posterior brachial cutaneous** innervates skin on the posterior aspect of the arm below the deltoideus. The **medial brachial cutaneous,** from the medial cord, is distributed to the posteromedial aspect of the lower third of the arm. The arm receives additional cutaneous innervation from the **intercostobrachial,** the lateral cutaneous branch of the second thoracic (intercostal) nerve, supplying the posteromedial surface of the arm from the axilla to the olecranon process.

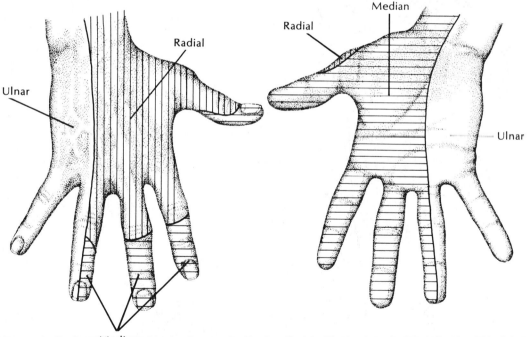

Figure 2-7. *Dermatomes of hand.*

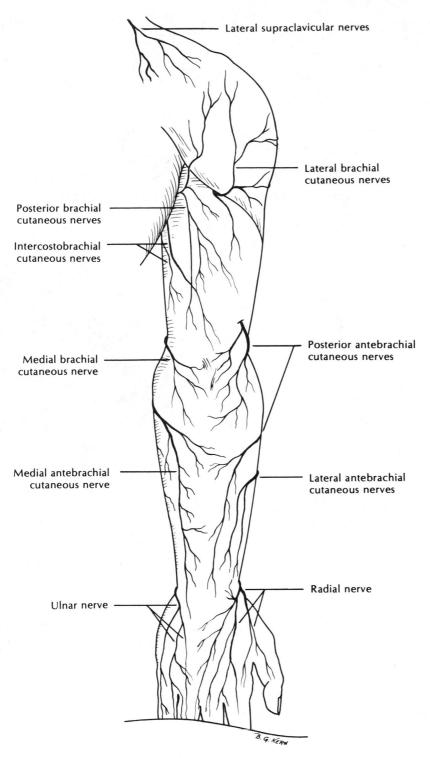

Lateral supraclavicular nerves

Lateral brachial cutaneous nerves

Posterior brachial cutaneous nerves

Intercostobrachial cutaneous nerves

Medial brachial cutaneous nerve

Posterior antebrachial cutaneous nerves

Medial antebrachial cutaneous nerve

Lateral antebrachial cutaneous nerves

Radial nerve

Ulnar nerve

B. G. KERN

Figure 2-8. Cutaneous innervation of the dorsal aspect of the superior extremity.

Cutaneous branches to the forearm include the **lateral antebrachial cutaneous,** a continuation of the musculocutaneous nerve, giving anterior and posterior branches to the radial half of the forearm; the **posterior antebrachial cutaneous,** from the radial nerve, supplying an upper branch to the distal half of the anterolateral aspect of the arm and a lower branch to the mid-dorsum of the forearm, and the **medial antebrachial cutaneous,** a branch of the medial cord, giving anterior and posterior branches to the medial aspect of the forearm.

The ulnar, median, and radial nerves all contribute to the cutaneous innervation of the hand. The cutaneous branches of the **ulnar nerve** supply both surfaces of the hand and fingers medial to a line passing through the midline of the ring finger. The **median nerve** gives cutaneous branches to the remainder of the palmar surface of the hand and fingers, and to the dorsal surface of the fingers distal to the middle phalanx. Cutaneous branches from the **radial nerve** supply the remainder of the dorsal surface of the hand and fingers, as well as a small area over the lateral aspect of the thenar eminence. It is interesting to note that the ring finger receives cutaneous innervation from all three nerves that supply the hand, namely, the radial, ulnar, and median.

Dermatome

By definition, a dematome is the area of skin supplied by a spinal nerve. For example, in the upper extremity the medial aspect of the forearm and hand is supplied by the eighth cervical spinal nerve. However, in the extremities specific areas of skin are also supplied by terminal branches of plexuses. Thus, distribution of the ulnar is limited to the medial aspect of the hand. By testing sensory loss of respective areas supplied by spinal nerves or plexus branches, the physician can make a tentative diagnosis of the site of a lesion to a nerve.

Venous Drainage

The superficial venous drainage of the arm, forearm, and hand begins as **palmar** and **dorsal digital veins** on the respective surfaces of the digits (Fig. 2-9). These veins join to form the **dorsal metacarpal veins,** which anastomose to form the **dorsal venous arch** lying proximal to the heads of the metacarpal bones.

The **cephalic vein** is the lateral continuation of the dorsal venous arch. It ascends along the radial side of the forearm to the antecubital fossa. Above the elbow it continues in the lateral bicipital groove, then follows the interval between the deltoideus and the pectoralis major muscles into the deltopectoral triangle. It terminates by perforating the clavipectoral fascia and drains into either the axillary or subclavian vein.

The **basilic vein** is the medial continuation of the dorsal venous arch. It ascends on the ulnar side of the forearm, receiving tributaries from both the anterior and posterior surfaces. In the arm it ascends a short distance in the medial bicipital groove, then penetrates the brachial fascia and unites with the brachial vein to form the axillary vein.

In the antecubital fossa, the **median cubital vein** passes obliquely across the fossa connecting the cephalic and basilic veins. It may conduct the bulk of the blood from the cephalic to the basilic vein. The median cubital is the vein most commonly used for venipuncture.

The deep venous drainage of the upper extremity originates from **deep venous arcades** of the hand, which parallel the arterial arches. The radial, ulnar, and brachial arteries have **venae comitantes,** which receive blood from the areas supplied by these vessels. Deep and superficial veins communicate extensively with each other. The veins accompanying the radial and ulnar arteries join to form the venae comitantes of the brachial artery, which unite with the basilic vein to form the axillary vein.

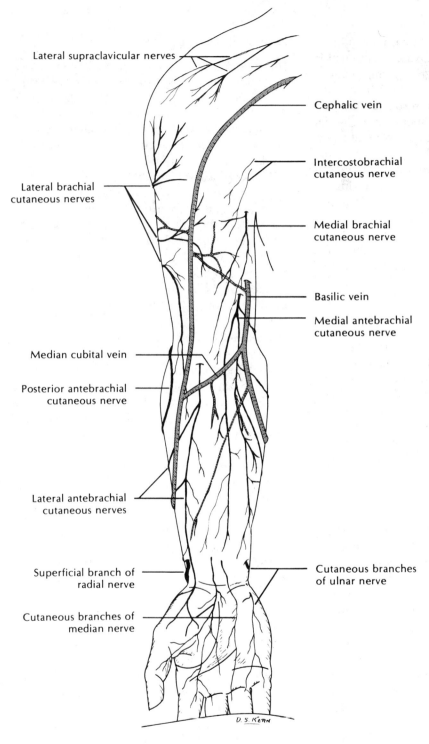

Lateral supraclavicular nerves

Cephalic vein

Intercostobrachial cutaneous nerve

Lateral brachial cutaneous nerves

Medial brachial cutaneous nerve

Basilic vein

Medial antebrachial cutaneous nerve

Median cubital vein

Posterior antebrachial cutaneous nerve

Lateral antebrachial cutaneous nerves

Superficial branch of radial nerve

Cutaneous branches of ulnar nerve

Cutaneous branches of median nerve

D. S. KERN

Figure 2-9. *Superficial venous drainage and cutaneous innervation of the volar aspect of the superior extremity.*

Lymphatic Drainage

The **superficial lymphatics** of the upper extremity begin as a meshwork around the fingers as the **digital lymphatic plexus.** It is drained by lymph vessels following the digital arteries that turn onto the dorsum of the hand where they form the **dorsal plexus.** The radial half of this plexus drains along the radial side, and the ulnar half along the ulnar side of the forearm. Lymphatics draining the palm pass to the sides of the hand to join either ulnar or radial channels, or may drain along channels that ascend in the midline of the volar aspect of the forearm. One or two **cubital lymph nodes,** located just above the medial epicondyle, are interposed in the ulnar channels. Their efferent vessels accompany the basilic vein to the **axillary lymph nodes.** The radial and posterior lymph channels of the forearm follow the cephalic vein to terminate in the **central nodes** of the axilla. Efferents of the central nodes form the subclavian lymph trunk. A **deltopectoral node** may be interposed before they reach the axilla.

The **deep lymphatic channels** parallel the arteries in the hand and forearm and drain into five or six small nodes in the cubital fossa. Efferent vessels accompany the brachial veins and ascend to drain into the **lateral and central groups of axillary nodes.**

Arm

Axilla

The axilla is a pyramidal-shaped area. It consists of four walls, an apex, and a base. The **anterior wall** is formed by the pectoralis major and minor muscles and the clavipectoral fascia; the **posterior wall** by the subscapularis, teres major, and latissimus dorsi muscles; the **medial wall** by the serratus anterior muscle, the first five ribs, and the intercostal muscles; and the **lateral wall** by the upper medial surface of the humerus. The **base** is formed by the axillary fascia. The truncated **apex,** directed superomedially, is bounded anteriorly by the clavicle, medially by the first rib, and posteriorly by the superior border of the scapula. Axillary vessels and the divisions of the brachial plexus traverse the apex of the axilla in their course from the neck to the upper extremity. They are surrounded by an extension of scalene (prevertebral) fascia, the cervicoaxillary sheath. Infections may follow this tubular elongation in passing between the neck and axilla. The axilla contains the axillary artery and vein, most of the brachial plexus, axillary lymph nodes, fat, and connective tissue.

Brachial Plexus Lesions

The upper-arm birth palsy (Erb–Duchenne paralysis) is the most common type of nerve injury during childbirth. It is caused by forcible widening of the angle between the head and the shoulder, which may occur from pulling on the head at birth or using forceps to rotate the fetus in utero. It may be caused in later life by falling on the shoulder. The site of injury is the junction of C_5 and C_6 as they form the upper trunk of the brachial plexus. This is called Erb's point. The injury causes paralysis of the abductors and lateral rotators of the shoulder, and the flexors of the elbow. Weakness is also noted in the adductors and medial rotators of the shoulder. The arm hangs at the side in internal rotation with the forearm pronated and the fingers and wrist flexed—the porter's tip hand.

Lower arm injuries (Klumpke's paralysis) occur when the arm is forcefully stretched upward, causing damage to the lower trunk (C_8 and T_1). It is also seen in scalenus anticus syndrome. Since these nerves supply most of the intrinsic muscles of the hand, the injury results in a claw-hand appearance, similar to an ulnar nerve injury.

Brachial Plexus

The **brachial plexus** is a network of nerves derived from the **ventral rami of the fifth through eighth cervical** (C₅ through C₈) and **first thoracic** (T₁) **nerves** (Fig. 2-10). It receives contributions from the fourth cervical (C₄) and second thoracic (T₂) nerves. The brachial plexus supplies muscular, sensory, and sympathetic fi-

bers to the upper extremity. The plexiform arrangement permits intermingling of nerve components from several segments of the spinal cord to form composite nerves that supply individual structures. It is composed of five roots, three trunks, six divisions, three cords, and sixteen named branches. It is located partly in the neck, under the clavicle, and in the axilla.

The **roots** of the brachial plexus are continua-

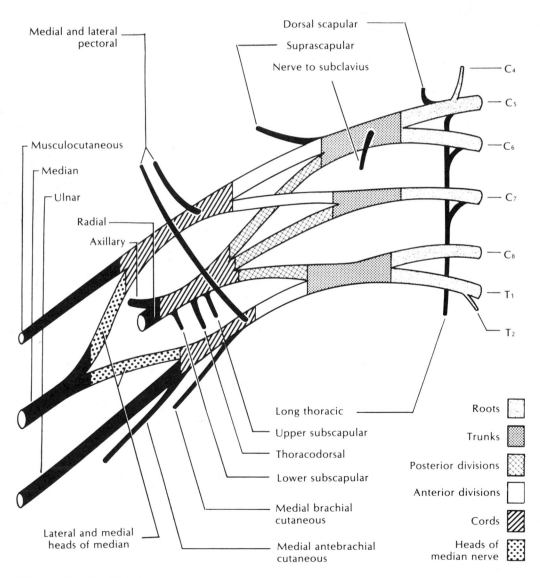

Figure 2-10. *Brachial plexus.*

tions of the ventral rami of its component spinal nerves, which emerge between the scalenus anterior and medius muscles in line with similar but more cranially situated roots constituting the cervical plexus. The roots of C_5 and C_6 unite to form the **upper trunk,** C_7 becomes the **middle trunk,** and C_8 and T_1 constitute the **lower trunk.** Each trunk divides distal to the scalene gap into **anterior** and **posterior divisions,** which anastomose to form lateral, medial, and posterior cords, so named for their relation to the second part of the axillary artery. The **posterior cord** is formed by the union of the posterior divisions of all three trunks; the anterior divisions of the upper and middle trunks form the **lateral cord;** and the **medial cord** is the continuation of the anterior division of the lower trunk.

Two branches are derived from the roots, the dorsal scapular and the long thoracic. The **dorsal scapular** (C_5) pierces the scalenus medius to course parallel to the medial border of the scapula and deep to the levator scapulae and the rhomboids to supply these muscles. The **long thoracic** (C_5, C_6, and C_7) descends posterior to the roots of the plexus to ramify on and supply the serratus anterior muscle.

Two nerves are also derived from the upper trunk. The **nerve to the subclavius** (C_5 and C_6) descends anterior to the plexus and posterior to the clavicle to supply the subclavius muscle. The **suprascapular** (C_5 and C_6) crosses the posterior triangle of the neck to the scapular notch, where it divides into **supraspinatus and infraspinatus branches** to supply muscles in their respective scapular fossae.

Two nerves originate from the lateral cord. The **lateral pectoral** (C_5, C_6, and C_7) pierces the clavipectoral fascia to supply the pectoralis major. The **musculocutaneous** (C_5, C_6, C_7) enters the coracobrachialis, supplying it and the two other flexor muscles in the arm, then continues into the skin on the lateral side of the forearm as the **lateral antebrachial cutaneous nerve.** The **lateral head of the median nerve** is also derived from the lateral cord. It joins with the **medial head** from the medial cord to form the **median nerve**

(C_5, C_6, C_7, C_8, and T_1), which supplies flexor muscles in the forearm, most of the short muscles of the thumb, and skin of the lateral two-thirds of the palm of the hand and the fingers.

In addition to the medial head of the median nerve, the medial cord gives rise to four nerves. The **medial pectoral** (C_8 and T_1) passes through and supplies the pectoralis minor as well as the overlying pectoralis major. The **medial brachial cutaneous** (C_8 and T_1) supplies skin over the medial and posterior aspect of the distal third of the arm. The **medial antebrachial cutaneous** (C_8 and T_1) innervates skin of the medial and posterior aspect of the forearm. The largest branch, the **ulnar** (C_8 and T_1) is distributed to some of the flexors in the forearm, most of the intrinsic muscles of the hand, the skin of the medial side of the hand, all the skin of the little finger, and the skin of the medial half of the ring finger.

The posterior cord gives origin to the **upper subscapular nerve** (C_5 and C_6), which passes posteriorly to enter the subscapularis muscle; the **lower subscapular** (C_5 and C_6), which descends to supply the subscapularis and terminates in the teres major; and the **thoracodorsal** (C_5, C_6, and C_7), which descends anterior to the subscapularis and teres major to terminate in the latissimus dorsi. Other branches include the **axillary** (C_5 and C_6), which passes posteriorly through the quadrangular space and innervates the deltoideus and teres minor, and the **radial** (C_5, C_6, C_7, C_8, and T_1), which courses posterior to the axillary artery to innervate the extensors of the arm and forearm. It also supplies skin on the posterior aspect of the arm and forearm, the lateral two-thirds of the dorsum of the hand, and the dorsum of the lateral three and one-half digits over the proximal and intermediate phalanges.

Axillary Artery

As a continuation of the **subclavian,** the **axillary artery** extends from the lateral border of the first rib to the lower border of the teres major, where it becomes the **brachial artery** (Fig. 2-11). For descriptive purposes, it is subdivided into **three**

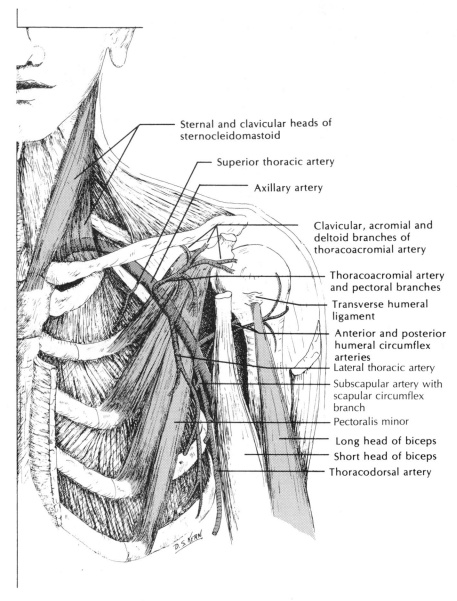

Sternal and clavicular heads of sternocleidomastoid

Superior thoracic artery

Axillary artery

Clavicular, acromial and deltoid branches of thoracoacromial artery

Thoracoacromial artery and pectoral branches

Transverse humeral ligament

Anterior and posterior humeral circumflex arteries

Lateral thoracic artery

Subscapular artery with scapular circumflex branch

Pectoralis minor

Long head of biceps

Short head of biceps

Thoracodorsal artery

Figure 2-11. *Axillary artery.*

parts by the overlying pectoralis minor muscle. The first portion lies proximal, the second deep, and the third distal to the muscle.

Six branches originate from the axillary artery, one branch from the first segment, two from the second, and three from the third. The single branch from the first part, the **superior thoracic** (highest thoracic), is distributed to superficial fascia over the first intercostal space.

The branches from the second portion are the thoracoacromial artery and the lateral thoracic. The **thoracoacromial artery** pierces the clavipectoral fascia to divide into four branches. The **acromial branch** passes laterally over the coracoid process, gives twigs to the deep surface of the deltoideus, and finally ramifies around the acromial process. The **deltoid branch** courses in the deltopectoral triangle to supply the deltoideus and pectoralis major. The **pectoral branches** pass between and supply both pectoral muscles and, in the female, distribute deep branches to the mammary gland, and the **clavicular branch** supplies the sternoclavicular joint and the subclavius muscle. The **lateral thoracic artery** arises near and descends along the lower border of the pectoralis minor, supplying the two pectoral muscles and the serratus anterior. In the female, it sends branches to the breast.

Branches from the third part of the axillary artery include the smaller **anterior** and the larger **posterior humeral circumflex arteries,** which encircle the neck of the humerus to supply the deltoideus and anastomose with the ascending branch of the profunda brachii artery. The largest branch of the axillary, the **subscapular artery,** arises opposite the lower border of the subscapularis muscle. It divides into the **scapular circumflex artery,** which passes through the triangular space to supply muscles on the dorsum of the scapula, and the **thoracodorsal branch,** which continues along the lateral border of the scapula to supply the latissimus dorsi.

Table 2-7
Muscles of the Arm

Muscle	Origin	Insertion	Action	Nerve
Biceps brachii	Long head, supraglenoid tubercle; short head, tip of coracoid process	Tuberosity of radius and antebrachial fascia via the bicipital aponeurosis	Flexes forearm and arm; supinates hand	Musculocutaneous
Coracobrachialis	Tip of coracoid process	Middle third of medial surface of humerus	Flexes and adducts arm	Musculocutaneous
Brachialis	Distal two-thirds of anterior surface of humerus	Coronoid process and tuberosity of ulna	Flexes forearm	Musculocutaneous and small branch of radial
Triceps brachii	Long head, infraglenoid tubercle; lateral head, posterior surface and lateral border of humerus; medial head, posterior surface of distal half of humerus	Posterior aspect of olecranon process of ulna	Extends forearm; long head aids in extension and adduction of arm	Radial
Anconeus	Lateral epicondyle of humerus	Lateral aspect of olecranon and upper fourth of posterior surface of ulna	Acts as weak extensor of forearm	Radial

Collateral Circulation

Collateral circulation of the axillary and brachial arteries will depend upon the level of ligation of the artery. For example, if the ligation of the axillary artery is distal to the humeral circumflex and subscapular branches, the blood flow in the limb is re-established through an anastomosis between these branches and the profunda brachii artery. If the ligation of the brachial artery is distal to the profunda brachii and the superior ulnar collateral arteries, these vessels will establish circulation distally, with the inferior ulnar collateral, radial, ulnar, and interosseous recurrent arteries.

Muscles of the Arm (Table 2-7)

The four muscles of the arm, three flexors and one extensor, are located in flexor (anterior) and extensor (posterior) compartments delineated by the lateral and medial intermuscular septa (Fig.

2-12). In the flexor compartment the fusiform **biceps brachii** originates by two heads: a short head (in common with the medially situated **coracobrachialis muscle**) from the coracoid process, and a long head from the supraglenoid tubercle of the scapula. Lying in the intertubercular (bicipital) groove, the tendon of the long head passes deep to the transverse humeral ligament, where it acquires a synovial sheath as it traverses the joint cavity of the shoulder to its origin on the scapula. The biceps is a powerful supinator as well as a flexor of the forearm. The **brachialis**, the strongest flexor of the forearm, covers the lower anterior half of the humerus and the capsule of the elbow joint. It forms a bed for the more superficially placed biceps.

The **triceps brachii** fills the posterior compartment. Its long and lateral heads obscure the more deeply placed medial head. Both biceps and triceps cross the elbow and shoulder joints, and hence, act upon both. The small triangular **anconeus muscle** is located superficially at the lateral aspect of the elbow.

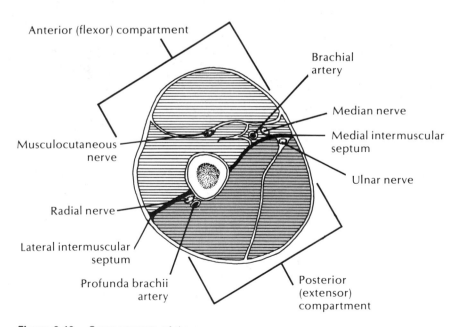

Figure 2-12. *Compartments of the arm.*

Arteries and Nerves in the Arm
(Fig. 2-13; Tables 2-8 and 2-9)

The **brachial artery,** the continuation of the axillary, begins at the lower border of the teres major, passes obliquely from a medial to a midline position in the arm, and terminates in the antecubital fossa by dividing into the radial and ulnar arteries. It may be palpated throughout its entire course.

Proximally it lies on the muscular septum and medial head of the triceps; distally it is located on the brachialis muscle medial to the tendon of the biceps and lateral to the median nerve.

Arterial Blood Pressure
Arterial blood pressure is routinely determined by using a sphygmomanometer (an inflatable cuff), a pressure recorder, a ma-

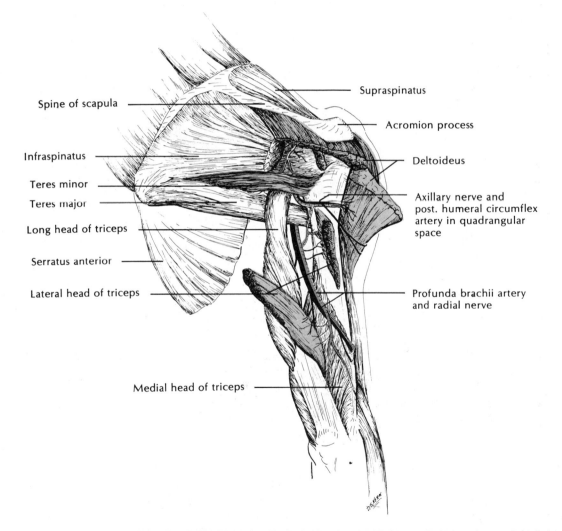

Spine of scapula

Infraspinatus

Teres minor

Teres major

Long head of triceps

Serratus anterior

Lateral head of triceps

Medial head of triceps

Supraspinatus

Acromion process

Deltoideus

Axillary nerve and post. humeral circumflex artery in quadrangular space

Profunda brachii artery and radial nerve

Figure 2-13.   *Innervation and arterial supply to the muscles on the posterior aspect of the arm.*

Table 2-8
Nerve Distribution to the Arm

Nerve	Origin	Course	Distribution
Lateral brachial cutaneous (upper)	Axillary	From deep to deltoid, winds around posterior border of muscle to reach the skin	Skin over the deltoid
Lateral brachial cutaneous (lower)	Radial	Pierces lateral intermuscular septum to reach skin of lower portion of arm	Skin over lateral and posterior aspect of arm below deltoid
Posterior brachial cutaneous	Radial	From axilla penetrates deep fascia to reach skin	Skin on posterior aspect of proximal third of arm
Medial brachial cutaneous	Medial cord of brachial plexus	From axilla penetrates deep fascia to reach skin	Skin on medial aspect of distal third of arm
Intercostobrachial cutaneous	Lateral cutaneous branch of second intercostal nerve	From intercostal space passes through axilla to reach skin	Skin on medial aspect of proximal portion of arm
Axillary	Posterior cord of brachial plexus	Passes through quadrangular space to reach deep surface of deltoid	Deltoid and twigs to teres minor
Musculocutaneous	Lateral cord of brachial plexus	Pierces coracobrachialis and emerges between biceps and brachialis to continue as lateral antebrachial cutaneous	Flexor muscles in the arm—coracobrachialis, biceps, and brachialis
Radial	Posterior cord of brachial plexus	Follows radiospiral groove between lateral and medial heads of triceps; pierces lateral intermuscular septum and continues to forearm	In arm, supplies triceps, brachioradialis, extensor carpi radialis longus, anconeus, and twigs to brachialis

nometer, and a stethoscope. The cuff is placed around the midarm and inflated with air, which compresses the brachial artery against the humerus, thus occluding the vessel. A stethoscope is placed over the artery below the cuff and the air is slowly released. At the first sound of blood flowing through the artery the pressure level is noted on the manometer as the systolic pressure. As the cuff is further deflated a murmur develops. The pressure just before this sound disappears is the diastolic pressure.

Four named branches arise from the brachial artery. The **profunda brachii** (deep brachial), accompanies the radial nerve in the radiospiral groove of the humerus. Posteriorly this vessel divides into an ascending branch that courses proximally between the lateral and long heads of the triceps, and a descending branch that runs with the medial head of the triceps to the posterior aspect of the elbow. The small **nutrient artery** of the humerus arises about the middle of the arm to enter the nutrient canal on the anteromedial aspect of the bone. The **superior ulnar collateral branch** begins about the middle of the arm,

Table 2-9
Arterial Supply to the Arm and Forearm

Artery	Origin	Course	Distribution	Anastomoses
Brachial	Continuation of axillary at distal border of teres major	Passes along medial aspect of the humerus to antecubital fossa	Musculature of the arm	No direct anastomoses
Profunda brachii	Brachial	Curves around humerus in radiospiral groove	Triceps and deltoideus	Anterior and posterior humeral circumflex; radial and common interosseous recurrent
Nutrient	Brachial	Through nutrient foramen, shaft of humerus	Marrow cavity of humerus	No direct anastomoses
Superior and inferior ulnar collaterals	Brachial	Extend distally to cross elbow joint	Muscles crossing elbow joint	Around elbow joint
Radial	Brachial at antecubital fossa	Courses deep to brachioradialis to reach the wrist	Extensor muscles in forearm	Ulnar artery in the hand
Ulnar	Brachial at antecubital fossa	Passes through forearm between flexor carpi ulnaris and flexor digitorum profundus	Flexor muscles in forearm	Radial artery in the hand
Recurrent interosseous	Common interosseous	Passes proximally to cross elbow joint	Muscles crossing elbow joint	Around elbow joint
Radial recurrent	Radial	Passes proximally to cross elbow joint	Muscles crossing elbow joint	Around elbow joint
Anterior and posterior ulnar recurrents	Ulnar	Pass proximally on respective aspects of elbow joint	Muscles crossing elbow joint	Around elbow joint
Common interosseous	Ulnar	Short branch terminates as anterior and posterior interossei	Divides into anterior and posterior interossei	No direct anastomoses
Anterior and posterior interossei	Common interosseous	Pass to respective aspects of interosseous membrane	Supply deep flexor and deep extensor muscles, respectively	With each other; carpal branches of radial and ulnar
Carpals	Both radial and ulnar arteries proximal to wrist joint	Form carpal arch	Tendons crossing wrist	With each other

pierces the medial intermuscular septum, accompanies the ulnar nerve, and sends branches to either side of the medial epicondyle. The **inferior ulnar collateral artery** arises about an inch proximal to the medial epicondyle. It divides into a posterior branch that pierces the intermuscular septum to descend deep to the triceps, and an anterior branch that passes inferiorly between the biceps and the brachialis muscles.

The **median nerve** arises from the brachial plexus lateral to the axillary artery, descends, crossing the brachial artery in the midarm, then proceeds into the cubital fossa where it then lies medial to the artery. The **ulnar nerve,** the terminal branch of the medial cord, courses distally medial to the brachial artery. In the midarm it pierces the medial intermuscular septum. At the elbow the nerve passes between the olecranon process and the medial epicondyle where it is superficial and easily palpable. Impingement upon the nerve at this point gives rise to the tingling sensation interpreted as arising from the "crazy bone" of the elbow. The median and ulnar nerves have no branches in the arm.

From the lateral cord the **musculocutaneous nerve** courses lateral to the axillary artery, pierces the coracobrachialis muscle, and continues distally between the biceps and the brachialis to innervate all three muscles. Crossing to the lateral side of the arm between the biceps and the brachialis, it pierces the deep fascia above the biceps tendon to continue distally as the **lateral antebrachial cutaneous nerve.**

The **axillary** (circumflex) **nerve,** arising from the posterior cord at the lower border of the subscapularis tendon, passes posteriorly through the quadrangular space with the posterior humeral circumflex artery. It supplies the teres minor and ramifies on the deep surface of the deltoideus muscle. It gives a branch, the **lateral brachial cutaneous,** which supplies skin over the deltoideus.

The **radial nerve,** the largest branch of the brachial plexus and the continuation of the posterior cord, initially lies posterior to the axillary artery. It passes distally between the teres major and the long head of the triceps, and spirals around the posterior aspect of the humerus in the radiospiral groove, between the lateral and medial heads of the triceps. In the groove it is accompanied by the profunda brachii artery. It pierces the lateral intermuscular septum and follows the interval between the brachialis and brachioradialis muscles into the antecubital fossa. Along its course it supplies the triceps and, just proximal to the elbow, sends branches to the brachialis, brachioradialis, extensor carpi radialis longus, and anconeus muscles.

Fractures of the Humerus

Fractures of the humerus are often serious because of injury to closely related nerves and blood vessels. The surgical neck is a common fracture site, which often involves damage to the axillary nerve, thus limiting abduction of the arm. A fracture in the middle third of the humerus may involve the radial nerve where it lies in the radiospiral groove. Such damage causes wrist drop since the nerve supply to the extensor muscles of the hand is lost.

Antecubital Fossa

The **antecubital fossa** is a triangular depression at the anterior aspect of the elbow joint (Fig. 2-14). Its base is formed by a line passing through the epicondyles of the humerus. Its apex is directed distally, with the brachioradialis muscle forming the lateral side and the pronator teres muscle the medial side. The floor of the fossa is formed by the brachialis and the supinator muscles. The fossa is covered by deep and superficial fasciae and skin. The **median antecubital vein** (used in drawing blood and for intravenous injection) crosses the fossa obliquely, superficial to the deep fascia. The tendon of the biceps brachii descends through the middle of the fossa to insert onto the radial tuberosity. It sends a secondary insertion as

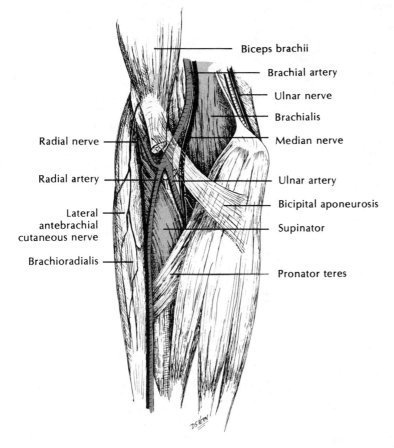

Figure 2-14. *Antecubital fossa.*

a fibrous expansion, the **bicipital aponeurosis,** into the deep fascia over the flexor muscles. The brachial artery passes through the fossa, medial to the biceps tendon, to bifurcate into the radial and ulnar arteries. The median nerve descends between the brachioradialis and brachialis muscle to enter the fossa medial to the brachial artery.

The Median Antecubital Vein

The median antecubital vein, the most common site for venipuncture, links the cephalic and basilic veins in the antecubital fossa. The tough, bicipital aponeurosis separates the deeper median nerve and the brachial artery from this superficial vein. This is an important relationship inasmuch as the fibrous band protects against the accidental injection of drugs into the artery, as well as possible damage or irritation to the median nerve.

Forearm

Muscles (Table 2-10)

The flexor muscles of the forearm are subdivided into superficial and deep groups (Fig. 2-15). The superficial group, the **palmaris longus, prona-**

Table 2-10
Muscles of the Forearm

Muscle	Origin	Insertion	Action	Nerve
Palmaris longus	Medial epicondyle of humerus	Flexor retinaculum and palmar aponeurosis	Flexes hand and tenses palmar aponeurosis	Median
Pronator teres	Humeral head, medial epicondyle; ulnar head, coronoid process of ulna	Middle of lateral surface of radius	Pronates hand	Median
Flexor carpi radialis	Medial epicondyle of humerus	Bases of second and third metacarpals	Flexes hand and elbow; slightly pronates and abducts hand	Median
Flexor carpi ulnaris	Humeral head, medial epicondyle; ulnar head, medial border of olecranon process and posterior border of ulna	Pisiform, hook of hamate, and base of fifth metacarpal	Flexes and adducts hand	Ulnar
Flexor digitorum superficialis	Humeral head, medial epicondyle; ulnar head, coronoid process; radial head, anterior border of radius	Palmar surface and sides of middle phalanges of fingers	Flexes middle phalanges; continued action flexes proximal phalanges and hand; aids in flexion of elbow	Median
Flexor digitorum profundus	Medial and anterior surface of ulna and adjacent interosseus membrane	Bases of distal phalanges of fingers	Flexes terminal phalanges; continued action flexes proximal phalanges and hand	Ulnar and median
Flexor pollicis longus	Anterior surface of radius, adjacent interosseus membrane, and coronoid process of ulna	Base of distal phalanx of thumb	Flexes thumb	Median
Pronator quadratus	Anterior surface of distal fourth of ulna	Anterior surface of distal fourth of radius	Pronates hand	Median
Brachioradialis	Lateral supracondylar ridge of humerus and lateral intermuscular septum	Lateral side of base of styloid process of radius	Flexes forearm	Radial
Extensor carpi radialis longus	Lateral supracondylar ridge of humerus	Posterior surface of base of second metacarpal	Extends and abducts hand	Radial
Extensor carpi radialis brevis	Lateral epicondyle of humerus	Posterior surface of base of third metacarpal	Extends and abducts hand	Radial
Extensor digitorum	Lateral epicondyle of humerus	Extensor expansion on fingers	Extends fingers and hand	Radial

(Continued)

Table 2-10   (Continued)

Muscle	Origin	Insertion	Action	Nerve
Extensor carpi ulnaris	Lateral epicondyle and posterior border of ulna	Base of fifth metacarpal	Extends and adducts hand	Radial
Extensor digiti minimi	From extensor digitorum and interosseus membrane	Extensor expansion on proximal phalanx of fifth digit	Extends fifth digit	Radial
Supinator	Lateral epicondyle of humerus, ligaments of elbow joint, and supinator crest and fossa of ulna	Lateral surface of upper third of radius	Supinates hand	Radial
Abductor pollicis longus	Posterior surface of ulna and middle third of posterior surface of radius	Base of first metacarpal	Abducts thumb and hand	Radial
Extensor pollicis longus	Middle third of posterior surface of ulna and adjacent interosseus membrane	Base of distal phalanx of thumb	Extends distal phalanx and abducts hand	Radial
Extensor pollicis brevis	Posterior surface of middle third of radius	Base of proximal phalanx of thumb	Extends and abducts hand	Radial
Extensor indicis	Posterior surface of ulna	Extensor expansion on index finger	Extends index finger	Radial

tor teres, **flexor carpi radialis,** and **flexor carpi ulnaris,** form the muscle mass at the medial side of the proximal forearm. The superficial group and the **flexor digitorum superficialis** originate, in part, from a common tendon attached to the medial epicondyle of the humerus. In the deep group, the **flexor pollicis longus** and **flexor digitorum profundus** pass superficially to the more distally placed **pronator quadratus.** With the exception of the flexor carpi ulnaris and ulnar half of the flexor digitorum profundus, which are supplied by the **ulnar nerve,** all the flexors are supplied by the **median nerve.** The flexors are separated from the extensor muscles by the **interosseous membrane** passing between the radius and ulna.

The extensor muscles are also divided into groups (Figs. 2-15 and 2-16). The superficial group, consisting of the **brachioradialis, extensor carpi radialis longus** and **brevis, extensor digitorum, extensor digiti minimi,** and the **extensor carpi ulnaris,** forms the muscle mass at the lateral aspect of the proximal portion of the forearm. Two of these muscles, the brachioradialis and the extensor carpi radialis longus, extend above the elbow joint, taking origin from the lateral supracondylar ridge of the humerus; the remainder arise, in part, by a common tendon attached to the lateral epicondyle of the humerus. The **brachioradialis,** while grouped with the extensors and innervated by the radial nerve, is functionally a flexor of the forearm in mid-supination. The deep extensor group (the **abductor pollicis longus,** the **extensor pollicis brevis** and **longus,** and the **extensor indicis**) are distal to the **supinator** and lie parallel to each other in the

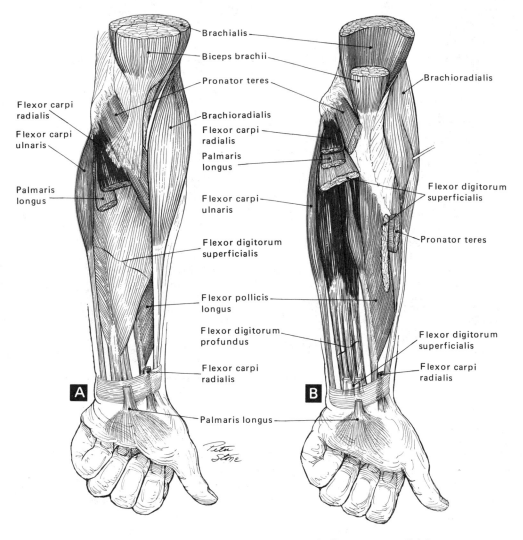

Figure 2-15. *Muscles of the anterior aspect of the forearm.* **A.** *The more superficial muscles are shown here, and two are cut to expose the flexor digitorum superficialis.* **B.** *Deeper-lying muscles, as well as tendons at the wrist.*

above order. They extend obliquely across the interosseous space. All the extensors are supplied by the deep branch (posterior interosseous) of the radial nerve.

Tennis Elbow

The medical term for this condition is lateral epicondylitis, which means an inflammation of the lateral epicondyle of the hu-

merus or the tissues surrounding it. Recall that most of the extensors of the fingers and wrist originate from a common tendon that attaches to the lateral epicondyle. Repeated strenuous contraction of these muscles, as occurs in the backhand stroke in tennis, causes a strain on the tendon, and results in exquisite tenderness and pain around the epicon-

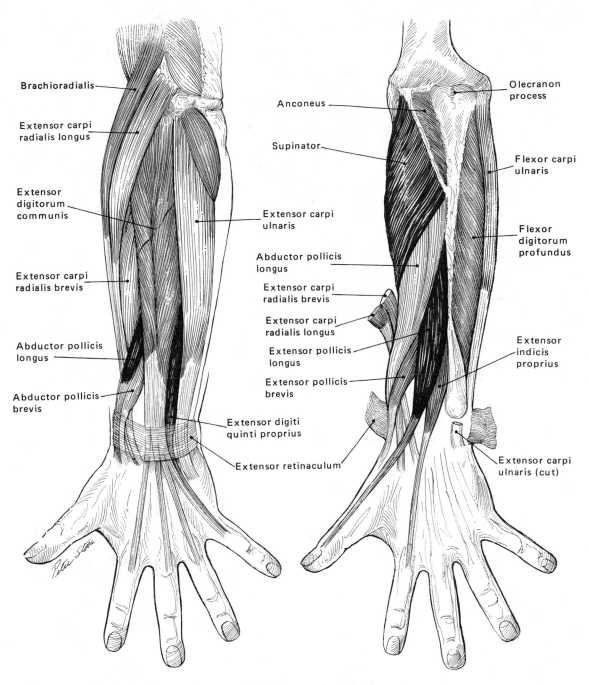

Brachioradialis

Extensor carpi
radialis longus

Extensor
digitorum
communis

Extensor carpi
radialis brevis

Abductor pollicis
longus

Abductor pollicis
brevis

Anconeus

Supinator

Extensor carpi
ulnaris

Abductor pollicis
longus

Extensor carpi
radialis brevis

Extensor carpi
radialis longus

Extensor pollicis
longus

Extensor pollicis
brevis

Extensor digiti
quinti proprius

Extensor retinaculum

Olecranon
process

Flexor carpi
ulnaris

Flexor
digitorum
profundus

Extensor
indicis
proprius

Extensor carpi
ulnaris (cut)

Figure 2-16. *Muscles on the posterior aspect of the forearm. Superficial muscles are shown at the left, more deeply placed muscles on the right.*

dyle. Rest eliminates the causative factor, and recovery usually follows.

Arteries and Nerves (see Table 2-9)

One of the terminal branches of the brachial artery, the **radial artery,** courses distally in the forearm between the brachioradialis and the pronator teres, and passes to the anterior aspect of the radius to become superficial at the wrist where the pulse is usually taken. It gives a **radial recurrent branch** in the antecubital fossa, which courses proximally anterior to the lateral epicondyle to anastomose with the profunda brachii artery. Muscular branches of the radial artery are distributed to the superficial extensor muscles. At the wrist it gives **palmar** and **dorsal carpal branches,** which aid in the formation of the **carpal arches,** which supply the wrist and carpal joints.

The larger of the two terminal branches of the brachial, the **ulnar artery,** passes deep to the pronator teres and the superficial flexors to lie on the surface of the flexor digitorum profundus, where it is overlapped distally by the flexor carpi ulnaris. After giving **palmar** and **dorsal carpal** branches to the **carpal arches,** it passes into the hand deep to the palmar carpal ligament. At the origin of the ulnar artery, **anterior** and **posterior ulnar recurrent branches** arise and course proximally to pass respectively anterior and posterior to the medial epicondyle, and anastomose with superior and inferior ulnar collateral branches of the brachial artery. The largest ulnar branch, the **common interosseous,** arises about 3 cm distal to the bifurcation of the brachial artery. A **recurrent branch** of the common interosseous artery turns proximally, deep to the supinator, to join the anastomosis around the elbow joint. The common interosseous artery terminates by dividing into the **anterior** and **posterior interosseous branches,** which pass to their respective sides of the interosseous membrane to supply the deep flexor and deep extensor muscles. The **anterior interosseous artery,** extend-

ing further distally than the posterior branch, pierces the interosseous membrane in the distal third of the forearm to supply the muscles on the dorsum of the forearm before it anastomoses with the carpal arches.

The **median nerve** leaves the antecubital fossa by passing between the heads of the pronator teres to gain a position deep to the flexor digitorum superficialis, where it continues distally between this muscle and the flexor digitorum profundus (Table 2-11). In the forearm it supplies the pronator teres, the pronator quadratus, and all the flexors, except the flexor carpi ulnaris. It shares the innervation of the flexor digitorum profundus with the ulnar nerve. At the wrist the median nerve becomes superficial to lie at the ulnar side of the tendon of the palmaris longus muscle, then passes deep to the flexor retinaculum to enter the hand.

Median Nerve Injuries

When the median nerve is injured at the wrist, sensation is lost on the palmar surface of the hand lateral to the midline of the ring finger. It is also absent on the dorsal aspect of corresponding distal phalanges. Paralysis occurs in the muscles of the thenar eminence, and the ball of the thumb flattens out as the muscles atrophy. Furthermore, the thumb cannot be placed in opposition to the other fingers and its pinching action is lost.

If the injury is at or above the elbow, additional paralysis is noted. For example, there is loss of pronation, and loss of the digital flexors, except for the little and ring fingers, which have an additional innervation from the ulnar nerve. Therefore, in making a fist the thumb will be extended and adducted; the index and middle fingers will be in extension, and the ring and little fingers flexed (papal sign).

From its position between the olecranon process and the medial epicondyle, the **ulnar nerve** enters the forearm by passing between the heads

Table 2-11
Nerve Distribution to Forearm

Nerve	Origin	Course	Distribution
Medial antebrachial cutaneous	Medial cord of brachial plexus	Follows brachial artery and in antecubital fossa pierces deep fascia to course in superficial fascia along medial aspect of forearm	Skin over medial aspect of forearm
Lateral antebrachial cutaneous	Continuation of musculocutaneous nerve	At antecubital fossa pierces deep fascia and courses in superficial fascia along lateral aspect of forearm	Skin of lateral aspect of forearm
Posterior antebrachial cutaneous	Radial nerve	Pierces lateral intermuscular septum in distal portion of arm to course in superficial fascia on posterior aspect of forearm	Skin on posterior aspect of forearm
Ulnar	Medial cord of brachial plexus	Follows brachial artery and at elbow passes along ulnar groove of medial epicondyle; passes through forearm between flexor carpi ulnaris and flexor digitorum profundus	Flexor carpi ulnaris and ulnar half of flexor digitorum profundus
Median	Lateral and medial cords of brachial plexus	Follows brachial artery; at antecubital fossa passes between heads of pronator teres to course through forearm deep to the flexor digitorum superficialis	Supplies all flexors in forearm not supplied by ulnar
Deep branch of radial (posterior interosseous)	Radial in proximal portion of forearm	Penetrates supinator to reach posterior compartment	Supplies all extensors in forearm not supplied by radial nerve proper

of the flexor carpi ulnaris. It continues distally in the forearm deep to this muscle, lying on the flexor digitorum profundus. In the distal half of the forearm it parallels the course of the ulnar artery to pass to the lateral side of the tendon of the flexor carpi ulnaris and enters the hand deep to the palmar carpal ligament. In the forearm it supplies the flexor carpi ulnaris and the ulnar half of the flexor digitorum profundus.

Ulnar Nerve Injury
Ulnar nerve injury may result in extensive motor and sensory loss to the hand. After complete interruption of the ulnar nerve above the elbow, the following changes in the hand are noted: 1) spaces between the tendons become sunken, called "guttering," due to atrophy of the interossei, third and fourth lumbricales, and the adductor pollicis; 2) partial (ulnar) *claw hand* involves hyperextension of the fourth and fifth digits at the metacarpophalangeal joints and flexion of the interphalangeal joints; 3) flattening of the hypothenar eminence occurs from atrophy of the palmaris brevis muscle and the intrinsic muscles of the little finger; and 4) persistent abduction of the little finger results due to loss of action of the third palmar interosseous with unopposed action of extensor digiti quinti.

Clinical tests for ulnar nerve damage include: 1) difficulty in making a fist because the fourth and fifth fingers cannot flex at the distal interphalangeal joint; 2) adduction of the thumb is lost—when the patient attempts to grasp a piece of paper between the thumb and the index finger he accomplishes it by strong flexion of the thumb at the interphalangeal joint (Froment's sign); and 3) abduction and adduction of the fingers are weakened and difficult to perform.

After entering the antecubital fossa, the **radial nerve** divides into superficial and deep branches. The **superficial branch** passes distally, deep to the brachioradialis, to the dorsum of the wrist. Here it divides into medial and lateral branches that supply cutaneous innervation to the dorsum of the hand. The **deep branch** of the radial (posterior interosseous) nerve pierces the supinator muscle to be distributed to all the extensor muscles within the forearm.

Radial Nerve Injury

In a complete radial nerve injury there is loss of extension with "wrist drop." The patient is unable to extend the wrist against gravity. If the lesion is in the distal third of the arm, extension of the elbow by the triceps, weak supination by the brachioradialis, and weak extension by the extensor carpi radialis longus will be preserved. In lesions at either level sensory loss will occur on the lateral aspect of the dorsum of the hand.

Hand

Objectives

At the completion of the study of the hand the student should be able to

▶ *List the intrinsic muscles of the hand and give their location, action and innervation*

▶ *Define the radial bursa, ulnar bursa, digital tendon sheaths; and give their locations and associated tendons*

▶ *Describe the extensor expansion and the attachments of the long flexor tendons to the phalanges*

▶ *Draw the cutaneous nerve distribution of the hand*

▶ *Draw and label the formation and branches of the arterial arches of the hand*

Muscles (Table 2-12)

The intrinsic muscles of the hand are divided into three groups: the thenar, the hypothenar, and the interossei and lumbricales (Fig. 2-17).

The **thenar eminence** (ball of the thumb) is formed by the laterally placed superficial **abductor pollicis brevis,** the intermediate **opponens pollicis,** and the more deeply placed **flexor pollicis brevis.** The latter is divided into a deep and superficial portion by the tendon of the flexor pollicis longus. The **hypothenar eminence** (heel of the hand) is composed of the medially situated **abductor digiti minimi,** the more lateral **flexor digiti minimi brevis** lying superficial to, and usually blending with, the more deeply placed **opponens digiti minimi.**

The seven interossei are disposed in two layers and are composed of three palmar and four dorsal muscles. These muscles are located in the spaces between the metacarpal bones. The unipennate **palmar interossei** arise from single metacarpals, while the bipennate **dorsal interossei** arise from two adjacent metacarpals. All the interossei insert into bases of proximal phalanges. The dorsal interossei act as abductors and the palmar interossei act as adductors of the digits. Abduction and adduction of the digits are relative to a line passing through the central axis of the middle finger. The relatively extensive **adductor pollicis** muscle is in the same plane as the palmar interossei muscles and arises by two heads. It is sometimes referred to as the fourth palmar interosseous.

The four **lumbricales** consist of muscular slips, which originate from each of the four tendons of the flexor digitorum profundus. They

Table 2-12
Muscles of the Hand

Muscle	Origin	Insertion	Action	Nerve
Palmaris brevis	Medial aspect of flexor retinaculum	Skin of palm	Wrinkles skin of palm	Ulnar
Abductor pollicis brevis	Flexor retinaculum, scaphoid, and trapezium	Lateral side of base of proximal phalanx of thumb	Abducts thumb; assists in flexion of proximal phalanx	Median
Flexor pollicis brevis	Flexor retinaculum and trapezium	Base of proximal phalanx of thumb with sesamoid interposed	Flexes thumb; assists in apposition	Median
Opponens pollicis	Flexor retinaculum and trapezium	Entire length of lateral border of first metacarpal	Draws first metacarpal toward center of palm	Median
Abductor digiti minimi	Pisiform and tendon of flexor carpi ulnaris	Medial side of base of proximal phalanx of fifth digit	Abducts fifth digit	Ulnar
Flexor digiti minimi brevis	Flexor retinaculum and hook of hamate	Medial side of base of proximal phalanx of fifth digit	Flexes proximal phalanx of fifth digit	Ulnar
Opponens digiti minimi	Flexor retinaculum and hook of hamate	Medial border of fifth metacarpal	Draws fifth metacarpal forward in cupping of hand	Ulnar
Adductor pollicis	Oblique head from capitate and bases of second and third metacarpals; transverse head from anterior surface of third metacarpal	Medial side of base of proximal phalanx of thumb	Adducts thumb; assists in apposition	Ulnar
Lumbricales (4)	Tendons of flexor digitorum profundus	Extensor expansion distal to metacarpophalangeal joint	Flex metacarpophalangeal and extend interphalangeal joints	Two lateral muscles by median; two medial by ulnar
Dorsal interossei (4)	By two heads from adjacent sides of metacarpal bones	Lateral sides of bases of proximal phalanges of index and middle fingers; medial sides of bases of middle and ring fingers	Abduct index, middle, and ring fingers; aid in extension of interphalangeal and flexion of metacarpophalangeal joints	Ulnar
Palmar interossei (3)	Medial side of second metacarpal; lateral sides of fourth and fifth metacarpals	Base of proximal phalanx in line with its origin	Adduct index, ring, and little fingers; aid in extension of interphalangeal and flexion of metacarpophalangeal joints	Ulnar

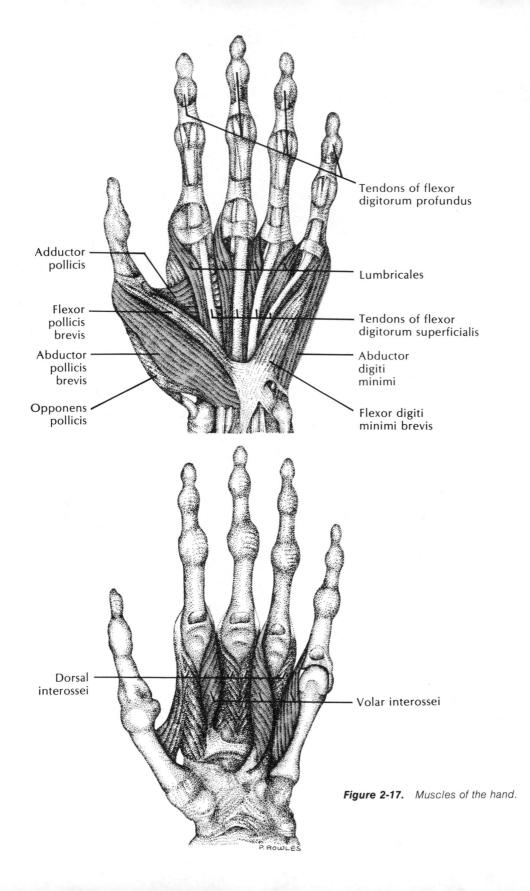

Adductor
pollicis

Flexor
pollicis
brevis

Abductor
pollicis
brevis

Opponens
pollicis

Tendons of flexor
digitorum profundus

Lumbricales

Tendons of flexor
digitorum superficialis

Abductor
digiti
minimi

Flexor digiti
minimi brevis

Dorsal
interossei

Volar interossei

P. ROWLES

Figure 2-17. *Muscles of the hand.*

course to the radial side of the metacarpophalangeal joints to insert into the extensor expansion. Their action is to flex the metacarpophalangeal joints and extend the interphalangeal joints, thus placing the fingers at right angles to the palm of the hand.

Synovial Tendon Sheaths

At the wrist and in the hand, lubricating synovial sheaths surround tendons as they extend from the forearm into the hand (Fig. 2-18). The **ulnar** **bursa** is a large synovial sac invaginated on the radial side by tendons of the flexor digitorum profundus and superficialis muscles. It extends from approximately 2 cm to 3 cm above the flexor retinaculum to the middle of the palm. Around the tendons of the fifth finger it continues to the distal phalanx. Separate from the ulnar bursa, the **digital synovial sheaths** invest the digital parts of the deep and superficial flexor tendons to the second, third, and fourth digits. These sheaths extend from the heads of the metacarpals to the insertions of the long flexor tendons.

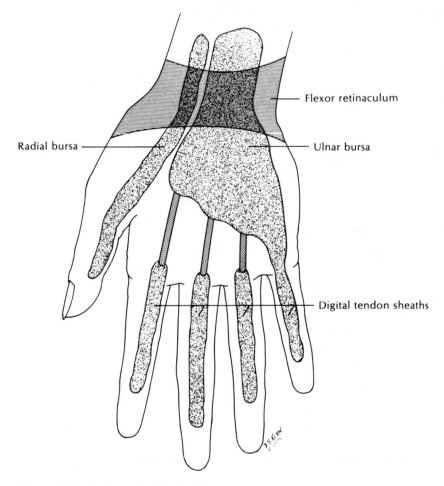

Flexor retinaculum

Radial bursa

Ulnar bursa

Digital tendon sheaths

Figure 2-18. *Radial and ulnar bursae.*

Tendosynovitis

Tendosynovitis is an infection of a synovial tendon sheath. Digital sheath infections are common and are often caused by a nonsterile penetrating wound from a thorn, needle, knife, or splinter, which carries pathogenic organisms into the closed synovial digital sheath. Cardinal signs of such an infection are exquisite tenderness over the sheath, sustained flexion of the infected digit, and upon extension, severe pain over the proximal end of the finger. The closed chamber of the digital sheath swells with edematous fluid and pus. The increasing pressure on the sheath can cause exquisite pain and necrosis of the flexor tendons. Tendosynovitis of the digital sheath is an emergency situation. The sheath must be opened and drained early to prevent ischemic necrosis of the tendons and a serious loss of hand function.

The **radial bursa** surrounds the flexor pollicis longus tendon, extending from about .2 cm to 3 cm proximal to the flexor retinaculum to its insertion. A separate synovial sheath surrounds the tendon of the flexor carpi radialis from a point proximal to the flexor retinaculum to its insertion into the base of the second metacarpal bone.

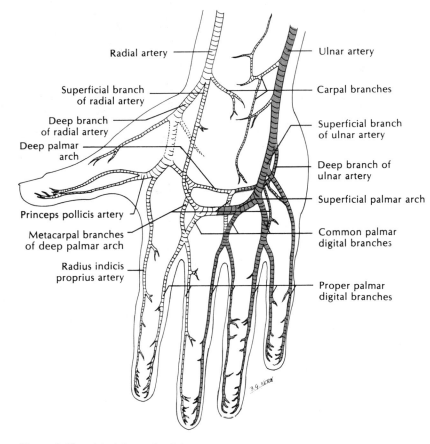

Figure 2-19. *Arterial supply of the hand.*

Synovial sheaths surround extensor tendons within six osseofibrous compartments or tunnels located deep to the extensor retinaculum. The sheaths, enclosing the tendons of the extensor carpi radialis longus and brevis and the extensor carpi ulnaris, extend from the proximal border of the extensor retinaculum to the insertions of these tendons. Sheaths of the digital extensors, including the extensors of the thumb and the abductor pollicis longus, terminate at about the middle third of the metacarpal bones.

Arteries and Nerves (Figs. 2-19 and 2-20; Table 2-13)

After passing deep to the palmar carpal ligament, the **ulnar artery** enters the hand by passing to the radial side of the pisiform bone to give muscular branches to the hand (Fig. 2-19). It terminates by dividing into a **deep branch** that joins with a similar branch of the radial artery to form the **deep palmar arch. A superficial branch** joins with a corresponding branch of the radial to complete the **superficial palmar arch.** The latter, predominately from the ulnar artery, lies transversely in the palm immediately under the palmar aponeurosis in line with the fully extended thumb. It gives a **proper digital branch** to the medial side of the little finger and three **common palmar digital branches,** which subsequently divide into **proper digital branches** to supply contiguous sides of the four fingers.

Passing from the anterior surface of the radius medial to the styloid process, the **radial artery** enters the hand deep to the tendons of the abductor pollicis longus and the extensor pollicis longus and brevis muscles. It pierces the first dor-

Table 2-13
Arterial Supply to the Hand

Artery	Origin	Course	Distribution	Anastomoses
Superficial palmar arch	Superficial branches of radial and ulnar; principally ulnar	Lies deep to the palmar aponeurosis; at the level of fully extended thumb	Primary blood supply to the hand	Deep palmar arch
Deep palmar arch	Deep branches of radial and ulnar; principally radial	Lies 3 cm proximal to superficial arch	To dorsum of hand and communicates with superficial palmar arch	Superficial palmar arch
Common palmar digitals	Superficial arch	Pass to webbings of fingers	Terminate as proper digitals	Deep palmar arch
Proper digitals	Common palmar digitals	Along lateral side of respective digits	Supplies tendons and skin to the digits	Dorsal digitals
Princeps pollicis	Radial	In space between the first dorsal interosseous and adductor pollicis	Primary blood supply to the thumb	Dorsal digitals
Radius indicis proprius	Radial	Passes along the radial side of index finger	Radial side of index finger	Dorsal digitals
Dorsal arch	Radial and ulnar	Superficial fascia on dorsum of hand	Dorsum of hand and dorsal aspects of digits	Palmar digitals

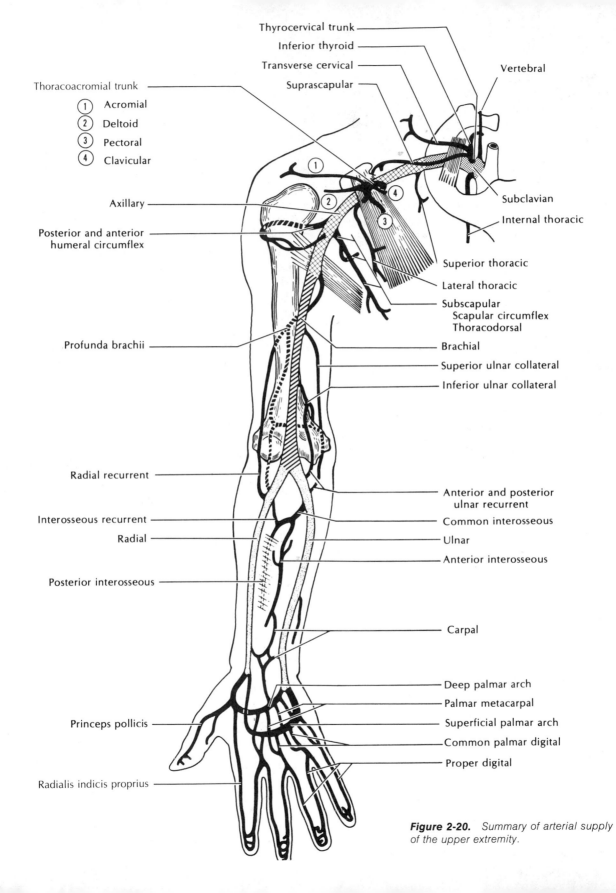

Thyrocervical trunk

Inferior thyroid

Transverse cervical

Suprascapular

Vertebral

Thoracoacromial trunk
1 Acromial
2 Deltoid
3 Pectoral
4 Clavicular

Subclavian

Internal thoracic

Axillary

Posterior and anterior
humeral circumflex

Superior thoracic

Lateral thoracic

Subscapular
Scapular circumflex
Thoracodorsal

Profunda brachii

Brachial

Superior ulnar collateral

Inferior ulnar collateral

Radial recurrent

Anterior and posterior
ulnar recurrent

Interosseous recurrent

Common interosseous

Radial

Ulnar

Anterior interosseous

Posterior interosseous

Carpal

Deep palmar arch

Palmar metacarpal

Princeps pollicis

Superficial palmar arch

Common palmar digital

Proper digital

Radialis indicis proprius

Figure 2-20. *Summary of arterial supply of the upper extremity.*

sal interosseous to lie between this muscle and the adductor pollicis. Branches of the radial artery include the **superficial palmar,** which with a similar branch of the ulnar, completes the **superficial palmar arch.** The **princeps pollicis branch** and the **indicis proprius branch** both arise after the radial artery pierces the first dorsal interosseous muscle. They supply, respectively, both sides of the thumb, and the radial side of the index finger. The **deep branch** of the radial artery forms most of the **deep palmar arch,** which lies on the carpal bones, a thumb's breadth proximal to the superficial arch. Branches of the deep arch are the **palmar metacarpal vessels,** which join the **common palmar digital arteries** of the superficial arch to supply the fingers.

On entering the hand deep to the flexor retinaculum, the **median nerve** gives a **motor (lateral) branch** to the muscles of the thenar eminence, namely, the opponens pollicis, the abductor pollicis brevis, the superficial head of the flexor pollicis brevis, and the two lateral lumbricales. Terminally the nerve supplies cutaneous innervation to the central area of the palm, palmar surface of the thumb, index, middle, and lateral half of the ring fingers, and all of the skin on the distal phalanges of these digits (see Fig. 2-7).

Carpal Tunnel Syndrome

Carpal tunnel syndrome is a painful condition owing to compression of the median nerve by the transverse carpal ligament, which forms the palmar aspect of the carpal tunnel. Median nerve compression at the wrist causes a burning sensation or numbness over the lateral three and one-half digits, and weakness or atrophy of the thenar muscles. The compression may be due to inflammation from chronic irritation of the transverse carpal ligament and is often found in patients with rheumatoid arthritis. Dividing the transverse carpal ligament decompresses the carpal tunnel and the median nerve and, after several months, usually results in almost complete return of median nerve function.

At the wrist the **ulnar nerve** divides into a **superficial branch** that supplies the palmaris brevis and the cutaneous innervation to both surfaces of the hand medial to a line passing through the midline of the ring finger, and a **deep branch** that supplies all the intrinsic muscles of the hand except those supplied by the median nerve. The deep branch pierces the opponens digiti minimi muscle and passes around the hamulus of the hamate bone to cross the palm in company with the deep palmar arch.

The **radial nerve** supplies no muscles in the hand. Its terminal distribution is the cutaneous innervation to a small area on the lateral surface of the thenar eminence and the dorsum of the hand lateral to a line passing through the midline of the ring finger, except for skin over the distal phalanges (Table 2-14).

Fasciae

The **external investing layer** of **deep fascia** completely ensheathes the upper extremity. In the arm it sends distinct **lateral** and **medial intermuscular septa** to attach to the humerus, which separate the musculature of the arm into extensor and flexor compartments. At the wrist this layer of deep fascia condenses and thickens to form a distinct band surrounding the wrist as the flexor and extensor retinacula (Fig. 2-21). The **extensor retinaculum** has deep attachments to the lateral side of the dorsum of the radius, the pisiform and triquetrum bones, and the styloid process of the ulna. It sends septa to attach to ridges on the dorsum of the radius and forms **six osseofibrous compartments** for the extensor tendons passing from the forearm into the hand. The six compartments transmit tendons, from the radial to the ulnar side respectively, as follows: the tendons of the abductor pollicis longus and extensor pollicis

Table 2-14
Nerve Distribution to the Hand

Nerve	Origin	Course	Distribution
Median	Lateral and medial cords of brachial plexus	Enters hand by passing deep to flexor retinaculum; distally divides into motor branch and palmar digital (cutaneous) branches	Motor branch supplies thenar muscles, lateral lumbricales (2), opponens, and short flexor and abductor of thumb; cutaneous branches supply lateral $2/3$ of palm, palmar aspect, and all skin of distal phalanges of lateral $3\frac{1}{2}$ digits
Ulnar	Medial cord of brachial plexus	Enters hand by passing deep to the palmar carpal ligament; distal to pisiform bone divides into superficial and deep branches of hand	Superficial branch supplies skin medial to a line bisecting the ring finger; deep branch supplies all intrinsic muscles of hand not supplied by the median nerve
Radial	Superficial branch—the continuation of radial after deep branch is given off at elbow	In forearm, courses deep to brachioradialis; at wrist, turns onto dorsum of the hand	Skin on dorsum of hand, except for distal phalanges, lateral to a line bisecting ring finger

brevis traverse the **first compartment,** located on the lateral border of the styloid process of the radius; the **second compartment,** situated on the radial side of the tubercle of the radius, contains the tendons of the extensor carpi radialis longus and brevis muscles; the **third,** on the ulnar side of the tubercle of the radius, conveys the tendon of the extensor pollicis longus muscle. The **fourth** and largest compartment, covering the ulnar third of the dorsum of the radius, transmits the four tendons of the extensor digitorum communis and the extensor indicis proprius; the **fifth,** located over the distal radioulnar articulation, contains the tendon of the extensor digiti minimi; and the **sixth,** located on the head of the ulnar bone, affords passage for the tendon of the extensor carpi ulnaris.

The **flexor retinaculum** (transverse carpal ligament) stretches across the concavity formed by the articulated carpal bones. It attaches to the tuberosity of the scaphoid and trapezium bones laterally, and the hamate and pisiform medially to form the carpal tunnel. The median nerve, tendon of the flexor pollicis longus, and tendons of the flexors digitorium superficialis and profundus traverse the carpal tunnel. The tendons of the superfi-

cialis to the third and fourth digits lie superficially, those to the second and fifth digits lie intermediately, and the deeply placed tendons of the profundus lie side by side in the compartment. Superficial to the flexor retinaculum the palmaris longus muscle passes distally in the midline, and the ulnar nerve and artery course to its ulnar side between the retinaculum and the **palmar carpal ligament.**

Dupuytren's Contracture

Dupuytren's contracture involves an insidious fibrosis of the palmar fascia, which forms subcutaneous nodules that attach first to the flexor sheath of the ring finger and then progressively involves the little, middle and index finger in that order. The progressive retraction of the flexor tendons pulls the fingers into continuous flexion which is like the sign of "papal benediction" with the ring and little fingers flexed into the palm and the thumb, index, and middle fingers extended. With increasing disability of the hand from these flexed fingers, surgical excision of this overgrowth of palmar fi-

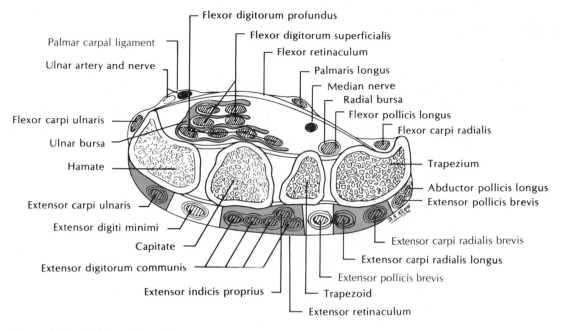

Figure 2-21. *Relationships of flexor and extensor retinacula.*

brous tissue may be necessary to restore the hand to normal function.

Deep fascia is also specialized on the palmar aspect of the hand to form compartments. The **thenar compartment** is formed by fascia that surrounds the small muscles of the thumb. Structures within this space include the abductor, flexor, and opponens muscles of the thumb, the muscular branch of the median nerve, and the superficial palmar branch of the radial artery. The **hypothenar compartment** is formed by fascia that surrounds the small muscles of the little finger. It contains the abductor, flexor brevis, and opponens muscles of the fifth digit, and the deep branches of the ulnar artery and nerve. The **central,** or intermediate, **compartment** (midpalmar space) is bounded superficially by the **palmar aponeurosis,** medially and laterally by fascia covering hypothenar and thenar muscles, and dorsally by the interosseous adductor compartment. The central space is incompletely divided by septa separating flexor tendons and associated lumbri-

cale muscles. It contains the long flexor tendons, the lumbricales, median and ulnar nerves, and the superficial palmar arch. The **interosseous adductor compartment** is bounded by dorsal and palmar interosseous fasciae and encloses the interossei muscles and the metacarpal bones. It contains the interossei muscles, metacarpal bones, the adductor of the thumb, the deep palmar arch, the deep branch of the ulnar nerve, and the arterial arch on the dorsum of the hand.

Surgical Repair of Flexor Tendons
Surgical repair of flexor tendons in the midpalmar space is often unsuccessful because of the limited blood supply to the tendons and the great difficulty in reestablishing an efficient sliding movement of the tendons within the damaged synovial sheaths. Because fully functional repair of these sheaths is almost impossible, the midpalmar space has been called the "no man's land" of tendon surgery.

Joints of the Superior Extremity

Shoulder Joint

The multiaxial, **ball-and-socket** shoulder joint, consisting of the **head of the humerus** articulating with the much smaller, relatively flat **glenoid fossa** of the scapula, has the greatest freedom of movement of any joint of the body (Fig. 2-22). The bony configurations result in a mobile, but inherently unstable, joint. Such anatomic instability is compensated for by the presence of an **articular (rotator) cuff of muscles and tendons,** which holds the humeral head in place and reinforces the joint capsule. A fibrocartilaginous rim around the glenoid cavity, the **glenoidal labrum,** deepens the articular fossa. The **articular capsule** attaches proximally to the margins of the glenoid fossa and distally at the anatomic neck of the humerus (Fig. 2-23). The joint cavity, lined by synovial membrane, is traversed by the **tendon of the long head of the biceps** ensheathed by a synovial membrane. The tendon continues distally in the **intertubercular groove,** which is bridged by the **transverse humeral ligament** to hold the tendon in place. The **glenohumeral ligaments** are three bands of tissue that blend with and strengthen the articular capsule. The **coracohumeral ligament** passes from the coracoid process to blend with the upper posterior part of the capsule and attaches to the anatomic neck and greater tubercle of the humerus. The **coracoclavicular ligament** forms the strongest union between the scapula and the clavicle. Passing from the coracoid process of the scapula, its lateral **trapezoid portion** attaches to the trapezoid ridge of the clavicle, while medially the **conoid portion** passes to the conoid tubercle of the clavicle.

Dislocation of the Shoulder Joint
Dislocation of the shoulder joint occurs frequently. The joint stability is largely dependent on the strong tendons (rotator cuff) of the scapular muscles. Because the rotator cuff is not present at the inferior aspect of the joint capsule, most dislocations occur in this area and are directed downward and forward. Such a dislocation may damage the axillary nerve. Superior dislocation seldom occurs because of the additional protection afforded by the acromial and coracoid processes and the coracohumeral and coracoclavicular ligaments that form an arch or cradle for the head of the humerus.

Elbow Joint

The elbow is a compound **hinge (ginglymus) joint** (Fig. 2-24). The spool-shaped **trochlea** of the humerus articulates with the **trochlear notch** of the ulna, and the **capitulum** of the humerus articulates with the **head of the radius.** A single articular capsule encloses the joint. Anteriorly and posteriorly the capsule is relatively thin and membranous, but is reinforced laterally and medially by strong **radial** and **ulnar collateral ligaments.** The synovial membrane lining the joint capsule is extensive and passes distally under the **annular ligament** to permit free rotation of the radius.

Radioulnar joints form the proximal and distal articulations between the radius and ulna, which permit rotation of the radius in pronation and supination of the forearm. The **superior radioulnar** articulation is enclosed in the articular capsule of the elbow joint and shares its synovial membrane. The **radial notch** of the ulna and the annular ligament of the radius form a ring in which the radial head rotates. The **annular liga-**

Text continues on page 94

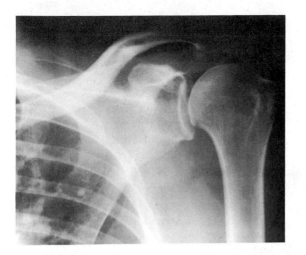

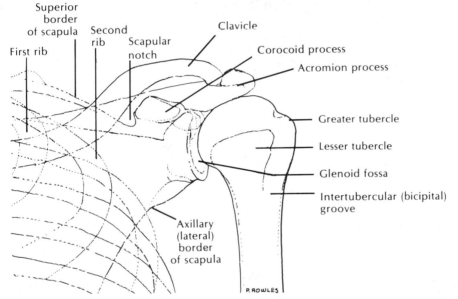

Figure 2-22. *Anteroposterior radiograph and schematic of right shoulder joint.*

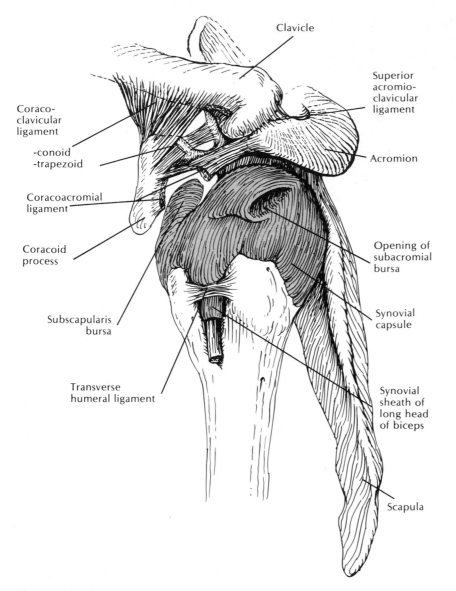

Clavicle

Superior
acromio-
clavicular
ligament

Coraco-
clavicular
ligament

-conoid
-trapezoid

Acromion

Coracoacromial
ligament

Coracoid
process

Opening of
subacromial
bursa

Synovial
capsule

Subscapularis
bursa

Transverse
humeral ligament

Synovial
sheath of
long head
of biceps

Scapula

Figure 2-23. *Synovial capsule of shoulder joint with important bursae and ligaments.*

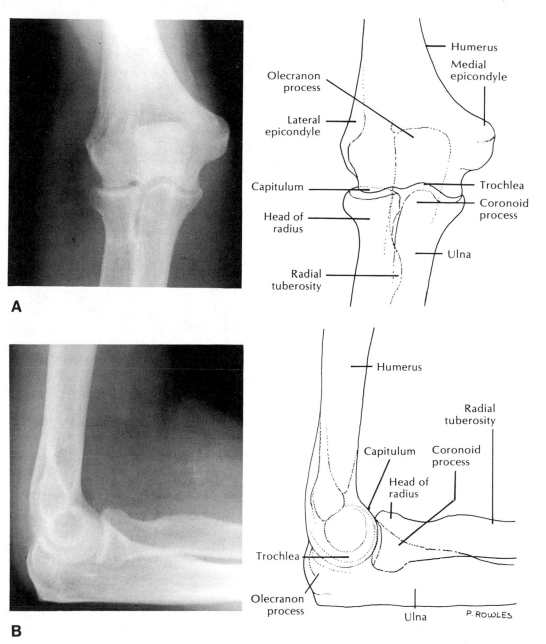

Figure 2-24. **A.** *Anteroposterior radiograph and schematic of right elbow joint.* **B.** *Lateral radiograph and schematic of right elbow joint.*

ment forms four-fifths of the ring and attaches to the margins of the radial notch of the ulna, cupping the head of the radius and blending with the articular capsule and the **radial collateral ligament.** The **quadrate ligament** forms a loose band extending from the distal border of the radial notch of the ulna to the medial surface on the neck of the radius.

The **interosseous membrane,** a strong tendinous sheet connecting the shafts of the radius and ulna, separates the extensor from the flexor compartment of the forearm and gives attachment to deep muscles of both groups. It extends proximally to within 2 cm to 3 cm of the tuberosity of the radius and distally to the inferior radioulnar articulation.

Wrist Joint

Colles' Fracture

When one falls on the outstretched arm, the radius bears the brunt of force transmitted through the hand. If a fracture occurs, it is usually a transverse break of the lower end of the radius about 3 cm proximal to the radiocarpal joint. In this type of injury, called Colles' fracture, the hand is displaced backward and upward. A tentative diagnosis can be made by the characteristic deformity of the wrist, resembling an overturned eating fork.

The **inferior radioulnar articulation** is a **pivot-type joint** lying between the head of the ulna and the ulnar notch of the radius. An **articular disc** of fibrocartilage, attaching to the medial edge of the distal end of the radius and the internal surface of the styloid process of the ulna, is interposed between the ulna and the proximal row of carpal bones. The articular capsule is relatively weak.

A **condyloid joint** is present between the distal end of the radius, the articular disc, and the proximal row of carpal bones, namely, the **scaph-**

oid (navicular), **lunate,** and **triquetrum** (triangularis). The articular capsule enclosing the joint is strengthened by **dorsal** and palmar radiocarpal ligaments and the **radial** and **ulnar collateral ligaments.**

Carpal Joints

The intercarpal articulations are **gliding joints** between the two transverse rows of carpal bones, with **dorsal** and **palmar intercarpal ligaments** passing transversely, and interosseous **intercarpal ligaments** linking the lateral borders of the proximal row to the distal row of bones (Fig. 2-25). The **midcarpal joint,** between the proximal and distal rows, allows a considerable range of movement. The central portion of this articulation forms a limited ball-and-socket type of joint, with the **scaphoid** and **lunate** forming the socket proximally and the **capitate** and **hamate** forming the ball distally. The lateral portions are **gliding joints** between the **trapezium** and **trapezoid** articulating with the **scaphoid** laterally and the **hamate** with the **triquetrum** medially. **Collateral ligaments** of the radial and ulnar borders connect the scaphoid with the trapezium and the triquetrum with the hamate. The **pisiform** articulates with the **triquetrum,** which is surrounded by a separate articular capsule.

Joints of the Hand

The **carpometacarpal joint** of the thumb forms a **saddle-shaped articulation** between the **trapezium** and the **first metacarpal** and is surrounded by an articular capsule. The remaining carpometacarpal joints have a common synovial cavity between the **intercarpal** and **intermetacarpal joints** linked by **dorsal, palmar,** and **interosseous ligaments.**

The **intermetacarpal joints** formed by contiguous sides of the bases of **second through**

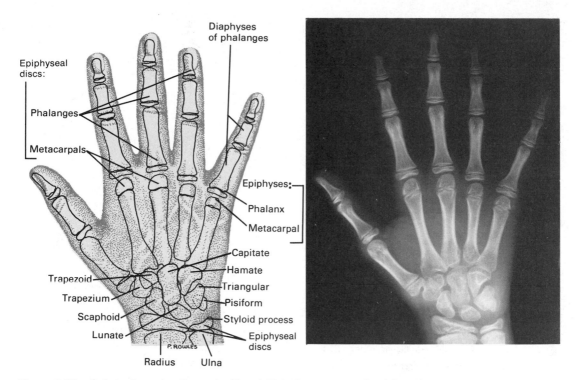

Figure 2-25. *Schematic and radiograph of hand. Note the presence of epiphyses.*

fifth metacarpals are joined by **dorsal, palmar, and interosseous ligaments.** The **deep transverse metacarpal ligaments** connect the heads of the **second, third,** and **fourth metacarpals** on the palmar aspect and thereby limit the abduction of these bones.

The condyloid **metacarpophalangeal joints** are formed by the rounded **head of the metacarpal bones** articulating with the concavities of the **proximal phalanges.** These joints are linked by articular capsules that are reinforced dorsally by the extensor tendons. **Palmar ligaments** bridge the joints on the palmar aspect of the hand and are continuous laterally with the strong cordlike **collateral ligaments,** which attach proximally to the tubercle and distally to the lateral aspect of the base of the phalanx.

The **interphalangeal joints** are **hinge-type joints,** structurally the same as the metacarpopha-

langeal articulations, with a **palmar ligament** and two **collateral ligaments** that are reinforced dorsally by the extensor expansion.

Bursae

Synovial bursae of the upper extremity are associated with the joints and sometimes communicate with the articular synovial cavity. The **subacromial (subdeltoid) bursa,** about 2 cm to 3 cm in diameter, is located deep to the deltoid between the tendon of the supraspinatus and the joint capsule. This bursa sends an extension deep to the acromial process and the coracoacromial ligament and may communicate with the joint cavity. Between the tendon of the subscapularis and the neck of the scapula the **subscapular bursa** usu-

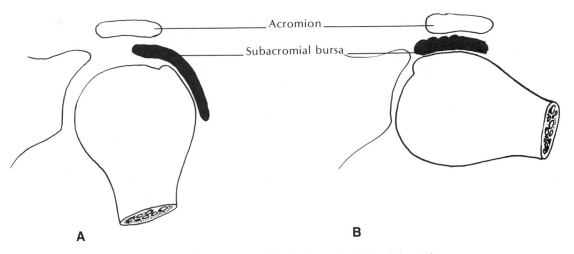

Figure 2-26. *Subacromial bursitis test.* **A.** *Arm at side.* **B.** *Arm abducted to right angle.*

ally communicates with the cavity of the shoulder joint. The **olecranon bursa** lies in the subcutaneous tissue spaces over the olecranon process, and the **subtendinous olecranon bursa** is located between the tendon of the triceps and the olecranon process. The small, but constant, **bicipitoradial bursa** is between the tendon of the biceps and the radial tuberosity.

Subacromial (Subdeltoid) Bursitis Test

The acromion process is separated from the tuberosities of the humerus by a single large bursa variously called the subdeltoid or subacromial. During examination, if the arm is at the side, pressure upon an inflamed bursa will elicit pain (Fig. 2-26A). However, in right-angled abduction of the arm, no pain will be felt from pressure because the bursa has moved under the shelter of the acromion (Fig. 2-26B). This observation (Dawborn's sign) aids in distinguishing subacromial bursitis from other lesions of the shoulder joint.

M A J O R   A N A T O M I C   A N D   C L I N I C A L   P O I N T S

Superficial Back and Shoulder

- □ The primary blood supply to the superficial muscles of the back and shoulder is derived from the transverse cervical and suprascapular arteries that course through the posterior triangle of the neck.
- □ The scapula is essentially embedded in muscle (supraspinatus, infraspinatus, subscapularis); muscle attachments to the vertebral border are insertions, and to the axillary border are origins.
- □ A mnemonic for the insertion of the "rotator cuff" muscles beginning on the greater tubercle is "SITS":
  - S—Supraspinatus
  - I —Infraspinatus
  - T—Teres minor
  - S—Subscapularis

- The deltoid muscle functions in all movements at the shoulder joint
- Muscular injections are placed in the fleshy belly of the deltoid.
- A "winged scapula" results from damage to the long thoracic nerve owing to denervation of the serratus anterior.
- A "shoulder droop" can result from a superficial wound in the posterior triangle of the neck that damages the spinal accessory nerve.
- Because of the strength of the fused tendons forming the rotator cuff, the protection of the acromion process, and the laxity of the joint capsule inferiorly, most shoulder dislocations occur inferiorly.

Arm

- The lateral and medial intermuscular septa divide the arm into anterior (flexor) and posterior (extensor) compartments.
- The radial nerve supplies all extensor muscles of the upper extremity distal to the shoulder joint.
- The musculocutaneous nerve supplies muscles in the anterior compartment of the arm.
- Both the biceps and triceps brachii are two-joint muscles.
- Blood pressure is usually taken by placing a pressure cuff so that it will occlude the brachial artery; this is the optimal site to apply a tourniquet for control of bleeding in the upper extremity.
- Fracture of the midshaft of the humerus endangers the radial nerve.
- Trauma to the radial nerve paralyzes the extensor muscles and produces a "wrist drop."
- A crushing injury of the elbow joint will usually damage the ulnar nerve and result in an "ulnar claw hand"; adduction and abduction of the fingers will also be impaired.
- In a dislocation of the shoulder the head of the humerus displaces inferiorly. This may damage the axillary nerve as it passes through the quadrangular space, and may impair the function of the deltoideus.

Pectoral Region and Axilla

- The breast is a mass of fat surrounding the mammary gland, which is located in the superficial fascia.

- The brachial plexus innervates all muscles of the upper extremity except the trapezius, sternocleidomastoid, and omohyoid.
- Cords of the brachial plexus are named from their relationship to the axillary artery.
- The apex of the axilla extends into the posterior triangle of the neck (boundaries: 1st rib, clavicle, superior border of scapula).
- The axillary artery is divided into three parts by the pectoralis minor muscle; one branch of the axillary artery arises from the first part, two from the second part, three from the third part.
- In fractures of the clavicle the subclavius muscle has a protective role to the underlying axillary vessels and brachial plexus.
- Inflammation from an infectious process in the neck can pass into the axilla along the cervico–axillary sheath.
- Metastasis from the breast, especially the lateral half, involves the axillary lymph nodes.
- Malignant invasion of the suspensory ligaments (of Cooper) of the breast may result in their shortening and in concomitant dimpling of the skin of the breast.
- Injury to the upper trunk of the brachial plexus, "Erb's point" results in a "waiter's tip hand."
- Trauma to the lower trunk of the brachial plexus, "Klumpke's paralysis" results in a "true claw hand."

Forearm

- The muscle mass on the medial aspect of the elbow is made up of flexors; on the lateral aspect, extensors. The tendons of the flexors sweep to the anterior aspect of the wrist; extensor tendons to the dorsal aspect.
- In the forearm the ulnar nerve supplies the flexor carpi ulnaris and the ulnar half of the flexor digitorum profundus. The median nerve supplies the remainder of the flexors.
- Three nerves supply cutaneous innervation to the forearm: the lateral antebrachial is a continuation of the musculocutaneous; the medial antebrachial arises from the medial cord of the brachial plexus; and the posterior antebrachial arises from the radial.
- Structures passing superficial to the flexor retinaculum are the tendon of the palmaris longus, the ulnar nerve, and the ulnar artery.

□ The median cubital vein is the vein most frequently used for intravenous injection and for drawing blood.

□ The pulse is usually taken by pressing the radial artery against the radius at the wrist.

□ In pronation of the hand the radius crosses over the ulna.

□ In falling on the hand the impact forces pass from the hand to the radius, across the interosseous membrane to the ulna, and from the ulna to the humerus.

□ Entrapment can occur as the major nerves enter the forearm where the radial passes between the two heads of the supinator, the median between the heads of the pronator teres, and the ulnar between the heads of the flexor carpi ulnaris.

Hand

□ Three major nerves supply cutaneous innervation to the hand. The ulnar supplies both surfaces medial to a line passing through the long axis of the ring finger. The median supplies the remainder of the palm and the dorsum of the distal phalanges on the lateral 3½ digits. The radial supplies the remainder of the dorsum of the hand.

□ The radial nerve supplies no muscles in the hand.

□ The median nerve supplies the thenar muscles; A mnemonic for this distribution is "LOAF":
  L—Lateral (2) lumbricales
  O—Opponens pollicis
  A—Abductor pollicis brevis
  F—Flexor pollicis brevis (superficial head)

□ The remaining intrinsic muscles are supplied by the ulnar.

□ The actions of the thumb are at right angles to those of the fingers.

□ Abduction and adduction of the fingers is relative to a line passing through the long axis of the middle finger.

□ The action of the lumbricales is flexion of the metacarpophalangeal joints and extension of the interphalangeal joints.

□ All intrinsic muscles of the hand are supplied by the first thoracic segment of the spinal cord.

□ An assessment of damage to the three major nerves can be made at the hand based on the following signs: In the radial, "wrist drop" and loss of sensation over the dorsum of the first interosseous space; in the median, loss of opposition and loss of sensation over the palmar aspect of the first interosseous space; and in the ulnar, weakness of adduction and abduction of the fingers and loss of sensation along the medial border of the hand.

□ The carpal tunnel syndrome is caused by swelling in the area between the flexor retinaculum and the carpal bones; it results in damage to the median nerve.

□ In surgical procedures the midpalmar space ("no man's land") is avoided if possible to lessen chances of infection of the ulnar bursa.

□ A "frozen" finger is caused by adhesions of the digital tendon sheath.

□ The collateral circulation is excellent in the hand owing to contribution to superficial and deep arches by both the radial and ulnar arteries.

QUESTIONS FOR REVIEW

1. The long thoracic nerve arises from cervical spinal nerves (C_5, C_6, C_7) and supplies which of the following muscles?

  A. Latissimus dorsi
  B. Teres major
  C. Teres minor
  D. Serratus anterior
  E. Rhomboideus major

2. All of the following statements are true of the human breast except

  A. Blood supply to the breast is derived principally from lateral thoracic and internal thoracic arteries
  B. The breast lies embedded in superficial fascia
  C. The deep lymphatics of the breast drain

primarily into parasternal and epigastric lymph nodes

    D. The breast is innervated by intercostal and supraclavicular nerves

    E. The structure of the breast includes fatty tissue, glandular tissue, and dense connective tissue septa

3. The brachial plexus gives rise to nerves that supply the upper extremity. The fifth and sixth cervical roots of the plexus (ventral rami) join to form the

    A. Lower (inferior) trunk

    B. Lateral cord

    C. Medial cord

    D. Middle trunk

    E. Upper (superior) trunk

4. The posterior cord of the brachial plexus gives rise to the axillary nerve. The median nerve is derived from which of the following?

    A. Only the lateral cord of the brachial plexus

    B. Only the medial cord of the brachial plexus

    C. Both the lateral and medial cords of the brachial plexus

    D. Only the posterior cord of the brachial plexus

    E. The middle trunk of the brachial plexus

5. The structure coursing along the bicipital (intertubercular) groove of the humerus is

    A. The tendon of the long head of the biceps

    B. The tendon of the long head of the triceps

    C. The radial nerve

    D. The profunda brachii artery

    E. The ulnar nerve

6. The musculotendinous rotator cuff

    A. Does not appreciably strengthen the shoulder joint, and therefore, has little functional significance

    B. Consists of the latissimus dorsi, teres major, supraspinatus, and subscapularis

    C. Consists of the subscapularis, supraspinatus, infraspinatus, and teres minor

    D. Lies within the articular capsule of the shoulder joint

    E. Consists of the glenohumeral ligaments, the subscapularis, and the long head of the biceps

7. The median nerve innervates all of the following except the

    A. Flexor carpi radialis

    B. Opponens pollicis

    C. Flexor carpi ulnaris

    D. Palmaris longus

    E. Flexor digitorum superficialis

8. Both pronators and the supinator insert on the

    A. Radius and ulna, respectively

    B. Ulna and radius, respectively

    C. Ulna only

    D. Radius only

    E. Humerus

9. Blood supply to the elbow joint is provided by all of the following except the

    A. Radial recurrent artery

    B. Ulnar recurrent artery

    C. Thoracodorsal artery

    D. Profunda brachii artery

    E. Superior ulnar collateral artery

10. The chief contents of the cubital fossa include all of the following except

    A. Biceps tendon

    B. Brachial artery

    C. Median nerve

    D. Ulnar nerve

    E. Bicipital aponeurosis

11. Wasting confined to the thenar eminence indicates a lesion of which of the following

    A. Radial nerve

    B. Ulnar nerve

    C. C_5 ventral ramus

    D. Median nerve

    E. Musculocutaneous nerve

12. Which of the following is true of muscles of the hand?

    A. The dorsal interossei adduct the fingers

    B. The median and ulnar nerves innervate an equal number of intrinsic muscles

    C. The lumbricales extend the metacarpophalangeal joints

    D. Both the thumb and the little finger have opponens muscles

    E. The flexor pollicis brevis flexes as well as abducts the thumb

13. The ulnar nerve innervates all of the following except the

    A. Hypothenar muscles
    B. Lumbricales of the index and middle fingers
    C. Flexor digitorum profundus to ring and little fingers
    D. Palmar interossei muscles

14. The most precise method for testing function of the median nerve in the hand is

    A. Sensation in the palmar aspect of the index finger and function of the opponens pollicis
    B. Sensation of the long, ring, and little fingers
    C. Extension of the fingers at the metacarpo-phalangeal joints
    D. Function of the first dorsal interosseous muscle

15. Which of the following are mismatched

    A. Quadrangular space : axillary nerve
    B. Radiospiral groove : radial nerve
    C. Triangular space : posterior humeral circumflex artery
    D. Pronator teres : median nerve
    E. Pronator quadratus : median nerve

16. When a patient has a shoulder injury, the best way to assess suprascapular nerve status or function is to observe the

    A. Flexion of the shoulder
    B. Adduction of the shoulder
    C. Abduction of the shoulder
    D. Extension of the shoulder
    E. Medial and lateral rotation of the shoulder

17. The muscle immediately deep to the belly of the biceps brachii is a strong
    _____ of the forearm and inserts onto the _____ .

    A. Flexor, ulna
    B. Supinator, radius
    C. Extensor, humerus
    D. Flexor, radius
    E. Pronator, ulna

18. The muscle that runs parallel to, and shares a common origin with, the short head of the biceps brachii muscle

    A. Is innervated by the ulnar nerve
    B. Flexes the arm

C. Receives all of its blood supply from the subscapular artery
D. Initiates abduction of the arm
E. Laterally rotates the humerus

19. The muscle immediately deep to the area of the skin supplied by the posterior brachial cutaneous nerve functions primarily in

    A. Flexing the arm
    B. Extending the arm and forearm
    C. Flexing the forearm
    D. Flexing the hand
    E. Adducting the hand and fingers

20. The muscle found immediately deep to the acromion process receives its motor innervation from a branch of the

    A. Median nerve
    B. Posterior cord of the brachial plexus
    C. Suprascapular nerve
    D. Cervical plexus
    E. Thoracodorsal nerve

21. Cutting the posterior cord of the brachial plexus would probably interrupt motor innervation to the _____ muscle.

    A. Serratus anterior
    B. Infraspinatus
    C. Biceps brachii (long head only)
    D. Pronator teres
    E. Teres minor

22. The roof of the antecubital (cubital) fossa is formed, in part, by the

    A. Bicipital aponeurosis
    B. Supinator
    C. Brachialis
    D. Interosseous membrane
    E. Radius and ulna

23. How many divisions are there in the brachial plexus?

    A. Three
    B. Four
    C. Five
    D. Six
    E. Seven

24. The muscle of the pectoral girdle that is used to divide the axillary artery into three parts is the _____ muscle. It normally receives its motor innervation from the _____ .

    A. Pectoralis minor, medial pectoral nerve

B. Subclavius, nerve to the subclavius
C. Pectoralis minor, medial and lateral pectoral nerve
D. Pectoralis major, lateral pectoral nerve
E. Pectoralis minor, lateral pectoral nerve

25. Medial brachial and medial antebrachial cutaneous nerves normally arise directly from

   A. Cord levels T_1 and T_2
   B. The medial cord of the brachial plexus
   C. The lower (inferior) trunk of the brachial plexus
   D. The median nerve
   E. The ulnar nerve

26. Which of the following is not classically a branch of the thoracoacromial trunk?

   A. Pectoral artery
   B. Lateral thoracic artery
   C. Acromial artery
   D. Deltoid artery
   E. Clavicular artery

27. The medial wall of the axilla is formed in part by the _____, while the _____ forms a portion of the anterior wall.

   A. Serratus anterior, pectoralis major
   B. Pectoralis minor, latissimus dorsi
   C. Humerus, teres major
   D. Coracobrachialis, subscapularis

28. Which of the following structures bounds the apex of the axilla?

   A. Lateral border of first rib
   B. Humerus
   C. Serratus anterior
   D. Anterior axillary fold
   E. Proximal border of the pectoralis minor

29. There is a single bony articulation between the upper extremity and the axial skeleton. This articulation is the

   A. Shoulder joint
   B. Acromioclavicular joint
   C. Sternoclavicular joint
   D. Sternocostal joint

30. One of the branches of the posterior cord of the brachial plexus is the _____ nerve.

   A. Medial brachial cutaneous
   B. Ulnar
   C. Thoracodorsal

D. Musculocutaneous
E. Long thoracic

31. The suprascapular nerve

   A. Innervates the infraspinatus muscle
   B. Is a branch of the posterior cord of the brachial plexus
   C. Innervates the teres minor
   D. Is a branch of the lower trunk of the brachial plexus

32. Superficial venous drainage from the ulnar side of the forearm and arm usually reaches the brachial vein, by way of the

   A. Basilic vein
   B. Saphenous vein
   C. Axillary vein
   D. Ulnar vein
   E. Cephalic vein

33. All of the following are usually direct branches of the brachial artery except the

   A. Profunda brachii artery
   B. Anterior interosseus
   C. Superior ulnar collateral artery
   D. Radial artery
   E. Ulnar artery

34. Hemorrhage from a superficial stab wound in the deltopectoral triangle would be primarily from the

   A. Axillary vein
   B. Basilic vein
   C. Subclavian vein
   D. Cephalic vein
   E. Deltopectoral vein

35. A powerful extensor of the elbow is the

   A. Latissimus dorsi muscle
   B. Anconeus muscle
   C. Biceps brachii muscle
   D. Triceps brachii muscle
   E. Brachioradialis muscle

36. Muscles inserting adjacent to each other along the intertubercular groove include all of the following except the

   A. Latissimus dorsi
   B. Coracobrachialis
   C. Pectoralis major
   D. Teres major

37. The costal surface of the scapula receives the insertion of a muscle whose function is to hold the scapula to the chest wall. When

the nerve to this muscle is destroyed, the result is a "winged" scapula. The muscle is the

A. Subscapularis
B. Supraspinatus
C. Infraspinatus
D. Teres major
E. Serratus anterior

38. The muscle that does not have its primary insertion on the scapula is the

A. Rhomboideus major
B. Rhomboideus minor
C. Latissimus dorsi
D. Trapezius
E. Levator scapulae

39. Inferior dislocation (subluxation) of the shoulder joint with impingement on a nerve would most probably impair the function of the _____ muscle.

A. Supraspinatus
B. Infraspinatus
C. Subscapularis
D. Deltoideus
E. Latissimus dorsi

40. A spiral fracture at the midshaft of the humerus would most likely damage the

A. Axillary nerve
B. Radial nerve
C. Musculocutaneous nerve
D. Median nerve
E. Ulnar nerve

41. The deep extensor muscles of the forearm are supplied by the

A. Ulnar artery
B. Radial artery
C. A branch of the common interosseus artery
D. The profunda brachii (deep brachial) artery

42. The ulnar artery

A. Enters the forearm by passing posterior to the medial epicondyle of the humerus
B. Lies along the radial border of the flexor carpi radialis muscle at the distal end of the radius
C. Passes through the wrist deep to the flexor retinaculum

D. Lies between the flexor digitorum superficialis and flexor digitorum profundus muscles in the forearm

43. The proximal boundary (base) of the antecubital fossa is delineated by the

A. Brachioradialis muscle
B. Brachialis muscle
C. Line drawn between the medial and lateral epicondyles of the ulna
D. Pronator teres muscle
E. Line drawn between the medial and lateral epicondyles of the humerus

44. In the classic dermatome distribution, pain around the base of the nail of the ring finger would be felt by means of the

A. Ulnar and radial nerves
B. Ulnar nerve
C. Median and ulnar nerves
D. Radial nerve
E. Lateral antebrachial cutaneous nerve

45. The lumbricales muscles, because of their attachment to the extensor hood of the digits, function primarily by

A. Abducting the digits
B. Adducting the digits
C. Flexing the metacarpophalangeal joints and extending the interphalangeal joints
D. Extending the metacarophalangeal joints and flexing the interphalangeal joints
E. Opposing the first and fifth digits

46. The _____ muscle, a flexor of the forearm in the midprone position, is innervated by the radial nerve and forms the lateral boundary of the _____

A. Brachioradialis, antecubital fossa
B. Pronator teres, antecubital triangle
C. Triceps brachii, quadrilateral space
D. Biceps brachii, triangular space
E. Pronator quadratus, carpal tunnel

47. All of the following statements concerning the forearm are true except

A. All of the extensor muscles are innervated by the radial nerve
B. All of the flexors (including pronator) are innervated by the median nerve
C. All of the tendons of flexor digitorum profundus pass deep (posterior) to the flexor retinaculum to reach the hand

D. The pronator quadratus is a flexor compartment muscle

E. The supinator is an extensor compartment muscle

48. The lateral antebrachial cutaneous nerve

A. Is the distal continuation of the musculo-cutaneous nerve

B. Is the direct continuation of the axillary nerve

C. Arises directly from the lateral cord of the brachial plexus

D. Arises directly from the lateral aspect of the medial cord of the brachial plexus

E. Carries sensory impulses from the lateral half of the hand and the first three digits

49. Which of the following nerves passes immediately deep to the deltoid muscle and innervates the teres minor muscle?

A. Musculocutaneous

B. Median

C. Radial

D. Ulnar

E. Axillary

50. One nerve that is not derived from the brachial plexus but innervates muscles of the upper extremity is the

A. Long thoracic

B. Thoracodorsal

C. Dorsal Scapular

D. Axillary

E. Spinal accessory

THREE

Thorax

Thoracic Cage

The thoracic cage, formed by the sternum, the ribs, and the thoracic vertebrae, gives protection for the lungs and heart and affords attachment for muscles of the thorax, upper extremity, back, and diaphragm.

Objectives

At the completion of the study of the thoracic wall the student should be able to

▶ *Palpate the surface landmarks on the thoracic wall*

▶ *List the anatomic features of a typical rib*

▶ *List the boundaries of the thorax and give the contents of an intercostal space*

▶ *Follow the origin, course, and distribution (drainage) of the intercostal artery, vein, and nerve*

▶ *Explain the role of the intercostal muscles and thoracic diaphragm in the movement of the lung in respiration*

Skeleton of the Thorax

The **sternum** is a flat bone consisting of three parts, the manubrium, the body and the xiphoid process. The **manubrium,** united with the body by fibrocartilage at the **sternal angle (of Louis),** presents superiorly a **suprasternal** (jugular) **notch** and articulates laterally with the clavicle and the first costal cartilage. At its junction with the body of the sternum, lateral demifacets are

present for articulation with the second costal cartilage. The **body,** forming the bulk of the sternum, articulates with the second through seventh costal cartilages. It is united with the xiphoid process inferiorly at the **xiphisternal junction.** The **xiphoid process,** the smallest part of the sternum, is thin, flattened, elongated, and presents a demifacet for the seventh costal cartilage. In youth the xiphoid process consists of hyaline cartilage, which is gradually replaced with bone.

There are twelve pairs of ribs. The first seven are **true ribs** in that they articulate by way of their costal cartilages with the sternum. The remaining five pairs are called **false ribs** because either their attachment anteriorly is through the costal cartilage of the rib above, rather than directly to the sternum; or in the case of the eleventh and twelfth ribs, they have no anterior attachment and are called **floating ribs.** Ribs one through seven increase progressively in length, while the remaining ribs decrease. The first rib has the greatest curvature and thereafter the curvature diminishes.

A **typical rib** (third through ninth) can be divided into three parts: head, neck, and shaft (Fig. 3-1). The **head** is wedge-shaped with a crest at the apex presenting two demifacets that articulate with the numerically corresponding thoracic vertebra and the vertebra above. At the **junction** of the **neck** and **shaft,** a **tubercle** on the posterior surface articulates with the transverse process of the numerically corresponding vertebra. The

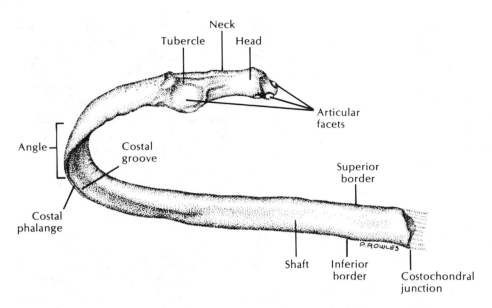

Figure 3-1. *A typical rib.*

long, thin **shaft,** or body, is curved, and twisted on its long axis. It forms the **angle** of the rib at its point of maximal curvature. The **costal groove,** on the inferior border of the rib, gives protection to intercostal nerves and vessels.

Thoracocentesis

Pus in the thorax, an excess of fluid in pleural effusion, or bleeding into the pleural cavity may necessitate draining this material from the thorax. In this procedure, called thoracocentesis, insertion of a large bore needle close to the superior border of a rib avoids injury to the vessels and nerves in the costal groove.

The first, second, tenth, eleventh, and twelfth ribs are classified as **atypical.** The **first rib** is short, broad, and flat. It has a long neck and only one facet on its head. There is a prominent tubercle on this rib for articulation with the transverse process of the first thoracic vertebra. The tubercle is separated by a shallow groove, formed by the subclavian artery and the brachial plexus, from the **scalene tubercle.** The latter affords attach-

ment for the anterior scalenus muscle. The **second rib** is appreciably longer than, but similar to, the first. Its shaft is not twisted but is still strongly curved, and the costal groove is indistinct. Its roughened external surface gives partial origin for the serratus anterior muscle. The tenth through twelfth ribs are atypical in that their heads have only single facets for articulation with the bodies of corresponding vertebrae. In addition, the **eleventh rib** is short, presents a slight angle, but has no neck or tubercle, while the **twelfth rib** is very short and lacks neck, tubercle, angle, and costal groove.

The **costal cartilages** are bars of hyaline cartilage that prolong the ribs anteriorly. The upper seven cartilages articulate with the sternum, the next three with the costal cartilages immediately above; the last two terminate in abdominal wall musculature.

The ribs and intercostal spaces constitute the thoracic wall. Each **intercostal space** is filled by muscles between which arteries, veins, and nerves course. The intercostal muscles elevate the ribs and receive their nerve supply from corresponding intercostal or thoracoabdominal nerves.

Intercostal Muscles

The intercostal muscles are disposed in three layers (Table 3-1, Fig. 3-2). The most superficial, the **external intercostal muscle,** partially fills the intercostal space from the vertebral column to the costochondral junction. Here its muscle fibers are lost but its fascia continues to the sternum as the **anterior intercostal membrane.** The **internal intercostal muscle** forms the middle layer of muscle. Its muscle fibers extend from the sternum anteriorly to the angle of the ribs posteriorly, where only its fascia continues to the vertebral column as the **posterior intercostal membrane.** The third muscular layer is discontinuous and is described as three individual muscles— anteriorly the **transversus thoracis** (sternocostalis), laterally the **innermost intercostals** (sometimes considered as a splitting of the internal intercostal), posteriorly the **subcostales.** The nerves, arteries, and veins within the intercostal space course between the second and third layers of muscles.

Intercostal Arteries and Nerves

Arteries passing in the intercostal space supply intercostal muscles, send twigs to the overlying pectoral muscles, mammary gland, and skin, and, at the lower intercostal spaces give branches to the diaphragm (Fig. 3-3; Table 3-2). The **internal thoracic** (internal mammary) **artery,** arising in the root of the neck as a branch of the first part of the subclavian, descends vertically at the side of the sternum to divide behind the sixth intercostal space into **musculophrenic** and **superior epigastric branches.** The **anterior intercostal arteries** supplying the first six intercostal spaces are branches of the internal thoracic artery. The remaining spaces are supplied by similar branches of the musculophrenic artery. The **posterior intercostal arteries** to the first two intercostal spaces are branches of the **superior intercostal,** a branch of the costocervical trunk from the subclavian. The remaining intercostal spaces receive their posterior (aortic) intercostal branches directly from the descending aorta. Additional branches from the posterior intercostals pass posteriorly to supply the deep muscles of the back, the vertebral column, and the spinal cord. The **subcostal artery** is the most inferior paired branch of the thoracic aorta. It arises in line with the posterior intercostals and passes inferiorly to supply structures in the abdominal wall.

The **intercostal nerves** are ventral rami of thoracic spinal nerves. The **typical intercostal nerves** (third through sixth thoracic) course along the posterior intercostal membrane to reach the **costal groove** on the inferior border of

Table 3-1
Muscles of the Thorax

Muscle	Origin	Insertion	Action	Nerve
External intercostals	Inferior border of rib	Superior border of rib below origin	Elevate rib	Segmentally by intercostals
Internal intercostals	Superior border of rib	Inferior border of rib above origin	Elevate rib	Segmentally by intercostals
Innermost intercostals	Variable in extent, sometimes considered as deep portion of internal intercostals, being separated by intercostal nerves and vessels			
Transversus thoracis	Posterior surface of xiphoid process and lower third of sternum	Inner surface of costal cartilage of second to sixth ribs	Draws costal cartilages inferiorly	Segmentally by intercostals
Subcostales	Inner surface of lower ribs near their angle	Inner surface of second or third rib below rib of origin	Draw adjacent ribs together	Segmentally by intercostals

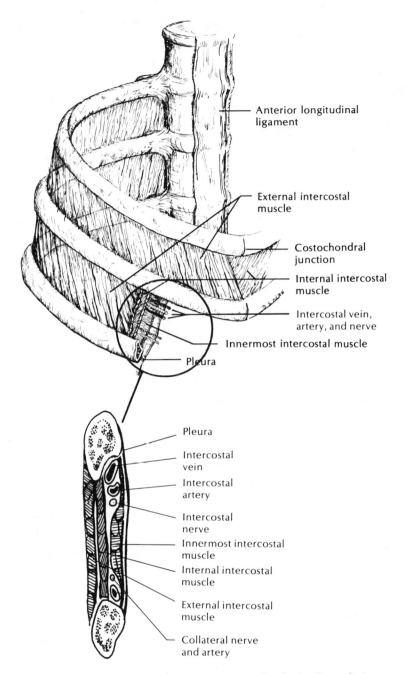

Figure 3-2. *Intercostal muscles. (Lower figure after Crafts, Roger C: A Textbook of Human Anatomy, 2nd ed. New York, John Wiley & Sons, 1979)*

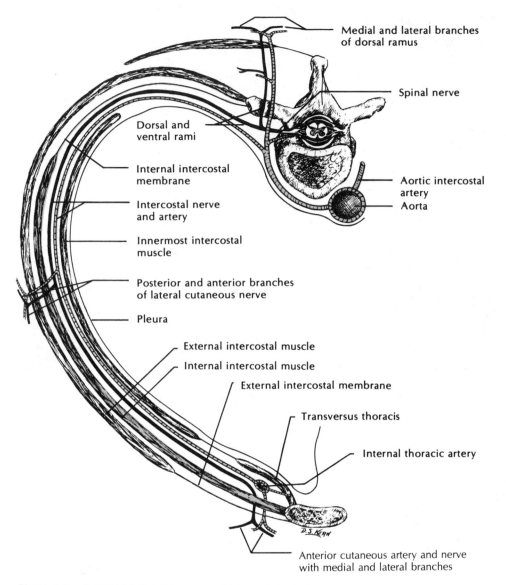

Medial and lateral branches
of dorsal ramus

Spinal nerve

Dorsal and
ventral rami

Internal intercostal
membrane

Intercostal nerve
and artery

Innermost intercostal
muscle

Posterior and anterior branches
of lateral cutaneous nerve

Pleura

Aortic intercostal
artery
Aorta

External intercostal muscle

Internal intercostal muscle

External intercostal membrane

Transversus thoracis

Internal thoracic artery

Anterior cutaneous artery and nerve
with medial and lateral branches

Figure 3-3. *Typical intercostal nerve and intercostal arteries.*

Table 3-2
Arterial Supply to the Thoracic Wall

Artery	Origin	Course	Distribution	Anastomoses
Internal thoracic	Subclavian	Passes inferiorly in a parasternal line between internal intercostal and sternocostalis muscles	Anterior intercostals to first six intercostal spaces	No direct anastomoses
Musculophrenic	Internal thoracic	Passes along infracostal margin	Anterior intercostals to lower five intercostal spaces; diaphragm	Superior and inferior phrenics
Anterior intercostals	Internal thoracic or musculophrenic	Pass between second and third layer of muscles in intercostal spaces	Muscles of intercostal space and overlying skin	Posterior intercostals
Superior intercostal	Costocervical trunk	Descends from costocervical trunk to terminate as posterior intercostals to upper two spaces	Intercostal muscles of first and second spaces	No direct anastomoses
Posterior intercostals	Superior intercostal and aorta	Pass between second and third layer of muscles in intercostal spaces	Intercostal muscles and overlying skin	Anterior intercostals
Subcostal	Aorta	Courses along inferior border of twelfth rib	Anterolateral abdominal wall muscles	Superior epigastric

the rib between the internal and subcostalis muscles and inferior to the intercostal vessels. They supply all the intercostal muscles and, in the midaxillary line give off **lateral cutaneous nerves,** which send **anterior** and **posterior branches** to supply the skin of the lateral chest wall. The intercostal nerves then proceed anteriorly to penetrate the anterior body wall just lateral to the sternum as the **anterior cutaneous nerves,** which divide into **medial** and **lateral branches** and supply the skin on the anterior aspect of the thorax.

Atypical intercostal nerves are also ventral rami, but have several distinctive characteristics. The **first thoracic nerve** is short and thick and divides unevenly into two branches. The small inferior branch supplies structures in the first inter-

costal space, while the much larger superior branch joins the brachial plexus. The cutaneous innervation over the first intercostal space is not supplied by this nerve, but by the **medial** and **intermediate supraclavicular nerves.** The lateral cutaneous branch of the second thoracic nerve distributes with the brachial plexus as the **intercostobrachial branch** to supply cutaneous innervation to the floor of the axilla as well as to the posteromedial aspect of the arm.

At the point where the costal cartilages course upward, the ventral rami of the seventh through eleventh thoracic **(thoracoabdominal)** nerves pass obliquely anteroinferiorly. They pass between the internal abdominal oblique and transversus abdominis muscles and supply the anterolateral abdominal wall musculature, the rectus abdominis, and the overlying skin. The ventral ramus of the twelfth thoracic nerve is the **subcostal nerve.** It passes anterior to the quadratus lumborum muscle, then to the abdominal wall, where its distribution is similar to that of the lower intercostal nerves. The dermatome flanking the umbilicus is supplied by the tenth thoracic nerve.

Pleural Cavity

Objectives

At the completion of the study of the lung and pleural cavity the student should be able to

▶ *Describe the surface projection of the pleurae, lungs, liver, and heart*

▶ *Describe the pleural cavity and the designation of portions of the parietal pleura.*

▶ *Label on a diagram the subdivisions of the lung and tracheobronchial tree*

▶ *Define the hilum of the lung and list the structures forming the root of the lung*

Pleura

The thorax contains three serous cavities, the two laterally situated pleural sacs and the midline pericardial cavity. The two **pleural cavities** are completely closed, separate, and lined by a serous membrane, the **pleura.** This membrane is a continuous sheet in each cavity but is subdivided for descriptive purposes into parietal and visceral portions. The **parietal pleura** applied to the thoracic cage is designated as the **costal parietal pleura;** the portion fused to the diaphragm, the **diaphragmatic parietal pleura,** and the portion adjacent to the mediastinum, the **mediastinal parietal pleura.** The **visceral pleura** is intimately adherent to the lung, covering its entire surface and continuing deeply into its fissures. In the adult the lung lies free in the cavity, except at the root of the lung, where the visceral pleura reflects to become the parietal pleura. The **pulmonary ligament** extends inferiorly from the root of the lung as a fusion of two layers of mediastinal pleura. Near the diaphragm the ligament ends in a free falciform border. Anteriorly the costal parietal pleura reflects to become the mediastinal parietal pleura forming the **costomediastinal recess;** inferiorly it is reflected onto the diaphragm to form the **costodiaphragmatic (phrenicocostal) recess.** The lung does not extend into these recesses except in maximal inspiration.

Projection of the **pleural reflection** onto the surface of the chest is related to certain ribs or costal cartilages. The anterior border of the parietal pleura passes inferiorly from the **cupula** (pleura over the apex of the lung) to the sternoclavicular joint, continues obliquely deep to the manubrium, and proceeds inferiorly along the posterior aspect of the midline of the sternum from the level of the second to the fourth costal cartilage. Here the margin of the left pleural reflection moves laterally, then inferiorly at the cardiac notch to reach the sixth intercostal space. On the right side, the pleural reflection continues inferiorly in the midline, and at the sixth costal cartilage both angle laterally to reach the midclavicular line at the level of the eighth costal cartilage. Both reflections cross the tenth rib at the midaxillary line and the twelfth rib at the midscapular line.

Pneumothorax

Because of the extension of the pleural cavity and lung into the posterior triangle of the neck, a puncture wound from a knife, needle, or bullet superior to the clavicle could penetrate the pleura and the lung and produce a pneumothorax (air in pleural cavity).

Lungs

The lungs are organs of respiration. **Inspiration** is an active process that results from a decrease in the intrathoracic pressure, as the thoracic cavities increase their volume, by the activity of the chest muscles and the diaphragm. Normal **expiration** is passive. Relaxation of the chest muscles results in a resilient recoil of the ribs and lungs with a concomitant intrathoracic volume decrease and pressure increase. The most extensive movement of the lung occurs in the lateral, anterior, and inferior directions. Each lung is accurately adapted to the space in which it lies and, when hardened *in situ,* bears impressions of structures in contact with its surface.

The **shape of the lung,** described as a bisected cone, presents an apex, a base, costal and mediastinal surfaces, and anterior, inferior, and posterior borders (Fig. 3-4). The rounded **apex,** lying behind the middle third of the clavicle and rising about 3 cm to 4 cm above the anterior aspect of the first rib, is grooved by the subclavian

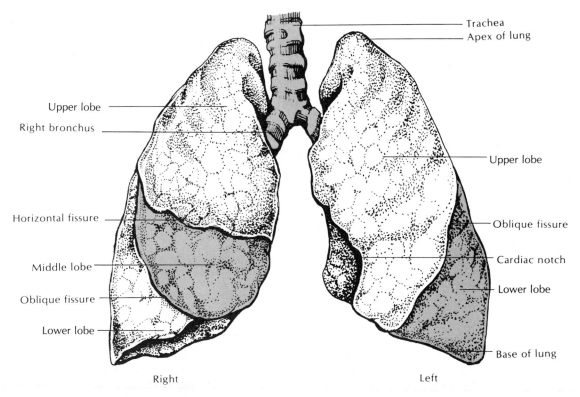

Figure 3-4. *Anterior view of trachea, bronchi, and lungs. (After Hollinshead, W Henry: Textbook of Anatomy, 3rd ed. Hagerstown, Harper & Row, 1974)*

artery and separated from structures in the neck by the **suprapleural membrane,** or **Sibson's fascia.** The concave semilunar **base** is adapted to the upper surface of the diaphragm. The convex and extensive **costal surface** follows the contour of the rib cage. The **mediastinal surface** is adjacent to, and marked by, structures in the mediastinum. It presents a large centrally depressed area, the **hilum,** through which bronchi, nerves, vessels, and lymphatics enter or leave the lung.

Owing to the high position of the right lobe of the liver, the **right lung** is shorter than the left. However, it has a greater volume than the left lung because of the displacement of the left lung by the position of the heart. It is related superiorly to the superior vena cava, the brachiocephalic and azygos veins, and, anterior to these structures, the ascending aorta. Posterior to the hilum, the right lung is grooved by the esophagus and more superiorly by the arch of the aorta, arch of the azygos vein, and the trachea.

The **left lung** displays a deep depression adjacent to the apex and left surface of the heart and is deeply grooved above and behind the hilum by the arch of the aorta and the descending aorta. Superiorly the left lung is grooved by the subclavian and common carotid arteries. The **anterior borders** of both lungs are comparatively short and thin and extend into the costomediastinal recesses. On the left lung the anterior border is displaced laterally by the heart to form the cardiac notch. The rounded, thick **posterior borders** form an indistinct confluence of costal and mediastinal surfaces occupying a deep hollow on either side of the vertebral column. Along its **inferior border** the mediastinal portion is blunt and related to the lower border of the pericardium. Anteriorly, laterally, and posteriorly this border is thin and sharp and extends into the costodiaphragmatic recess.

Each lung is partially transected by an **oblique fissure** from its surface to within a short distance of the hilum. On the left lung the fissure extends from the posterior border approximately 6 cm to 7 cm below the summit, runs obliquely and inferi-

orly to cross the fifth rib in the midaxillary line, and ends anteriorly opposite the sixth costal cartilage. On the right lung the oblique fissure begins slightly lower. The oblique fissure in each lung separates the **upper** (superoanterior) **lobe** from the **lower** (posteroinferior) **lobe.** The right lung is further subdivided by the **horizontal fissure,** which extends from the anterior border of the lung at the fourth intercostal space to follow the upper border of the fifth rib to the oblique fissure to form the wedge-shaped **middle lobe.** In the left lung the **lingula** is homologous to the middle lobe of the right lung and consists of a small tonguelike appendage between the cardiac notch and the oblique fissure.

Air Conduction System

The **trachea** is a wide tube about 10 cm in length. It is kept patent by a series of sixteen to twenty U-shaped horizontal cartilaginous bars embedded in its wall. Posteriorly the open cartilages of the tube are closed by the **trachealis muscle** and fibrous tissue. The trachea begins at the cricoid cartilage of the larynx, at the level of the sixth cervical vertebra. It terminates between the fourth and fifth thoracic vertebrae by dividing into right and left primary bronchi (Fig. 3-5). In inspiration the respiratory shift can carry the bifurcation as low as the seventh vertebra. The trachea lies anterior to the esophagus and inferior laryngeal nerves. It is posterior to the left brachiocephalic vein, left common carotid, and the arch of the aorta. Laterally, on the right, the trachea is related to the pleura, vagus nerve, and arch of the azygos vein, and on the left to the common carotid and subclavian arteries, phrenic and vagus nerves, and the arch of the aorta.

Tracheotomy

Tracheotomy is a life-saving procedure, often needed to relieve an upper respiratory obstruction from inhalation of a foreign object, which can cause spasm of the laryngeal muscles and airway obstruction. A tracheotomy can be per-

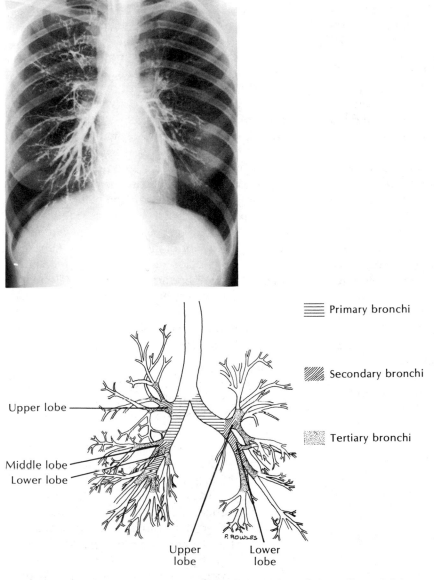

Primary bronchi

Secondary bronchi

Tertiary bronchi

Upper lobe

Middle lobe
Lower lobe

Upper
lobe

Lower
lobe

P. ROWLES

Figure 3-5. *Posteroanterior radiograph of the bronchial tree (refer to Figure 3-6 for tertiary bronchi).*

formed by making a midline skin incision from just above the cricoid cartilage to the jugular notch of the manubrium. The incision is deepened so that the third and fourth tracheal rings can be incised. A metal or plastic tracheal tube is then inserted into the trachea and breathing is re-established. Structures to be avoided include the inferior thyroid veins, the jugular venous arch, and the isthmus of the thyroid. To avoid these structures and some complications from the tracheal tube, some physicians prefer to open the trachea just inferior to the cricoid cartilage.

The **right primary bronchus,** a more direct continuation of the trachea, is shorter, straighter, and larger than the left bronchus; therefore, foreign bodies are more apt to pass into the right than into the left lung. The right primary bronchus enters the lung at the hilus, where it divides into **secondary bronchi** passing to the upper, middle, and lower lobes. The superior lobe bronchus passes above the pulmonary artery as the **eparterial bronchus.** Secondary bronchi give rise to **tertiary bronchi** named for the **bronchopulmonary segments** they supply. The secondary bronchus to the right upper lobe divides into apical, posterior, and anterior tertiary bronchi. The secondary bronchus to the right middle lobe divides into lateral and medial tertiary bronchi and the secondary bronchus to the right lower lobe divides into superior, medial basal, anterior basal, lateral basal, and posterior basal tertiary bronchi. Thus, ten bronchopulmonary segments are present in the right lung (Fig. 3-6).

The **left primary bronchus** is smaller in caliber, but roughly twice as long as the right. It passes initially superior to the pulmonary artery, but divides inferior to the artery into upper and lower lobe **secondary bronchi.** The secondary bronchus to the upper lobe subdivides into apicoposterior, anterior, superior, and inferior tertiary bronchi. The lower lobe bronchus subdivides into superior, anteromedial basal, lateral basal, and posterior basal tertiary bronchi. This results in eight bronchopulmonary segments in the left lung.

The **root of the lung** is formed by structures entering or leaving the lung. It consists of the bronchi, the pulmonary artery (carrying venous blood), pulmonary veins (carrying arterial blood), bronchial arteries supplying the bronchial tree, lymph vessels and nodes, and the pulmonary plexus of nerves. These structures are held together by connective tissue and are surrounded by visceral pleura reflecting from the surface of the lung to become the parietal pleura. The superior vena cava and right atrium are anterior to the root of the right lung, and the azygos vein arches above it. The root of the left lung lies anterior to the descending aorta and inferior to the arch of the aorta. The phrenic nerves, pericardiophrenic arteries and veins, and the anterior pulmonary plexuses are located anterior to both roots. The posterior pulmonary plexuses and the vagus nerves are posterior to both roots.

Bronchoscopy

The trachea, main bronchi, and the first segmental bronchi can be viewed directly through a bronchoscope. Such an instrument also enables the physician to remove foreign objects from these passageways. These objects usually fall into the right bronchus because it has a larger caliber and is more in line with the vertical trachea. Spraying a radiopaque oil into the bronchi outlines, on a chest x-ray, the deeper bronchi and the lung proper. The procedure, called a bronchiogram, helps localize bronchial lesions or obstructions.

Blood Vessels, Lymphatics, and Nerves (Table 3-3)

The **right** and **left pulmonary arteries** are branches of the **pulmonary trunk,** which, in turn, stems from the right ventricle. The **right**

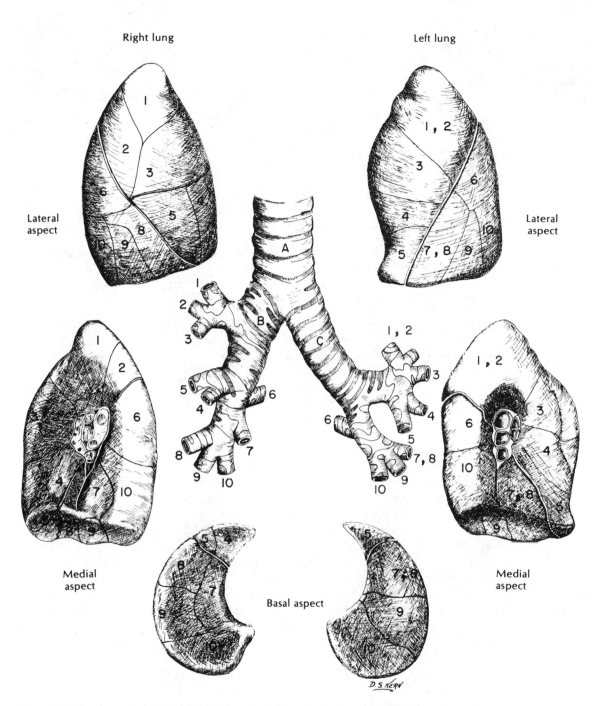

Figure 3-6. *Lung and tracheobronchial tree.* **A.** *Trachea.* **B.** *Right primary bronchus.* **C.** *Left primary bronchus. Tertiary bronchi and bronchopulmonary segments include:* (right lung). *1. apical, 2. posterior, 3. anterior, 4. medial, 5. lateral, 6. superior, 7. medial basal, 8. anterior basal, 9. lateral basal, 10. posterior basal; and* (left lung) *1. and 2. apicoposterior, 3. anterior, 4. superior, 5. inferior, 6. superior, 7. and 8. anteromedial basal, 9. lateral basal, 10. posterior basal.*

Table 3-3
Pulmonary Circulation

Artery or Veins	Origin	Course	Distribution	Anastomoses
Pulmonary trunk	Right ventricle	Ascends approximately two and a half inches; terminates as right and left pulmonary arteries	Transmits blood to lung to be oxygenated	No direct anastomoses
Pulmonary artery (right and left)	Pulmonary trunk	Passes to hilum of lungs	To respective right and left lungs	No direct anastomoses
Lobar	Pulmonary arteries	At hilum, three on right, two on left	Lobes of lungs	No direct anastomoses
Tertiary	Lobar arteries	Follow tertiary bronchi	Bronchopulmonary segments	No direct anastomoses
Right bronchial	Posterior intercostal of third or fourth space	Passes with bronchial tree	Bronchial and parabronchial tissue	No direct anastomoses
Left bronchial	Aorta	Passes with bronchial tree	Bronchial and parabronchial tissue	No direct anastomoses
Pulmonary venules	Begin at periphery of pulmonary lobule	Course in intersegmental planes	Coalesce to form 2–6 pulmonary veins, which terminate in left atrium	No direct anastomoses

branch, longer and of larger caliber than the left, passes posterior to the ascending aorta and superior vena cava and anterior to the right bronchus. It divides into two branches, one to the upper lobe and a second to the middle and lower lobes. The **left pulmonary artery** passes horizontally, anterior to the descending aorta and the left bronchus, and divides into branches to the upper and lower lobes. Further subdivisions of the pulmonary arteries correspond to the divisions of the bronchial tree.

Pulmonary veins begin at alveolar capillaries and coalesce at the periphery of the lung lobule. They maintain a peripheral relation in the bronchopulmonary segments running in intersegmental connective tissue. They drain adjacent, rather than single, segments. On the right side the middle lobe vein joins the lower lobe vein at the hilus, usually resulting in two veins from each lung draining into the left atrium.

The **bronchial arteries** are small vessels coursing along the posterior aspect of the bronchi to supply the bronchi and bronchioles to the level of the respiratory bronchiole. Classically, two **left bronchial arteries** arise from the anterior aspect of the descending aorta, while a single **right bronchial artery** arises from the first aortic intercostal or the upper left bronchial artery.

The **lymphatics** of the lungs are disposed as a **superficial plexus** lying immediately deep to the visceral pleura, and a **deep plexus** arising in the submucosa of the bronchi and peribronchial connective tissue. The efferent lymph vessels of the superficial plexus follow the surface of the lung to the bronchopulmonary trunks; those of the deep plexus follow the pulmonary vessels to the hilus. The **nodes** of the lung are disposed in five groups: the tracheal or paratracheal nodes along the trachea, the superior tracheobronchial

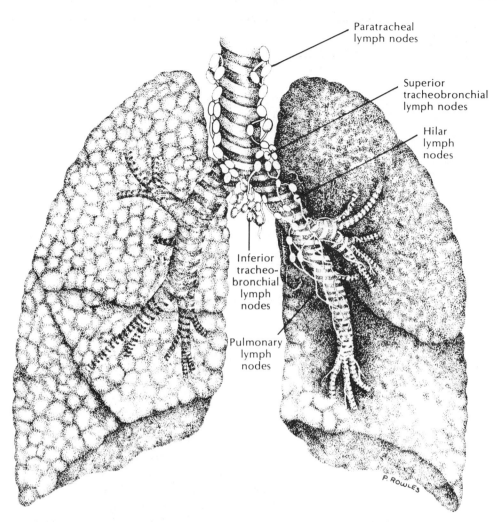

Paratracheal
lymph nodes

Superior
tracheobronchial
lymph nodes

Hilar
lymph
nodes

Inferior
tracheo-
bronchial
lymph
nodes

Pulmonary
lymph
nodes

P. ROWLES

Figure 3-7. *Lymphatics of lungs.*

nodes in the angle between the trachea and bronchi, the inferior tracheobronchial nodes in the angle between the bronchi, the bronchopulmonary nodes at the hilus of each lung, and the pulmonary nodes in the substance of the lung (Fig. 3-7).

Innervation to the lung is supplied by the vagus and thoracic sympathetic nerves distributed through the **anterior** and **posterior pulmonary nerve plexuses,** located on corresponding surfaces of the root of the lung. The cell bodies of the preganglionic sympathetic neurons are located in the upper three to five thoracic segments of the spinal cord.

Asthma
Efferent (motor) vagal stimuli cause smooth muscle contraction, thus narrowing the lumens of bronchi and bronchioles. Efferent sympathetic fibers are bronchiodilatory, therefore epinephrine (adrenalin) injections will relieve the bronchial spasms in asthma by blocking the vagal stimuli.

Pericardial Cavity

Objectives

At the completion of the study of the heart and pericardial cavity the student should be able to

▶ *Locate the oblique and transverse pericardial sinuses and describe the reflections of parietal to visceral pericardium that delineate these sinuses.*

▶ *Describe the surface anatomy of the heart and relate each chamber to the borders and surfaces of the heart*

▶ *List and define the anatomic features of each chamber of the heart*

▶ *Label on a diagram the arteries and veins of the coronary circulation*

▶ *Follow the pathway of both parasympathetic and sympathetic impulses along the conduction system of the heart*

▶ *Relate the valves of the heart to the skeleton of the heart and to blood flow through the heart*

The **pericardial cavity,** a conical fibroserous sac, surrounds the heart and proximal portions of the great cardiac vessels. The **pericardium** is disposed into three layers: an external strong **fibrous layer,** which blends with the adventitia of the great vessels, a **parietal layer** of serous pericardium adherent to the inner surface of the fibrous layer, and a **visceral layer** of serous pericardium that reflects onto and is intimately applied to the heart and great vessels. The proximal portions of the aorta, the pulmonary trunk, and the superior vena cava are surrounded by visceral pericardium. Near the median plane, the base of the pericardial cavity rests on and fuses with, the central tendon of the diaphragm. The

right posteroinferior aspect of the base is pierced by the inferior vena cava and posteriorly, at the junction of the upper part of the posterior and lateral surfaces, by the pulmonary veins.

Cardiac Tamponade

Inflammation of the pericardium (pericarditis) with an effusion of fluid or extensive bleeding into the pericardial sac, if untreated, is a life threatening condition. The tough, fibrous, inelastic pericardium will not stretch to accommodate the excess fluid. This results in a compression of the heart (cardiac tamponade) and circulatory failure.

Anteriorly the pericardium is separated from the sternum and the second to sixth costal cartilages by the lungs and pleura, except in the region of the left fourth through sixth intercostal spaces, where the left lung is deficient at the **cardiac notch.** In this area, known as the "bare area," the pericardium comes in contact with the anterior chest wall. **Superior** and **inferior sternopericardial ligaments** attach the pericardium to the sternum. Laterally the pericardium is adjacent to the mediastinal parietal pleura, with the phrenic nerve and pericardiophrenic vessels interposed between them. Posteriorly it is related to the descending aorta, the esophagus, and the bronchi.

The **transverse sinus** of the pericardium is a space within the pericardial cavity situated between the aorta and pulmonary trunk, and the superior vena cava. It is formed by the reflection of the serous pericardium. The **oblique sinus** of the pericardium, a boxlike diverticulum of the pericardial cavity within the irregularities of the pericardial reflection, is bounded by the pericardial reflection over the right pulmonary veins and the inferior vena cava on one side and the pericardial reflection over the left pulmonary veins on the other.

Heart (Table 3-4)

Anteriorly the heart presents a **sternocostal surface** formed superiorly by the right and left auricular appendages and by the infundibulum (Figs. 3-8 and 3-9). Inferiorly two-thirds of this aspect of the heart is formed by the right ventricle and one-third by the left ventricle. The **apex** of the heart is formed entirely by the left ventricle. The **lower border** of the sternocostal surface presents a sharp margin (margo acutis) formed almost entirely by the right ventricle. The **right border,** formed by the right atrium, is continuous superiorly with the superior vena cava and inferiorly

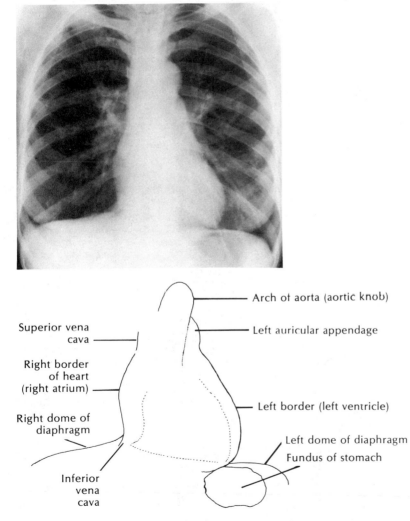

Figure 3-8. *Posteroanterior radiograph and diagram of heart.*

Table 3-4
Coronary Circulation (Arterial and Venous)

Artery or Vein	Origin	Course	Distribution	Anastomoses
Right coronary artery*	Right coronary sinus of aorta	Passes to right along atrioventricular sulcus	Right atrium, right ventricle, interventricular septum, and nodal tissue	Circumflex branch of left coronary
Nodal artery	Right coronary	Ascends towards sinuatrial node	Pulmonary trunk and sinuatrial node	No direct anastomoses
Marginal artery	Right coronary	Passes along inferior margin of heart	Right ventricle and apex	Interventricular branches
Posterior interventricular artery	Right coronary	Posterior interventricular sulcus	Right and left ventricles and interventricular septum	Anterior interventricular
Left coronary artery	Left coronary sinus	Passes to the left for half inch, divides into anterior interventricular and circumflex branches	Aorta and pulmonary trunk	No direct anastomoses
Anterior interventricular artery	Left coronary	Anterior interventricular sulcus	Right and left ventricles and interventricular septum	Posterior interventricular
Circumflex artery	Left coronary	Passes to the left in atrioventricular sulcus	Left atrium and left ventricle	Right coronary
Pericardiophrenic artery	Internal thoracic	Descends between mediastinal parietal pleura and parietal pericardium	Mediastinal parietal pleura and parietal pericardium	Superior phrenic
Great cardiac vein	Myocardial capillary plexuses	Ascends in anterior interventricular sulcus	Terminates in coronary sinus	Venae cordis minimae, middle cardiac
Coronary sinus	Continuation of great cardiac vein	Courses in atrioventricular sulcus	Drains left atrium, left ventricle and receives middle and small cardiac veins	No direct anastomoses
Middle cardiac vein	Myocardial capillary plexus	Ascends in posterior interventricular sulcus, terminates in coronary sinus	Drains right and left ventricular myocardium	Venae cordis minimae, great cardiac vein
Small cardiac vein	Myocardial capillary plexuses	Passes along inferior margin of heart to terminate in coronary sinus	Drains right ventricular myocardium	Venae cordis minimae, great cardiac vein
Anterior cardiac veins	Myocardial capillary plexuses	Anterior surface of right ventricle	Right ventricular myocardium directly into right atrium	Vena cordis minimae, great cardiac vein
Venae cordis minimae artery	Capillary plexuses of deep myocardium	Substance of myocardium	Drains myocardium of all chambers	No direct anastomoses

* Note that the right coronary artery supplies the sinuatrial and atrioventricular nodes.

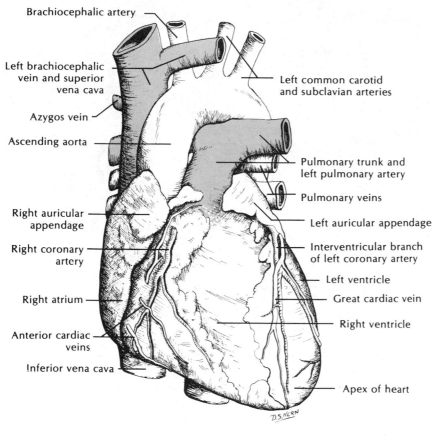

Brachiocephalic artery

Left brachiocephalic
vein and superior
vena cava

Azygos vein

Ascending aorta

Right auricular
appendage

Right coronary
artery

Right atrium

Anterior cardiac
veins

Inferior vena cava

Left common carotid
and subclavian arteries

Pulmonary trunk and
left pulmonary artery

Pulmonary veins

Left auricular appendage

Interventricular branch
of left coronary artery

Left ventricle

Great cardiac vein

Right ventricle

Apex of heart

D.S.HERN

Figure 3-9. *Anterior aspect of the heart.*

with the inferior vena cava. The more convex **left border** is formed mostly by the left ventricle. The **diaphragmatic surface** of the heart rests on the diaphragm and is formed entirely by the ventricles, with the greatest contribution from the left ventricle. The **base** of the heart, or **posterior surface,** is formed almost entirely by the left atrium. The **atrioventricular groove** completely encircles the heart and separates the atria, which lie superiorly and to the right, from the ventricles, which lie inferiorly and to the left.

The nutrient **coronary arteries** of the heart may be considered as greatly enlarged vasa vasorum (Fig. 3-10). They originate from the **aortic sinuses** (dilations of the aorta) immediately

above the aortic cusps. The **right coronary artery** arises from the anterior aortic sinus, courses forward between the pulmonary trunk and the right auricular appendage, then passes inferiorly to the atrioventricular groove, in which it runs posteriorly to anastomose with the left coronary. It gives small branches to the root of the aorta, pulmonary trunk, and the walls of the right atrium and ventricle. It supplies both the sinuatrial and the atrioventricular nodes. A **marginal branch** of the right coronary passes from right to left along the lower border and anterior aspect of the heart. A **posterior interventricular** (posterior descending) **branch** extends anteriorly on the diaphragmatic surface in the interventricular groove.

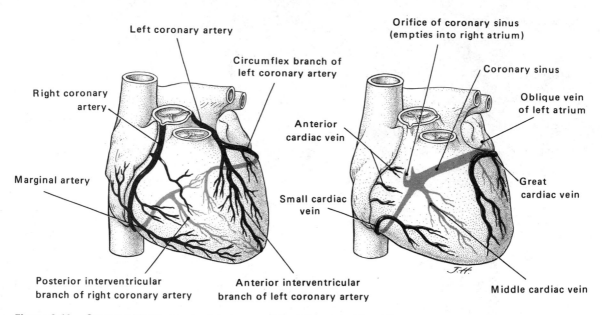

Figure 3-10. *Coronary blood vessels. Arteries are at the left, veins at the right.*

The large **left coronary,** originating from the left posterior aortic sinus, passes a short distance to the left, behind the pulmonary trunk, then between the pulmonary trunk and the left auricular appendage, where it divides into an anterior interventricular (anterior descending) branch and the circumflex branch. The **anterior interventricular branch** descends in the interventricular groove on the anterior aspect of the heart toward the apex, where it anastomoses with the posterior interventricular branch. The **circumflex artery** continues in the atrioventricular groove to the lower border of the base of the heart to anastomose with the right coronary.

Heart Attacks

An insufficient blood supply to the heart muscle (myocardial ischemia) may precipitate a heart attack, which causes the patient severe chest pains (angina pectoris) over the area of the heart (precordium). These severe chest pains are an example of referred pain because the pain usually extends from the chest wall, to the left shoulder and down the left arm. A common cause for the reduced blood supply is a marked reduction of the diameter of the lumen of one or more of the coronary arteries, caused by an accumulation of atherosclerotic plaques within the inner wall of the artery. Another cause is by occlusion of a coronary artery by a blood clot (thrombus).

If the circulatory deficiency is of sufficient duration, local tissue necrosis results, causing permanent loss of muscle fibers (myocardial infarction). If the damage covers a substantial area of the heart wall, the heart cannot contract and cardiac arrest results.

The largest vein draining the heart is the **coronary sinus.** It begins at the upper end of the anterior interventricular groove as a continuation of the **great cardiac vein,** the companion vein of the anterior descending branch of the left coronary artery. It runs posteriorly from left to right in the atrioventricular groove. The **middle cardiac**

vein, accompanying the posterior descending branch of the right coronary artery, empties into the coronary sinus near its termination. The **small cardiac vein** curves around the lower margin of the right border of the heart and also empties into the coronary sinus at its termination. In its course, it follows the marginal branch of the right coronary artery. The small **oblique vein of the left atrium** descends over the posterior surface of this chamber to drain into the midportion of the coronary sinus. **Valves** in the coronary sinus are usually present at the junction of the great and small cardiac veins, and at its termination. The **anterior cardiac veins** are small vessels on the right atrial and ventricular surfaces that empty directly into the right atrium near the atrioventricular groove. A small part of the blood is collected directly from the heart musculature by small veins, the **venae cordis minimae,** which arise within the muscle of the heart wall and drain directly into all chambers of the heart.

Nerves

Modifications of the intrinsic rhythm of heart muscle is produced through autonomic nerve plexuses. **Parasympathetic innervation,** from the vagus nerve, slows the rate and reduces the force of the heart beat. **Sympathetic innervation,** from the thoracic and cervical ganglia has the opposite effects. The **superficial cardiac plexus** lies in the concavity of the aortic arch, proximal to the ligamentum arteriosum, and receives the inferior cervical cardiac branch of the left vagus and the superior cervical cardiac branch of the left sympathetic trunk. The more extensive **deep cardiac plexus** is located on the base of the heart posterior to the aortic arch and anterior to the tracheal bifurcation. From the cervical sympathetic ganglia the plexus receives all the cardiac branches on the right side and the middle and inferior cardiac branches on the left side. From the right vagus the deep cardiac plexus receives the superior and inferior cervical branches, and from the left vagus, the superior cervical branch.

In the thorax additional vagal and sympathetic branches pass directly to the deep cardiac plexus.

Coronary plexuses, surrounding and accompanying the coronary arteries, are extensions from the cardiac plexus. Vagal fibers within the coronary plexus produce vasoconstriction, while sympathetic fibers cause vasodilatation of the arterioles. **Visceral afferent fibers** from the heart and coronary arteries end almost entirely in the first four thoracic segments of the spinal cord.

Conduction (Purkinje) System

The orderly sequence in which ventricular contraction follows atrial contraction and proceeds from the apex to the base of the heart is due to a specialized conduction system composed of two aggregates of nodal tissue and a band of specialized cardiac muscle (Fig. 3-11).

The **sinuatrial node** is a small collection of specialized myocardial tissue at the junction of the right atrium and the superior vena cava, at the superior end of the sulcus terminalis. This node, the **pacemaker,** appears to initiate the heartbeat, and the impulse is propagated over the atria.

Artificial Pacemaker of the Heart
An artificial pacemaker of the heart is a battery-operated device that supplies electric impulses to regulate the heartbeat in patients suffering from arrhythmias or bradycardia. The pacemaker is implanted subcutaneously over the upper anterior chest. A special electrode catheter is passed intravenously into the cephalic, external jugular, or internal jugular vein through the superior vena cava and into the right ventricle. The tip of the catheter is positioned in the endocardium, so that a small voltage through the wire from the pacemaker will cause ventricular contraction.

The **atrioventricular node** is a nodule of similar tissue located in the septal wall of the right atrium, immediately above the opening of the cor-

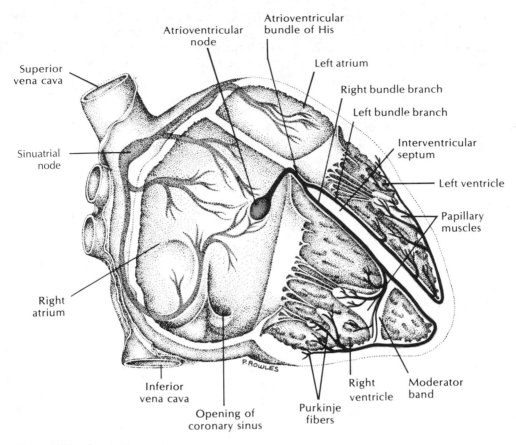

Figure 3-11. *Conduction system.*

onary sinus. Activity at this node is probably initiated by stimuli reaching it through the atrial musculature. The impulse is then directed toward the apical portions of the ventricles by way of the **atrioventricular bundle of His.** This structure is a strand of specialized myocardium (Purkinje fibers) passing from the atrioventricular node into the interventricular septum, where it divides into a right and left bundle within the septal musculature.

Heart or AV Block
Heart or AV block is an interruption of nervous impulses through the atrioventricular bundle (of His) causing the ventri-

cles to beat at a rhythm slower than the atria. Such a loss of synchronization causes the ventricles to contract whether or not they are filled. Thus, the efficiency of the pumping action of the heart is drastically reduced and cardiac arrest may follow.

The **right bundle branch** follows the endocardium along the septum of the right ventricle giving branches to the septum, then extends across the ventricular cavity, traversing the **moderator band** to reach the anterior papillary muscle of the right ventricle. The **left bundle branch** follows the endocardium of the left ventricle to the apex, where it is distributed to the anterior and posterior papillary muscles and the ventricular myocardium.

Ventricular Fibrillation

Ventricular fibrillation follows a complete functional breakdown in the conduction system. The impulse follows no regular pattern but seems to be stimulating all regions of the ventricle simultaneously. Such a loss of coordination of the ventricular contraction causes a loss of blood flow to the myocardium and low blood pressure. To stop ventricular fibrillation or to defibrillate the heart, an electric shock is delivered to the heart through external paddles. This usually stops the fibrillating ventricles. A normal heartbeat will usually follow defibrillation.

Interior of the Heart

The interior of the heart is divided into right and left atria, each with an auricular appendage, and right and left ventricles (Fig. 3-12).

The **right atrium** is described as having two parts, the sinus venarum and the atrial portion. The sinus venarum, a large quadrangular cavity with smooth walls, lies between the two venae cavae and is continuous inferiorly with the right ventricle. The auricle (proper) is a small conical muscular pouch that projects from the upper anterior part of the sinus venarum to overlap the root of the aorta. The **crista terminalis** is a vertical ridge extending from the orifice of the superior vena cava to the opening of the inferior vena cava. The **musculi pectinati** are parallel muscular ridges at right angles to the crista terminalis and extend into the auricular appendage.

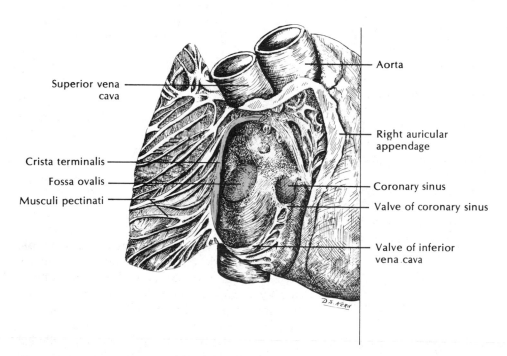

Figure 3-12. *Internal aspect of the right atrium.*

Openings into the right atrium include: (1) the superior vena cava, devoid of valves, opening into the superoposterior part of the chamber; (2) the inferior vena cava, with a thin rudimentary valve along the anterior margin between its orifice and the atrioventricular orifice, opening into the inferoposterior part; (3) the opening of the coronary sinus, guarded by a crescentic valve directly posterior to the atrioventricular orifice; and (4) the right atrioventricular aperture, closed by the tricuspid valve in the anterioinferior part of the atrium, opening into the right ventricle. The rudimentary valve of the inferior vena cava is continuous on its left with the **anulus ovalis** of the interatrial septum, which surrounds a wide shallow depression, the **fossa ovalis.** The upper end of the fossa is open in fetal circulation as the **foramen ovale,** which permits passage of blood from the right to the left atrium.

Auscultation of the Heart Sounds

To interpret cardiac valve sounds, one must be able to locate the sounds accurately as they are projected onto the anterior chest wall. The distance between the heart and the skin of the anterior thorax affects the intensity of the heart sounds— softer or lower intensity with increased distances, and louder or higher intensity with shorter distance. Using a stethoscope, the areas of greatest audibility for the various valves are: 1) left atrioventricular (mitral or bicuspid), over the apex of the heart in the fifth intercostal space in the midclavicular line; 2) right atrioventricular (tricuspid), midline over the inferior surface of the sternum, or the right fifth intercostal space at the sternal margin; 3) aortic semilunar, over the right second intercostal space at the sternal margin; and 4) pulmonary semilunar, over the left second intercostal space at the sternal margin.

The **right ventricular cavity** is triangular in shape. Its funnel-shaped superior portion, the **in-fundibulum,** or **conus arteriosus,** leads into the pulmonary trunk. The infundibulum has smooth internal walls, while the remainder of the cavity possesses projecting muscular bundles, the **trabeculae carneae. Papillary muscles** are conical projections from the wall into the cavity that are connected at their apices to the tricuspid valve by the **chordae tendineae.** The **moderator band,** a well-marked trabecula, projects from the interventricular septum to the base of the anterior papillary muscle and may act to prevent over-distention of the chamber. The **right atrioventricular orifice,** about 2 cm to 3 cm in diameter, is guarded by a valve with three triangular cusps, the **tricuspid valve** (Fig. 3-13). The anterior cusp intervenes between the atrioventricular orifice and the infundibulum, the septal cusp lies in relation to the septal wall, and the posterior cusp is adjacent to the inferior wall. The bases of the cusps are attached continuously to a fibrous ring, the **anulus fibrosus,** a deeply lying structure at the periphery of the atrioventricular orifice. The apex of each cusp extends into the ventricle and is attached by chordae tendineae to papillary muscles. The **pulmonary orifice,** at the apex of the infundibulum, is surrounded by a thinner fibrous ring to which the bases of three cuplike **pulmonary semilunar cusps** are attached. The center of the free margin of each semilunar cusp is thickened to form a **nodule** (corpus Arantii). The **lunulae** of the valves are the thinner crescentic regions at either side of the nodules. When the valve is closed, the three nodules meet at the center of the orifice, effecting complete occlusion.

The **left atrium** forms most of the base of the heart and consists of the atrium proper and its auricular appendage. Four **pulmonary veins,** two on each side, enter the chamber at its superolateral aspect. The internal surface of the wall of the left atrium is generally devoid of muscular ridges except in the auricular appendage. Anteroinferiorly the left atrioventricular orifice, guarded by the bicuspid valve, opens into the left ventricle.

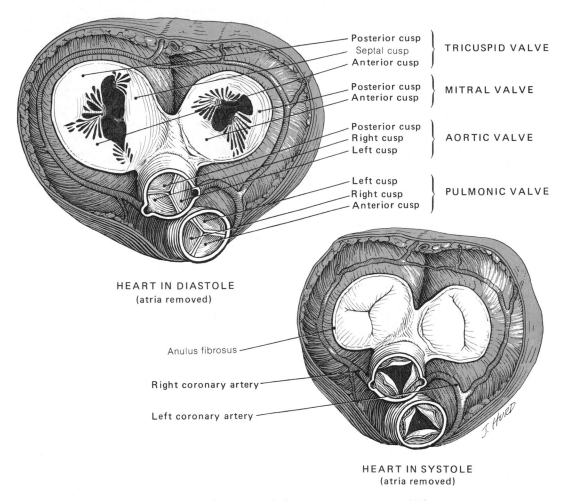

Posterior cusp
Septal cusp — TRICUSPID VALVE
Anterior cusp

Posterior cusp — MITRAL VALVE
Anterior cusp

Posterior cusp
Right cusp — AORTIC VALVE
Left cusp

Left cusp
Right cusp — PULMONIC VALVE
Anterior cusp

HEART IN DIASTOLE
(atria removed)

Anulus fibrosus

Right coronary artery

Left coronary artery

HEART IN SYSTOLE
(atria removed)

Figure 3-13. *Valves of the heart with atria removed. At the left, atrioventricular (AV) valves are open to permit blood into the ventricles. At the right, semilunar (aortic and pulmonic) valves are open to pass blood into the aorta and pulmonary trunk.*

Cardiac Malformations

Cardiac malformations usually arise from developmental defects in the formation of the heart valves, septa, or both. An example of an atrioseptal defect is a postnatal patent foramen ovale. If of sufficient size, such an abnormal opening allows venous blood from the right atrium to mix freely with the well-oxygenated blood in the left atrium. A ventricular septal defect usually occurs in the uppermost (membranous) portion of the septum. It is more serious than an atrial septal defect because the blood flow from the left to the right ventricle is much greater and heart failure occurs much sooner.

Valvular abnormalities may be either a stenosis (constriction) or an incompetence (leakage) of valves. In the latter the valve does not close completely, which allows a backflow (leakage) of blood and causes a sound called a heart murmur. Most valvular and septal defects are now amenable to surgical correction.

The cone-shaped cavity of the **left ventricle** is longer and narrower than the right and has much thicker walls. The internal surface is covered with a dense meshwork of muscular ridges, **trabeculae carneae,** which are finer and more numerous than those of the right ventricle. The **papillary muscles,** much larger and stronger than in the right ventricle, are usually two in number. The interventricular septum and the upper portion of the anterior wall are relatively free of muscular ridges. The **left atrioventricular orifice,** slightly smaller than the right, is closed by the **bicuspid (mitral) valve.** The cusps are set obliquely, with the larger anterior cusp to the right and the smaller posterior cusp to the left. The bases of the cusps are attached to the **anulus fibrosus** surrounding the left atrioventricular orifice. Their apices project into the cavity with **chordae tendineae** extending from each cusp to both papillary muscles. The **aortic vestibule** is that portion of the left ventricle immediately below the aortic orifice. Its walls, composed mainly of fibrous tissue, are quiescent during ventricular contraction so that outflow of blood from the left ventricle is not impeded. In the upper right posterior part of the cavity the **aortic orifice** is surrounded by an **anulus fibrosus** for the attachment of the bases of the three **semilunar aortic cusps** guarding the opening. This valve is similar to the previously described pulmonary valve. The musculomembranous **interventricular septum** separates the ventricles as well as separating the left ventricle from the inferiormost part of the right atrium. The

muscular portion is thickest near the apex and gradually thins to become membranous near the aortic opening.

Tetralogy of Fallot

Tetralogy of Fallot is a combination of septal and valvular defects. It has four (tetra-) consistent malformations: (1) stenosis of the pulmonary trunk, (2) ventricular septal defect (VSD), (3) overriding of the aorta over the VSD, and (4) hypertrophy of the right ventricle. Although it is one of the most crippling of the cardiac abnormalities, it was the first to be alleviated by surgery. Dr. Alfred Blalock of Johns Hopkins University devised an operation to anastomose the right subclavian artery to the right pulmonary artery and thus provide adequate circulation through the lungs and correct the severe hypoxemia, which caused all infants with tetralogy to be cyanotic or "blue" babies.

The **fibrous skeleton** of the heart consists of fibrous rings, the anuli fibrosi, surrounding the atrioventricular, aortic, and pulmonary orifices. The fibrous rings are connected by the right and left trigona fibrosa and the conus tendon. The **right trigonum fibrosum,** situated between the two atrioventricular orifices, sends an extension into the membranous portion of the interventricular septum. The **left trigonum fibrosum** lies between the left atrioventricular orifice and the root of the aorta. The **conus tendon** extends between the root of the aorta and the pulmonary artery. In addition to giving attachment to the bases of the valves of the heart, this fibrous skeleton also serves for the attachment of the various muscle bundles composing the myocardium.

Mediastinum

Objectives

After the completion of the study of the mediastinum the student should be able to

▶ *Delineate the subdivisions of the mediastinum; list and describe the contents of each portion*

▶ *Follow the aorta from its origin to the aortic hiatus; give its relationships and branches*

▶ *Label on a diagram the tributaries and components of the azygos system of veins*

▶ *List elements of the sympathetics that are present in the thorax*

▶ *Define the esophageal plexus; list the branches of the vagus nerve that appear in the thorax and give their distribution*

The **mediastinum** is the midline area between the two pleural cavities and contains all the thoracic structures except the lungs (Fig. 3-14). The heart occupies a central subdivision of this area, the middle mediastinum. The rest of the mediastinum is subdivided in relation to the heart (or pericardial cavity) into superior, anterior, and posterior portions (Fig. 3-15). The anterior, middle, and posterior mediastina are sometimes referred to collectively as the **inferior mediastinum.**

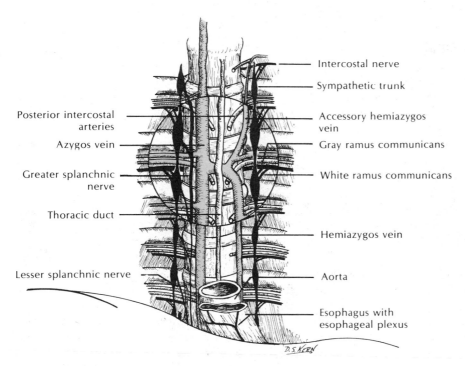

Posterior intercostal arteries

Azygos vein

Greater splanchnic nerve

Thoracic duct

Lesser splanchnic nerve

Intercostal nerve

Sympathetic trunk

Accessory hemiazygos vein

Gray ramus communicans

White ramus communicans

Hemiazygos vein

Aorta

Esophagus with esophageal plexus

D.S.KERN

Figure 3-14. *Structures within the posterior mediastinum.*

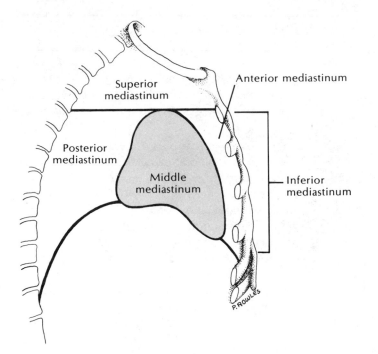

Figure 3-15. *Mediastinum.*

The **superior mediastinum** is the area above the fibrous pericardium. It is demarcated from the inferior mediastinum by a line passing from the sternal angle to the intervertebral disc between the fourth and fifth thoracic vertebrae. It contains all structures passing between the neck and the thorax, including the remnant of the thymus, brachiocephalic veins and superior vena cava, aortic arch and its three branches, thoracic duct, trachea, esophagus, phrenic, vagus, and cardiac nerves, and the sympathetic trunk. The **anterior mediastinum** is a limited area anterior to the pericardium and contains connective tissue with some remains of the thymus, fat, lymph nodes, and the internal thoracic vessels. The **middle mediastinum,** centrally located and limited by the fibrous pericardium, contains the heart and its eight great vessels, namely, aorta, pulmonary artery, superior and inferior venae cavae, and the four pulmonary veins. Laterally, to either side of the pericardium, the phrenic nerve and the peri-

cardiophrenic vessels course between the parietal pericardium and the mediastinal parietal pleura. The **posterior mediastinum** is posterior to the pericardium and the diaphragm. It contains the descending (thoracic) aorta, posterior intercostal vessels, azygos and hemiazygos veins, thoracic duct, vagus and splanchnic nerves, sympathetic trunk, and esophagus.

Aorta (Table 3-5)

In the thorax the **aorta** is divided, for descriptive purposes, into the ascending, arch, and descending parts. Originating at the aortic orifice, the **ascending aorta** passes superiorly to the level of the second costal cartilage, where it becomes the arch. At its origin the lumen is not uniform, owing to the three aortic sinuses, which are opposite the cusps of the aortic semilunar valves. The right and left coronary arteries originate from two of these dilatations. The **arch of the aorta** begins at the

Table 3-5
Aortic Arch Branches and Branches of the Anterior Aspect of the Aorta

Artery	Origin	Course	Distribution	Anastomoses
Ascending aorta	Left ventricle	Ascends approximately 5 cm to become the arch of the aorta	Gives origin to right and left coronary arteries	No direct anastomoses
Arch of the aorta	Continuation of ascending aorta; concavity of arch at level of disc between T_4 and T_5	Arches posteriorly and to the left over bifurcating pulmonary trunk	Gives origin to brachiocephalic, left common carotid and left subclavian	No direct anastomoses
Descending aorta	Continuation of arch of aorta	Courses initially alongside vertebral column; gradually shifts to the right to lie in midline at aortic hiatus	Gives origin to posterior intercostals, and visceral branches	No direct anastomoses
Esophageals	Anterior aspect descending aorta	Extend anteriorly to esophagus	Muscular wall of esophagus	Left gastric and short gastrics
Superior phrenics	Anterior aspect of the aorta	Arise at aortic hiatus to pass to superior aspect of diaphragm	Thoracic diaphragm	Inferior phrenic, pericardiophrenic, and musculophrenic
Bronchials	Anterior aspect of aorta or posterior intercostal	Course with the tracheobronchial tree	Bronchial and peribronchial tissue	No direct anatomoses

level of the second costal cartilage, ascends to the level of the middle of the manubrium, and then arches to the left and anteroposteriorly to reach the lower border of the fourth thoracic vertebra, where it becomes the descending or thoracic aorta. The **brachiocephalic, left common carotid,** and **left subclavian arteries** arise from the arch. The **ligamentum arteriosum,** the remnant of the ductus arteriosus, passes from the inferior aspect of the arch to the left pulmonary artery.

Aortic Aneurysm
The aorta has a higher incidence of aneurysms than any other artery, probably due to its curved shape and large size. In the thorax, aneurysms usually occur in the ascending or descending aorta but seldom in the aortic arch. Symptoms of thoracic aortic aneurysm usually depend on pressure exerted on adjacent structures.

For example, pressure on the inferior (recurrent) laryngeal nerve causes hoarseness and a brassy cough. Pressure on the esophagus may cause dysphagia (difficulty in swallowing). Dyspnea (difficulty in breathing) may follow pressure on the trachea, root of the lung, or phrenic nerve.

The **left inferior laryngeal (recurrent branch)** of the vagus nerve loops around the arch of the aorta adjacent to the ligamentum arteriosum to pass along the tracheoesophageal groove, where it ascends to reach the larynx. The **descending aorta** continues from the arch to pass through the posterior mediastinum, inclining to the right and then anteriorly to gain a position anterior to the vertebral column. It passes into the abdomen through the aortic opening in the diaphragm, opposite the lower border of the twelfth thoracic vertebra, to become the abdominal aorta. From the anterior aspect its branches include **two**

left bronchial and several small branches to the esophagus, mediastinum, diaphragm, and pericardium. From the posterior aspect, nine pairs of **posterior** (aortic) **intercostal** and the two **subcostal arteries** arise. The posterior intercostal arteries divide to send one branch anteriorly into the intercostal space and a second branch posteriorly to the deep muscles of the back, the vertebral column, and the spinal cord. The subcostal arteries follow the lower border of the twelfth rib to supply the abdominal wall musculature.

Azygos System of Veins (Table 3-6)

The **azygos system of veins** drains most of the blood from the thoracic wall and posterior wall of the abdomen. The **azygos vein** is formed by the junction of the right subcostal and the right ascending lumbar veins and ascends through the posterior and superior mediastina. It receives blood from the posterior intercostal veins as high as the level of the third intercostal space. Arching over the root of the right lung, it empties into the

Table 3-6
Azygos System of Veins

Vein	Origin	Course	Distribution	Anastomoses
Azygos	Junction of right ascending lumbar with right subcostal	Passes through posterior mediastinum, arches over root of right lung; terminates in superior vena cava	Drains posterior abdominal wall, wall of the right thoracic cavity, receives hemiazygos and accessory hemiazygos	Superior intercostal
Hemiazygos	Junction of left ascending lumbar with left subcostal	Ascends in posterior mediastinum; terminates in azygos	Drains posterior abdominal wall, the inferior third of left side of thoracic wall	Accessory hemiazygos
Accessory hemiazygos	Coalescence of posterior intercostals, middle third of thoracic wall	Crosses midline to drain into azygos vein	Drains middle third of left side of thoracic wall	Hemiazygos
Ascending lumbar	Formed by coalescence of lumbar veins draining posterior abdominal wall	Ascends to join subcostal on right to form azygos; subcostal on left to form hemiazygos	Drains posterior abdominal wall	Retroperitoneal veins and tributaries of superior and inferior epigastric
Superior intercostal	Coalescence of second, third, and fourth posterior intercostals	Descends to drain into azygos on right; brachiocephalic on left	Upper third of thoracic wall	Azygos and accessory hemiazygos
Highest intercostal	Continuation of posterior intercostal, first intercostal space	Terminates in brachiocephalic	Drains first intercostal space	Anterior intercostal, superior intercostal
Posterior intercostals	Capillary plexuses within intercostal space	Pass between second and third layer of muscles of intercostal space	Drains into azygos, hemiazygos, superior intercostals, or highest intercostal	Anterior intercostals
Anterior intercostals	Capillary plexuses within intercostal space	Pass between second and third layer of muscles of intercostal space	First six spaces into internal thoracic; lower five into musculophrenic	Posterior intercostal, lateral thoracic, pectorals

superior vena cava. During development it may deviate to the right, cleaving the developing right lung to demarcate an **azygos lobule.** The **hemiazygos vein,** formed by the junction of the left subcostal and the left ascending lumbar veins, receives the posterior intercostal veins draining the ninth, tenth, and eleventh left intercostal spaces. It crosses the midline to empty into the azygos vein. The **accessory hemiazygos** usually begins at the fourth intercostal vein on the left side, descends to the eighth or ninth space receiving the intervening intercostal veins to empty into the azygos or hemiazygos. The **superior intercostal veins** drain the second and third intercostal spaces. The right vessel joins the azygos, and the left vein drains into the left brachiocephalic vein. Each of the **highest intercostal veins** drain directly into the brachiocephalic vein of the same side.

Sympathetic Trunk, Phrenic and Vagus Nerves

The **sympathetic trunk,** or vertebral chain, is a series of ganglia (collections of nerve cell bodies) connected by interganglionic nervous tissue (Fig. 3-16). It extends from the **superior cervical ganglion** at the base of the skull to the **ganglion impar** at the coccyx. The cell bodies of the preganglionic fibers, passing to the chain ganglia, are located in the **intermediolateral cell column** of gray matter of the spinal cord from the **first thoracic to the second lumbar spinal cord segment.** The **preganglionic fibers** leave the central nervous system by way of the ventral root of a spinal nerve to pass to the ganglia by way of the **white rami communicantes.** The cell bodies of the **postganglionic neurons,** located within the chain ganglia, send their fibers by way of **gray rami communicantes** back to the spinal nerve to be distributed with its peripheral branches. Preganglionic fibers may also form **splanchnic nerves** by passing through the chain ganglia without synapsing and continue to para- or preaortic ganglia, where they synapse with the postganglionic cell bodies. In the thorax the **greater splanchnic nerve** is formed by filaments from the fifth through ninth ganglia, the **lesser splanchnic** from the tenth and eleventh, and the **least splanchnic** from the twelfth ganglion. These nerves descend anteromedially and pass through the diaphragm. They terminate at **preaortic ganglia** lying adjacent to the major branches of the abdominal aorta. Preganglionic fibers may also pass up or down the chain to synapse in ganglia at higher or lower levels, where gray rami communicantes carry the postganglionic fibers to the spinal nerves.

The **phrenic nerve,** which arises from the fourth cervical spinal nerve (with additional twigs from the third and fifth cervical spinal nerves), descends obliquely across the scalenus anterior muscle, passing anterior to the first part of the subclavian artery to enter the thorax. It descends through the thorax, between the mediastinal parietal pleura and the fibrous pericardium, to supply the diaphragm.

After traversing the neck, the **vagus nerves** enter the thorax anterior to the subclavian arteries. The right vagus gives an **inferior laryngeal (recurrent)** branch that loops around the corresponding subclavian artery, while the inferior laryngeal nerve from the left vagus loops around the arch of the aorta. Both vagi pass posterior to the root of the lung, giving branches to the **pulmonary plexus,** and thoracic cardiac branches to the **cardiac plexus,** after which they become plexiform around the esophagus. As the **esophageal plexus,** the vagi continue through the thorax to pass through the esophageal hiatus of the diaphragm, where they are reconstituted from their plexiform arrangement. The left vagus becomes the **anterior,** and the right the **posterior** trunks as they extend into the abdominal cavity (Table 3-7).

Esophagus

The **esophagus** originates at the level of the cricoid cartilage or sixth cervical vertebra as a continuation of the pharynx. It descends through the

Text continues on page 138

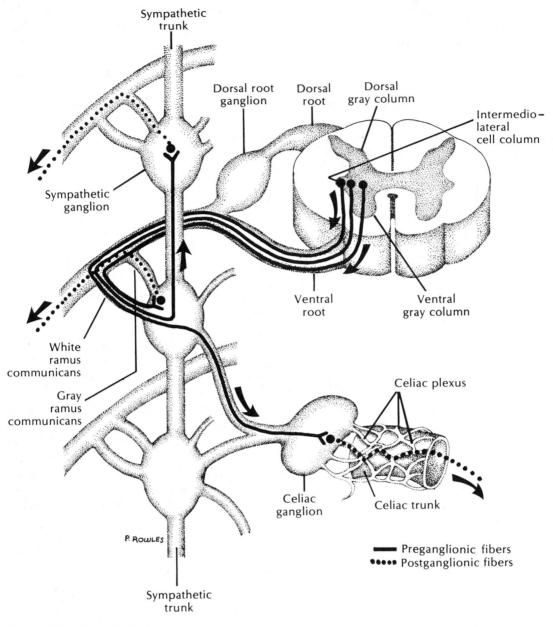

Figure 3-16. Sympathetic trunk.

Table 3-7
Nerve Distribution to Thorax

Nerve	Origin	Course	Distribution
Supraclaviculars	Third loop of cervical plexus	From midpoint posterior border of sternocleidomastoid pierces platysma to course in superficial fascia	Skin over shoulder, lower neck, and first two intercostal spaces
Phrenic	Third loop of cervical plexus	Passes on superficial aspect of scalenus anterior to thoracic inlet; in thorax, lies between mediastinal pleura and pericardium	Central portion of diaphragm
Intercostals	Ventral rami of thoracic nerves; first intercostal twig from T_1 as it passes to brachial plexus; ventral ramus T_{12} is subcostal	In intercostal spaces between second and third layer of muscles	Muscles of, and skin over, intercostal spaces; lower intercostals supply muscles and skin of anterolateral abdominal wall; dermatome T_{10} flanks umbilicus
Inferior (recurrent) laryngeal	Vagus nerve	On right, loops around subclavian, on left, around arch of aorta to ascend in tracheoesophageal groove	Intrinsic muscles of larynx (except cricothyroid); sensory below level of vocal folds
Cardiac plexus	Cervical and cardiac branches of vagus and sympathetic trunk	From concavity of arch of aorta and posterior surface of heart, fibers extend along coronary arteries and toward sinuatrial node	Impulses pass to S–A node; parasympathetics—slow rate, reduce force of heart beat, constrict coronary arteries; sympathetics, opposite action
Pulmonary plexus	Vagus and sympathetic trunk	Plexus forms on root of lung and extends along bronchial subdivisions	Parasympathetics constrict bronchioles; sympathetics opposite effect
Esophageal plexus	Vagus	Distal to tracheal bifurcation vagi become plexiform around esophagus; in abdomen reform as anterior and posterior vagal (gastric) trunks	Digestive tract—parasympathetic innervation (in general) increases peristalsis and secretions, and relaxes sphincters
Splanchnic nerves	Sympathetic trunk—greater (T_{5-9}), lesser (T_{10-11}), least (T_{12})	Penetrate thoracic diaphragm to terminate in ganglia adjacent to major branches of abdominal aorta	Digestive tract—sympathetic innervation (in general) decreases peristalsis, closes sphincters, and causes vasoconstriction

thorax, posterior to the trachea and anterior to the vertebral bodies, to pass through the **esophageal hiatus** of the diaphragm at the level of the tenth or eleventh thoracic vertebra.

Esophageal Constrictions

The esophagus has three distinct areas of normal narrowing. These areas are located in (1) the neck, at its junction with the pharynx, (2) the mediastinum, where the left main bronchus crosses the anterior surface of the esophagus, and (3) the diaphragm, at the esophageal hiatus. These three areas have clinical importance because disorders of the esophagus, such as (1) swallowed foreign bodies, (2) strictures following drinking of caustic substances, (3) sites of perforations, or (4) cancers often occur at these normal constrictions.

Thoracic Duct

The **thoracic duct** begins in the abdomen as a dilatation, the **cisterna chyli,** which receives lymph drainage from the abdomen, pelvis, and inferior extremities (see Fig. 1-11). Traversing the aortic hiatus of the diaphragm, it ascends in the thorax between the aorta and the azygos vein. In the neck it arches above the subclavian artery to empty into the junction of the subclavian and internal jugular veins. In addition to draining the lower half of the body, the thoracic duct receives the lymphatic drainage from the left side of the thorax, head, and left upper extremity. The **right lymphatic duct,** a short, 1 cm- to 2-cm-long vessel, is formed by the junction of the right jugular, subclavian, and mediastinal lymphatic trunks and empties at the origin of the right brachiocephalic vein. It drains the right side of the head and thorax, and the right upper extremity.

Thymus

The **thymus gland** is a prominent organ in the infant and reaches its greatest relative size at 2 years and greatest absolute size at puberty. It occupies the anterior mediastinum and the anterior part of the superior mediastinum. At puberty it begins to involute and is largely replaced by adipose tissue. It is a lymphoid organ whose principal function is lymphocyte and antibody production.

MAJOR ANATOMIC AND CLINICAL POINTS

Thoracic Wall

□ The sternoclavicular articulation is the only skeletal attachment of the upper extremity to the trunk.

□ The inferior extent of the neck and the superior extent of the thoracic cavity are demarcated by the first thoracic vertebra, first rib, and suprasternal (jugular) notch.

□ The true ribs (except for the first) have freely movable joints with both the sternum and the vertebral column.

□ Three layers of muscles are in the thoracic wall (intercostal space); vessels and a nerve course between the second and third layers.

□ Intercostal nerves are the ventral rami of spinal nerves; they supply both the thoracic and abdominal walls. The skin flanking the umbilicus is the dermatome of T_{10}.

□ Intercostal arteries enter both the anterior and the posterior aspect of an intercostal space.

□ The sternal angle, delineating the second costal cartilage/rib, is the reference point for accurately identifying rib number or intercostal space.

□ In thoracocentesis the needle is placed over the superior border of the rib to avoid damage to the vein, artery, or nerve that courses along the inferior border of the rib in the costal groove.

□ The anastomoses between anterior and posterior intercostals can provide collateral circulation in blockage of the descending aorta.

Lung and Pleural Cavity

□ The apex of the lung and the pleural cavity extend into the neck.

□ The posterior mediastinum is continuous with the retropharyngeal space of the neck.

□ The pleural and pericardial cavities are closed sacs; that is, the serous membrane of each is a continuous sheet.

□ Four lobes of the lungs (two on right [middle and lower]; two on left [upper and lower]) are contiguous to the diaphragm.

□ In inspiration both the vertical dimension and the circumference of the thoracic (pleural) cavity increase.

□ Primary bronchi pass to each lung; secondary bronchi pass to the lobes of the lung; tertiary bronchi pass to bronchopulmonary segments.

□ The anterior and inferior borders of the lungs move into the costomediastinal and costodiaphragmatic recesses only in deep inspiration.

□ A penetrating wound above the level of the clavicle could result in a pneumothorax by piercing the cupula of the pleural cavity.

□ The right bronchus is more in line with the trachea and larger in caliber than is the left. Hence, aspirated objects are more likely to become lodged in the right lung.

□ In a pleural effusion a fluid level may be visible in the costodiaphragmatic recess by fluoroscoping the patient in an upright position.

□ In intractable hiccups it may be necessary to transect one phrenic nerve.

Heart and Pericardial Cavity

□ The heart sits obliquely in the thoracic cavity with the right atrium to the right, left ventricle to the left, right ventricle anterior, and left atrium posterior.

□ The sequence of blood flow through the heart is venae cavae → right atrium → right ventricle → pulmonary trunk → lungs → pulmonary veins → left atrium → left ventricle → aorta.

□ The fossa ovalis and ligamentum arteriosum represent fetal structures through which blood by-passed the lungs.

□ The sequence of nerve impulse conduction to the heart is cardiac plexus → S/A node → atrial musculature → A/V node → A/V bundle → bundle branches → apex.

□ The bundle of His is the only "muscular connection" between the atria and the ventricles.

□ Thickened (fibrous) parietal pericardium is nondistensible. Therefore pericardial effusion or bleeding into the pericardial cavity is life threatening, the fluid displaces the heart (cardiac tamponade) and may result in cardiac arrest.

□ In a high interventricular septal defect the back-flow of blood may pass from the left ventricle into both the right ventricle and right atrium.

□ To visualize the left atrium on a radiograph it is necessary to take a lateral x-ray of the chest.

Mediastinum

□ The arch of the aorta spirals to the left and posteriorly as it crosses over the pulmonary trunk and is visible on a P/A radiograph.

□ The thoracic duct drains lymph from three-fourths of the body.

□ In the thorax the vagus nerve gives recurrent laryngeal, cardiac, pulmonary, and esophageal branches.

□ The azygos system drains both thoracic and abdominal walls and communicates with the vertebral plexus of veins.

□ In a barium swallow three normal constrictions of the esophagus may be visualized: at the origin (C_6), as the esophagus passes posterior to the arch of the aorta (T_4), and at its termination (esophageal hiatus T_{10}).

□ An aortic aneurysm may cause dysphagia (difficulty in swallowing) and/or hoarseness.

□ An hiatal hernia is a protrusion of abdominal viscera or the omentum through the esophageal hiatus into the thoracic cavity. It is usually accompanied by gastritis and epigastric pain.

□ The thoracic vertebral level of structures traversing the diaphragm is T_8 for the inferior vena cava, T_{10} for the esophagus, and T_{12} for the aorta. The anteroposterior position is inferior vena cava, esophagus, and aorta.

Q U E S T I O N S   F O R   R E V I E W

1. The muscles of the thoracic cavity proper include all of the following except the

    A. Internal intercostals
    B. External intercostals
    C. Subcostales
    D. Transversus thoracis (sternocostalis)
    E. Serratus anterior

2. Concerning intercostal nerves, all of the following are true except

    A. They carry both sensory and motor fibers
    B. They are ventral rami of thoracic spinal nerves
    C. They course along the superior border of the rib
    D. The lower intercostal nerves supply anterolateral abdominal wall muscles
    E. They provide sensory innervation to costal pleura

3. All of the following are characteristics or parts of a typical rib except

    A. Head
    B. Neck
    C. Articular tubercle
    D. Subcostal groove
    E. Broad and flat superior and inferior surfaces

4. All of the following are true concerning the internal thoracic artery except

    A. It is a branch of the subclavian
    B. It lies directly posterior to the sternum
    C. Its intercostal branches anastomose with intercostal branches of the aorta
    D. The superior epigastric is one of its terminal branches

5. How many bronchopulmonary segments are there in the right lung?

    A. Two
    B. Three
    C. Six
    D. Eight
    E. Ten

6. The lungs

    A. Lie inside the pleural cavities
    B. Have "mirror-image" configurations and are of equal size

    C. Are concave at the base and ride on the dome-shaped diaphragm
    D. Contact each other across the midline behind the heart
    E. Have root structures at their bases

7. The root of the lung includes all of the following except

    A. Bronchi
    B. Lymph vessels and nodes
    C. Trachea
    D. Bronchial arteries
    E. Pulmonary veins

8. Blood supply to large airways in the lung is derived mainly from the

    A. Intercostal arteries
    B. Internal thoracic arteries
    C. Pulmonary arteries
    D. Bronchial arteries
    E. Pericardiophrenic arteries

9. Chordae tendineae are present in

    A. Only the right ventricle
    B. The right and left ventricles
    C. The right and left ventricles and the right atrium
    D. The right and left atria
    E. All chambers of the heart

10. Borders of structures in the middle mediastinum normally seen on a posteroanterior (P/A) radiograph include all of the following except

    A. Aortic arch
    B. Left ventricle
    C. Right atrium
    D. Left atrium
    E. Superior vena cava

11. The crista terminalis is located in the

    A. Aortic arch
    B. Left ventricle
    C. Right atrium
    D. Right ventricle
    E. Superior vena cava

12. In the middle mediastinum, the ascending aorta gives rise to the

    A. Brachiocephalic artery
    B. Right coronary artery

C. Pericardial arteries
D. Bronchial arteries
E. Esophageal arteries

13. Which one of the following is found in the anterior mediastinum?

A. Trachea
B. Thymus
C. Vagus
D. Phrenic nerve
E. Esophagus

14. The esophagus

A. Presents four slight but distinct constrictions in its course
B. Lies anterior to the trachea
C. Begins at the level of the lower border of the cricoid cartilage
D. Passes anterior to the arch of the aorta
E. Has a very prominent sphincter where it joins the stomach that is formed by the external layer of muscle of the wall of the esophagus

15. All of the following are found in the posterior mediastinum except the

A. Thoracic duct
B. Arch of the aorta
C. Splanchnic nerves
D. Intercostal branches of the thoracic aorta
E. Hemiazygos vein

16. The azygos vein receives venous drainage from all of the following except the

A. Posterior intercostal veins
B. Left superior intercostal vein
C. Hemiazygos vein
D. Ascending lumbar vein

17. Which of the following arteries is the most usual source of the blood supply to the atrioventricular node?

A. The left coronary
B. The left circumflex
C. The right coronary
D. The left anterior descending
E. The marginal

18. Which one of the following pairs of structures is not correctly associated by location

A. The left coronary artery and the small cardiac vein
B. The posterior interventricular artery and the middle cardiac vein

C. The circumflex artery and the coronary sinus
D. The anterior interventricular artery and the great cardiac vein

19. Which of the following is not true concerning the right coronary artery? It

A. Is longer than the left coronary artery
B. Gives rise to a marginal branch
C. Shares a groove on the surface of the heart with coronary sinus
D. Supplies the A/V node but not the S/A node

20. The sinuatrial node is located

A. In the interatrial septum
B. Near the opening of the coronary sinus
C. Between the septal cusp and the coronary sinus
D. Near the junction of the crista terminalis and the superior vena cava
E. In the fossa ovalis

21. The foramen ovale, which usually closes at birth or shortly after, is located

A. In the medial wall of the right atrium
B. In the anterior wall of the right atrium
C. In the interventricular sulcus
D. Between the aorta and pulmonary artery
E. In the posterior wall of the right atrium

22. Which of the following structures is found in the membranous part of the interventricular septum?

A. Atrioventricular node
B. Sinuatrial node
C. Atrioventricular bundle (of His)
D. Septomarginal trabeculum (moderator band)

23. During ventricular systole (contraction) which pair of valves is open?

A. Right A/V and pulmonic
B. Right A/V and aortic
C. Right A/V and left A/V
D. Pulmonic and aortic
E. Left A/V and aortic

24. All of the following concerning the surface of the heart are correct except

A. The left ventricle forms most of the left border of the heart
B. The right ventricle forms a major portion of the anterior surface of the heart

C. The apex of the heart is roughly at the fifth intercostal space in the midclavicular line

D. The base of the heart is considered to be the diaphragmatic surface of the heart

E. The right atrium forms the right border of the heart

25. Reflections of the serous pericardium over the inferior vena cava, right pulmonary veins, and left pulmonary veins form the boundary of the

A. Oblique pericardial sinus
B. Transverse pericardial sinus
C. Sinus venosus
D. Coronary sinus
E. Pulmonary ligament

26. How many pairs of ribs articulate directly with the sternum?

A. Seven
B. Five
C. Nine
D. Six
E. Ten

27. The junction of the manubrium with the sternal body is known as the sternal angle (of Louis). Its significance is that it is where the costal cartilages of which of the following ribs articulate?

A. First
B. Second
C. Third
D. Fourth
E. No costal cartilages articulate at this junction

28. The inlet (superior extent) of the thoracic cavity demarcating the neck from the thorax is bounded by all of the following except the

A. First thoracic vertebra
B. First rib
C. Suprasternal (jugular) notch
D. Clavicle

29. With regard to the pleural cavities, all of the following statements are true except

A. They are normally potential or capillary spaces containing a small amount of serous fluid
B. They invest the lungs

C. They are closed serous sacs
D. They extend into the neck
E. They communicate with each other by means of the pulmonary ligaments

30. The costomediastinal recess is related to the

A. Posterior border of the lung
B. Anterior border of the lung
C. Inferior border of the lung
D. Pulmonary ligament

31. Anatomic surface projections are important to know in order to conduct a good physical examination. The anterior borders of both the right and left lungs follow similar courses (from behind the sternoclavicular joint) until the level of a costal cartilage where the anterior border of the left lung deviates laterally because of the cardiac notch. At which costal cartilage does this occur?

A. Second
B. Third
C. Fourth
D. Fifth
E. Sixth

32. The mediastinal surface of the right lung is in contact with all of the following except

A. Azygos vein
B. Heart
C. Thoracic aorta
D. Esophagus
E. Superior vena cava

33. The area homologous to the middle lobe of the right lung is the

A. Upper division of the left lung
B. Anterior segment of the upper left lung
C. Lingular division of the upper left lung
D. Superior segment of the lower left lung
E. Anteromedial basal segment of the lower left lung

34. Concerning the left primary bronchus, all of the following are true except

A. It is nearly twice as long as the right
B. It is smaller in diameter than the right
C. It passes anterior to the esophagus and thoracic (descending) aorta
D. It terminates by dividing into two secondary bronchi

E. Its first 2.5 cm is formed by U-shaped bars of cartilage

35. The intercostal nerves and vessels of the thoracic cage travel deep to the _____ and superficial to the _____ .

    A. Superficial fascia, external intercostal muscles
    B. Innermost intercostal complex, parietal pleura
    C. External intercostal muscles, internal intercostal muscles
    D. Internal intercostal muscles, innermost intercostal muscle complex

36. The cell bodies of origin of the nerve fibers that provide sensory innervation to the nipple of the male are found in which of the following?

    A. Spinal cord level C_4
    B. Spinal cord level T_4
    C. Dorsal root ganglion T_4
    D. Dorsal root ganglion C_4

37. Which of the structures listed below normally is not found in most of the intercostal spaces?

    A. A branch of the aorta
    B. A tributary to the azygos or hemiazygos system of veins
    C. A dorsal ramus of a spinal nerve
    D. A ventral ramus of a spinal nerve
    E. A branch of the internal thoracic (mammary) artery

38. The dermatome segment that includes the skin around the umbilicus sends its sensory innervation to spinal cord level

    A. T_6
    B. T_{10}
    C. L_1
    D. L_3
    E. L_5

39. The posterior intercostal arteries

    A. Are embedded in the parietal pleura
    B. Arise from esophageal branches of the aorta
    C. Supply blood to the visceral pleura of the lung
    D. Anastomose with branches of the internal thoracic artery
    E. Are not accompanied by veins

40. Which one of the following statements is true?

    A. The anterior intercostal arteries are all direct branches of the internal thoracic artery
    B. The anterior intercostal nerves interdigitate with the posterior intercostal nerves at the region of the costochondral junctions
    C. The first two posterior intercostal arteries are branches of a trunk arising from the subclavian artery
    D. The intercostal nerves are cutaneous nerves only
    E. All intercostal veins drain directly into the azygos vein

41. Indicate which of the following structures is not supplied by direct branches of the internal thoracic artery

    A. Diaphragm
    B. Pectoralis major
    C. Pericardium
    D. External intercostal muscles

42. The subcostalis muscle

    A. Spans two intercostal spaces
    B. Lies in the midaxillary line
    C. Receives its blood supply from the anterior intercostal artery
    D. Is not active in inspiration

43. The internal (posterior) intercostal membrane is an extension

    A. Of the fascia of the internal intercostal muscle from the costochondral junction to the sternum
    B. Of the fascia of the internal intercostal muscle from the rib angle to the vertebral column
    C. Of the fascia of the external intercostal muscle from the costochondral junction to the sternum
    D. Of the fascia of the external intercostal muscle from the rib angle to the vertebral column
    E. From the subcostal muscle to the vertebral column

44. The phrenic nerve accompanies the _____ en route to the diaphragm.

    A. Superior epigastric artery and veins

B. Mediastinal branch of the descending aorta
C. Pericardiophrenic artery and veins
D. Superior phrenic artery and veins
E. Inferior phrenic artery and veins

45. The right and left vagus nerves superior to the diaphragm form the

A. Hepatic and gastroduodenal branches
B. Major sympathetic nerve supply of the stomach
C. Celiac plexus
D. Esophageal plexus
E. Hypogastric plexus

46. The azygos vein:

A. Ascends through the anterior mediastinum
B. Arches over the root of the left lung
C. Ends in the inferior vena cava
D. Receives blood from the hemiazygos veins

47. The _____ carries lymph from most of the entire lower half and the left upper quadrant of the body to the venous system. It usually empties into the veins of the root of the neck near the junction of the _____ internal jugular and subclavian veins.

A. Right lymphatic duct, right
B. Jugular lymph trunk, left
C. Thoracic duct, right
D. Thoracic duct, left

48. The _____ artery is the first large branch of the aorta.

A. Left common carotid
B. Left subclavian
C. Brachiocephalic
D. Right common carotid
E. Right subclavian

49. Which structure most closely follows the esophagus through the posterior mediastinum?

A. Azygos vein
B. Accessory hemiazygos (when present)
C. Vagus nerves
D. Phrenic nerve

50. The superior mediastinum contains all of the following structures except the

A. Brachiocephalic artery
B. Superior vena cava
C. Left common carotid
D. Thoracic duct
E. Hemiazygos vein

FOUR

Abdomen

Abdominal Wall

Objectives

After completion of the study of the abdominal wall the student should be able to

▶ *Palpate surface landmarks and demarcate regions of the abdominal cavity*

▶ *Delineate the muscles and aponeuroses of the anterolateral abdominal wall*

▶ *Draw and label variations (by levels) of the rectus sheath*

▶ *Define the inguinal ligament and its extensions*

▶ *Relate the boundaries of the inguinal canal to muscles and aponeuroses of the abdominal wall*

▶ *Label the component parts of the lumbar plexus, list the branches of the plexus and give their distribution*

▶ *Give the origin, insertion, and major orifices of the thoracic diaphragm*

Surface Anatomy

The **linea alba** is a linear depression in the median plane extending from the xiphoid process to the pubic symphysis. It is formed by the fused insertions of the aponeuroses of the anterolateral muscles of the abdominal wall. The **umbilicus** (navel) is interposed in the linea alba and results from the closure of the umbilical cord shortly af-

ter birth, leaving a puckered, yet depressed scar. The **linea semilunaris** indicates the lateral extent of the rectus abdominis muscle and its sheath. The point where this line meets the right ninth costal cartilage indicates the position of the gallbladder. Transverse bands of connective tissue, the **tendinous inscriptions,** are interposed segmentally within the rectus abdominis and are visible as transverse lines in a muscular individual. The **xiphoid process** can be palpated in the midline at the thoracic outlet. The **crests of the ilia,** the **superior** and **inferior anterior iliac spines,** and the **pubic tubercle** are all palpable at the inferior extent of the abdominal wall. **Pubic hair** distribution differs in the two sexes. In the male it is dispersed in a diamond-shaped area and extends from the pubis to the umbilicus, while in the female it is triangular, with the base of the triangle above the mons pubis.

Muscles and Rectus Sheath

The anterolateral abdominal wall between the thoracic outlet and the innominate bone is composed of skin, superficial fascia, the three abdominal muscles (**external abdominal oblique, internal abdominal oblique,** and **transversus abdominis**) and their fascial envelopes. Anteriorly the **rectus abdominis** muscle and its sheath contribute to the abdominal wall, while posteriorly the muscular components of the wall are the **quadratus lumborum** and the **psoas major** and **minor** muscles (Table 4-1).

Table 4-1
Muscles of the Abdominal Wall

Muscle	Origin	Insertion	Action	Nerve
External abdominal oblique	External surface of lower eight ribs	Anterior half of iliac crest and linea alba	Compresses and supports abdominal viscera; rotates and flexes vertebral column	Lower five intercostals and subcostal
Internal abdominal oblique	Lateral two-thirds of inguinal ligament, iliac crest, and lumbodorsal fascia	Cartilages of lower three or four ribs, linea alba, and by conjoint tendon into pubis	Compresses and supports abdominal viscera; flexes and rotates vertebral column	Lower five intercostals, subcostal, and first lumbar
Transversus abdominis	Lateral one-third of inguinal ligament, iliac crest, lumbodorsal fascia, and cartilages of lower six ribs	Linea alba and by conjoint tendon into pubis	Compresses and supports abdominal viscera	Lower five intercostals, subcostal, and first lumbar
Rectus abdominis	Xiphoid process and fifth to seventh costal cartilages	Crest and symphysis of pubis	Tenses abdominal wall and flexes vertebral column	Lower five intercostals and subcostals
Pyramidalis	Front of pubis and anterior pubic ligament	Linea alba midway between umbilicus and pubis	Tenses linea alba	Subcostal
Quadratus lumborum	Lumbar vertebrae, lumbodorsal fascia, and iliac crest	Lower border of twelfth rib and transverse processes of upper four lumbar vertebrae	Draws last rib toward pelvis and flexes vertebral column laterally	Subcostal and first three or four lumbar
Psoas major	Transverse processes, intervertebral discs and bodies of all lumbar vertebrae	Lesser trochanter of femur	Flexes vertebral column	Second and third lumbar
Psoas minor	Bodies and intervening discs of twelfth thoracic and first lumbar vertebrae	Pectineal line and iliopectineal eminence of pelvis	Flexes vertebral column	First lumbar

The muscles of the anterolateral wall are disposed in three layers. The fibers of the outermost external oblique pass inferomedially, those of the intermediate internal oblique, inferolaterally, and those of the innermost transversus abdominis, transversely. Their insertions, by broad aponeuroses into the midline, envelop the rectus abdominis to form the rectus sheath (Fig. 4-1). Nerves and vessels to the anterolateral wall course between the internal oblique and transversus abdominis muscles. The posterior abdominal wall muscles, namely, the quadratus lumborum and the psoas major and minor, are the deepest lying structures in the small of the back. They extend from the twelfth rib to the iliac crest, with the psoas muscles continuing inferiorly into the iliac fossa.

The **rectus sheath** extends from the xiphoid

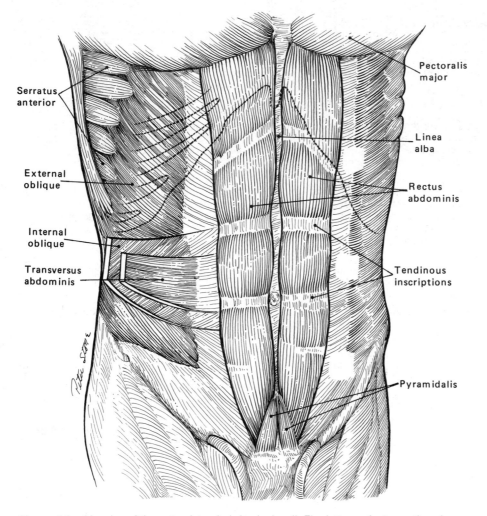

Figure 4-1. *Muscles of the anterolateral abdominal wall. The letters refer to sections in* Fig. 4-2.

process and adjacent costal cartilages to the pubic bone (Fig. 4-2). It holds the rectus abdominis muscle in place, but does not restrict its movement, owing to the presence of anterior and posterior fascial clefts. The **anterior wall** of the rectus sheath covers the muscle from end to end, but the **posterior wall** is incomplete superiorly and inferiorly. In the uppermost part of the abdomen the anterior wall of the sheath is formed by the aponeurosis of the external oblique; the posterior wall in this area is absent. From the lower margin of the thoracic outlet to the midpoint between the umbilicus and pubis, the aponeurosis of the internal oblique splits into two laminae. In this area the anterior wall of the sheath is formed by the aponeurosis of the external oblique and the superficial lamina of the aponeurosis of the internal oblique, while the posterior wall is composed of the deep lamina of the aponeurosis of the internal oblique and the aponeurosis of the transversus

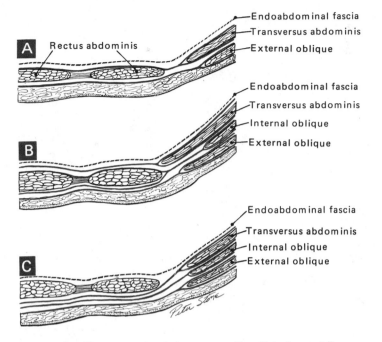

Figure 4-2. *The rectus sheath in cross-section. Note the contributions of the aponeuroses of the anterolateral abdominal muscles. Letters refer to levels indicated on Fig. 4-1.*

abdominis. Inferiorly the anterior wall is formed by the aponeuroses of all three abdominal muscles, whereas posteriorly the rectus abdominis rests directly on the **endoabdominal fascia,** which lines the entire abdominal cavity. The **arcuate line** (semicircular line of Douglas) is the lower edge of the aponeurotic components of the posterior wall of the rectus sheath. **Contents of the rectus sheath** include the rectus abdominis and pyramidalis muscles, terminations of the lower five intercostal and subcostal nerves, and the inferior and superior epigastric arteries and their venae comitantes.

Fasciae

The superficial fascia of the abdominal wall is divided into a **superficial (Camper's) layer,** with abundant adipose tissue, and a **deep membranous (Scarpa's) layer.** In the male, Scarpa's fascia extends over the pubic symphysis and thickens to form the **fundiform ligament** of the penis, which extends inferiorly to attach to the dorsum and sides of this structure. Camper's fascia extends over the inguinal ligament as the **superficial fascia of the thigh.** Over the scrotum the superficial layer loses its fat and fuses with the membranous layer to become the **dartos tunic** of the scrotum. This same combined layer, devoid of fat, elongates to ensheath the penis. Inferiorly Scarpa's fascia passes over the inguinal ligament and is continuous with the **fascia cribrosa** in the thigh, while over the perineum it becomes the **superficial perineal (Colles') fascia.**

Inguinal Region

Inferiorly the abdominal muscles contribute to the formation of the inguinal ligament and inguinal canal (Fig. 4-3 and Fig. 4-4). The **inguinal**

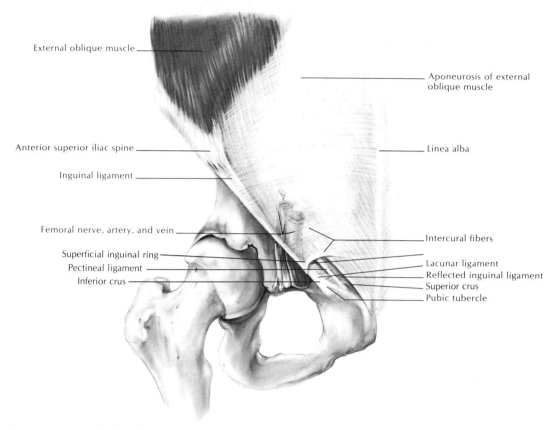

External oblique muscle

Aponeurosis of external oblique muscle

Anterior superior iliac spine

Linea alba

Inguinal ligament

Femoral nerve, artery, and vein

Intercural fibers

Superficial inguinal ring

Lacunar ligament

Pectineal ligament

Reflected inguinal ligament

Inferior crus

Superior crus

Pubic tubercle

Figure 4-3. *Inguinal region.*

(Poupart's) ligament is formed by the inferior free border of the aponeurosis of the external abdominal oblique muscle. The inguinal ligament attaches laterally to the **anterosuperior iliac spine** and medially to the **pubic tubercle** and adjacent area. As it attaches medially, it splits to form a triangular gap with a **superior** and an **inferior crus.** Intercrural fibers transform this triangular gap into the **superficial inguinal ring.** Extensions from the deep surface of the inguinal ligament pass to the superior pubic ramus to gain additional attachment and separate structures passing from the abdomen into the thigh.

The **lacunar (Gimbernat's) ligament** is a portion of the medial end of the inferior crus that rolls under the spermatic cord to attach to the **pectineal line** on the superior pubic ramus lateral to the pubic tubercle. The lacunar ligament is triangular in form and has a lateral crescentic base and an apex directed medially toward the pubic tubercle. The **pectineal ligament (of Cooper)** is a lateral prolongation from the lacunar ligament. It forms a strong narrow band that attaches along the **iliopectineal line.** A tendinous medial expansion is the **reflected** (reflex) **inguinal ligament,** continuous with the lacunar ligament. It sweeps beneath the superficial inguinal ring to reach the linea alba.

The **superficial inguinal ring,** easily palpa-

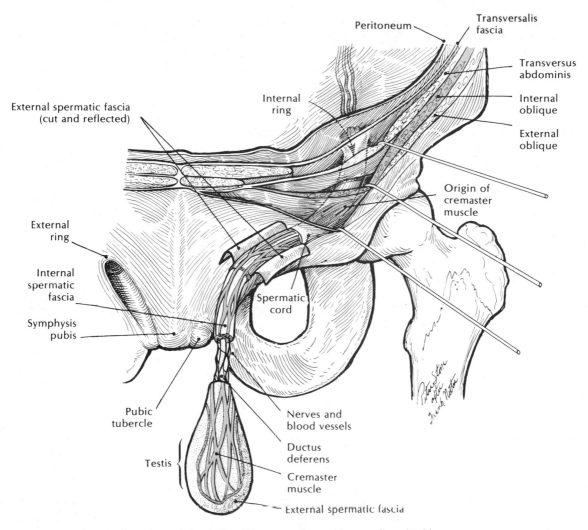

Figure 4-4. *Contributions from abdominal wall to spermatic cord covers. (Langley LL, Telford IR, Christensen JB: Dynamic Anatomy and Physiology, 5th ed. New York, McGraw-Hill, 1980)*

ble in the living male subject, allows passage of the spermatic cord in the male (round ligament of the uterus in the female) from the inguinal canal to the scrotum (labium majus). The **deep inguinal ring** is an opening in the endoabdominal fascia. It lies 1 cm to 2 cm above the midinguinal point and is immediately lateral to the inferior epigastric vessels. The **inguinal canal** is a narrow channel about 3 cm to 4 cm in length, passing between the deep and superficial rings. It is formed when the **processus vaginalis** evaginates the abdominal wall into the scrotum prior to the descent of the testis. The inguinal canal, directed inferiorly, medially, and anteriorly, has a floor, roof, and anterior and posterior walls. In the male it transmits the spermatic cord, cremasteric vessels, genital branch of the genitofemoral nerve, and cutaneous branches of the ilioinguinal nerve. In the female the round ligament of the uterus, with its vessels and nerves, traverse the canal.

The grooved surface of the inguinal ligament, reinforced medially by the lacunar ligament, forms the **floor** of the inguinal canal. The **anterior wall** is formed by the aponeurosis of the external abdominal oblique muscle. In its lateral half it is supplemented by the muscle fibers of the internal abdominal oblique muscle. The canal is **roofed** by the arching fibers of the internal abdominal oblique and its medial continuation, the cremaster muscle. The endoabdominal fascia covers the entire **posterior wall.** It is reinforced medially by the fusion of the aponeuroses of the internal oblique with the transversus abdominis muscle to form the **conjoint tendon** (falx inguinalis).

The wall of the inguinal canal and the superficial and deep rings are potential areas of weakness in the abdominal wall through which inguinal herniae pass. The anterior wall of the canal is strongest opposite the deep inguinal ring because of the presence of the muscle fibers of the internal oblique; similarly, the posterior wall is strongest opposite the superficial ring because of the presence of the conjoint tendon.

A **hernia** is an abnormal protrusion of any structure beyond its normal site. In the inguinal region, herniae are classified as direct or indirect, depending upon their course through the abdominal wall. A **direct inguinal hernia** bypasses the deep inguinal ring to penetrate the posterior wall of the inguinal canal medial to the inferior epigastric artery. This area of direct inguinal hernia, the **inguinal (Hesselbach's) triangle,** is formed medially by the rectus abdominis, laterally by the inferior epigastric artery, and inferiorly by the inguinal ligament. The direct inguinal hernia occurs more frequently in elderly people and is usually limited to the region of the superficial ring. The

wall of the hernial sac of a direct hernia is composed, from within outward, of peritoneum, extraperitoneal connective tissue, endoabdominal fascia, cremasteric muscle and fascia, external spermatic fascia, subcutaneous fascia, and skin.

In the male the **indirect inguinal hernia** is three times more common than the direct hernia. It traverses the deep inguinal ring (lateral to the inferior epigastric artery) and passes through the inguinal canal to exit through the superficial ring into the scrotum. Thus, the indirect inguinal hernia follows the same course as the spermatic cord. This type is also known as the **congenital hernia,** since failure of closure of the processus vaginalis predisposes to this type of hernia. The coverings of the indirect hernia are the same as the coverings of the spermatic cord. The position of the inferior epigastric artery, immediately medial to the deep inguinal ring, aids in distinguishing direct from indirect inguinal herniae.

The internal aspect of the anterior abdominal wall presents five longitudinal ridges of peritoneum. The single median ridge, the **median umbilical fold,** encloses a slender fibrous cord (median umbilical ligament), the remnant of the **urachus.** It passes from the apex of the bladder to the umbilicus between the peritoneum and the endoabdominal fascia. Two **medial umbilical folds,** representing the **obliterated umbilical arteries** (lateral umbilical ligaments) continue from the superior vesicular branch of the internal iliac artery to pass to the umbilicus. The **inferior epigastric arteries** and **veins** raise ridges of peritoneum (**lateral umbilical folds**) as they pass toward the rectus abdominis.

Arteries and Nerves (Tables 4-2 and 4-3)

The **inferior epigastric artery,** a branch of the external iliac, pierces the endoabdominal fascia to ascend behind the rectus abdominis, but within the rectus sheath. It anastomoses with the **superior epigastric artery** from the internal thoracic.

Table 4-2
Arterial Supply to the Abdominal Wall

Artery	Origin	Course	Distribution	Anastomoses
Superior epigastric	Internal thoracic	Descends in rectus sheath; deep to rectus abdominis	Rectus abdominis, anterolateral abdominal wall	Inferior epigastric, lumbars, lateral thoracic
Inferior epigastric	External iliac	Angles towards rectus sheath; forms lateral boundary of Hesselbach's triangle; ascends in sheath deep to rectus abdominis	Rectus abdominis, anterolateral abdominal wall	Superior epigastric, lumbars, lateral thoracic
Deep iliac circumflex	External iliac	Along deep aspect of inguinal ligament	Iliacus and inferior portion of anterolateral abdominal wall	Iliac branches of iliolumbar, superior gluteal, lateral femoral circumflex
Superficial iliac circumflex	Femoral	Follows inguinal ligament in superficial fascia	Superficial fascia and skin over lower portion anterolateral abdominal wall	Lateral thoracic
Superficial epigastric	Femoral artery	Angles towards umbilicus in superficial fascia	Superficial fascia and skin over hypogastric region	Lateral thoracic
Inferior phrenic	Abdominal aorta	Passes on inferior aspect of thoracic diaphragm	Thoracic diaphragm	Superior phrenic, left gastric, short gastrics, musculophrenic
Lumbars	Abdominal aorta	Pass (segmentally) in retroperitoneal position	Posterior abdominal wall	Superior and inferior epigastrics, iliolumbars
Iliolumbar	Posterior division internal iliac	Ascends along vertebral column; divides into lumbar and iliac branches	Iliacus, posterior abdominal wall and vertebral column	Iliac branch with deep iliac circumflex; lumbar branches with lumbar arteries

Branches of the superior and inferior epigastric arteries include muscular branches to the rectus muscle and cutaneous branches to overlying skin. Additional branches of the inferior epigastric artery include the **cremasteric** supplying the cremaster muscle, which passes through the deep inguinal ring into the inguinal canal, and **pubic branches** ramifying on the pubic bone. The **deep iliac circumflex artery,** the second branch of the external iliac, passes laterally to ramify on and supply the iliacus muscle in the iliac fossa. Four or five segmental branches, the **lumbar arteries,** arise from the abdominal aorta. They supply muscles of the posterior abdominal wall and, similar to the posterior intercostals, course between the second and third layers of muscle to supply anterolateral abdominal wall muscles and overlying skin. Additional blood supply, to the superficial fascia and skin, is derived from the **superficial iliac circumflex** and **su-**

Table 4-3
Nerve Distribution to Abdomen

Nerve	Origin	Course	Distribution
Thoracoabdominal nerves	Continuations of lower intercostal nerves	Pass between second and third layers of muscle	Anterolateral abdominal muscles and overlying skin (T_{10} dermatome flanks umbilicus); periphery of diaphragm
Iliohypogastric	Posterior divisions of lumbar plexus	Crosses posterior abdominal wall to pierce transversus abdominis	Skin over iliac crest and adjacent buttock and hypogastric regions
Ilioinguinal	Anterior divisions of lumbar plexus	Passes between second and third layers of muscle to reach and traverse inguinal canal	Skin of scrotum (labium majus), adjacent thigh and mons pubis
Segmental branches	Ventral rami of all lumbar nerves	Short twigs to successive segments of muscles of posterior abdominal wall	Quadratus lumborum and psoas muscles (iliacus by twigs from femoral)
Autonomic plexus	Formed around branches of abdominal aorta and their subsequent branches; parasympathetic (preganglionic) from vagi and pelvic splanchnics; sympathetic (postganglionic) by splanchnics from lower levels of sympathetic trunk	Follow arteriolar branchings to walls of digestive organs; myenteric and submucosal plexuses are locations of postganglionic parasympathetic nerve cell bodies	Parasympathetics increase peristalsis and secretion, relax sphincters; sympathetics opposite action

perficial epigastric branches of the femoral artery.

Innervation of muscles and skin of the anterolateral abdominal wall is derived from the ventral rami of spinal nerves from the seventh thoracic to the first lumbar, as thoracoabdominal, subcostal, iliohypogastric, and ilioinguinal nerves. The **thoracoabdominal** (lower intercostals) and **subcostal nerves** pass deep to costal cartilages and descend medially between the transversus abdominis and the internal abdominal oblique muscles. They supply abdominal muscles and terminate as lateral and anterior cutaneous branches in line with the more superior intercostal nerves.

The **iliohypogastric nerve** (T_{12} and L_1) emerges at the upper lateral border of the psoas major, crosses the quadratus lumborum to the crest of the ilium, where it pierces the transversus abdominis muscle. It gives **lateral cutaneous branches,** which supply the skin immediately above the iliac crest and the gluteal region, and **anterior cutaneous branches,** which pierce the aponeurosis of the external abdominal oblique 2 cm to 3 cm above the subcutaneous inguinal ring to innervate the skin of the hypogastric region. The **ilioinguinal nerve** (L_1) follows a course parallel but inferior to the iliohypogastric nerve, piercing the internal abdominal oblique and accompanying the spermatic cord through the superficial inguinal ring. It supplies skin over the upper medial part of the thigh, mons pubis, root of the penis or clitoris, and scrotum or labium majus.

Lumbar Plexus

The **lumbar plexus** is formed by the ventral rami of the first four lumbar spinal nerves with a contribution from the twelfth thoracic nerve (Fig. 4-5). Part of the fourth lumbar nerve joins with the fifth lumbar nerve to form the lumbosacral trunk, which contributes to the formation of the sacral plexus. The lumbar plexus differs from the brachial plexus in that no intricate interlacing of fi-

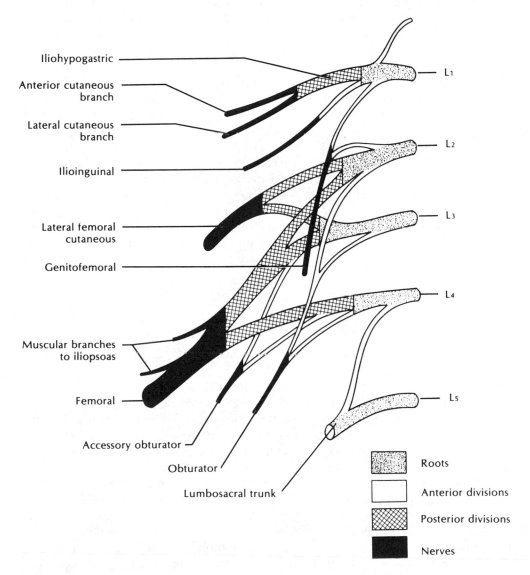

Figure 4-5. *Lumbar plexus.*

bers occurs. Leaving intervertebral foramina, the nerves pass obliquely outward behind the psoas major and anterior to the quadratus lumborum muscles to form their several branches.

The **iliohypogastric** and **ilioinguinal nerves**, described above, are the first two branches of the lumbar plexus. **Muscular branches,** arising segmentally as independent twigs, pass from all four lumbar spinal nerves to the quadratus lumborum, and from the second and third lumbar nerves to the psoas major and minor muscles. The **genitofemoral nerve** (L_1 and L_2) passes through the substance of the psoas major, emerges from its medial border close to the lower lumbar vertebrae to descend on the anterior surface of this muscle. It divides into two branches, **genital** and **femoral.** The former pierces the endoabdominal fascia to pass through the inguinal canal behind the spermatic cord to supply the cremaster muscle and the skin of the scrotum. The femoral branch descends on the external iliac artery to enter the femoral sheath. The **lateral femoral cutaneous nerve** (L_2 and L_3) emerges at the middle of the lateral border of the psoas major to cross the iliacus muscle obliquely

toward the anterior superior iliac spine. It courses deep to the inguinal ligament, an inch medial to the anterior superior iliac spine, to be distributed to the skin of the lateral aspect of the thigh. The **obturator nerve** (L_2, L_3, and L_4) descends through the substance of the psoas major to emerge from the medial border of this muscle near the brim of the pelvis. It passes behind the common iliac vessels, along the lateral wall of the pelvis superoanterior to the obturator vessels, to leave the pelvis through the obturator canal. Its distribution will be considered with the inferior extremity.

The **femoral nerve** (L_2, L_3, and L_4), the largest branch of the lumbar plexus, descends through the fibers of the psoas major to emerge from its lateral border. It gives a few filaments to the iliacus and psoas major and minor, then continues inferiorly passing deep to the inguinal ligament to enter the thigh. The ventral ramus of the fourth lumbar nerve divides into an upper and lower division. The upper division contributes to the lumbar plexus, the lower division joins the fifth lumbar to form the **lumbosacral trunk** passing to the sacral plexus.

Diaphragm

The **diaphragm** is a movable musculotendinous partition between the thoracic and the abdominal cavities. It forms the concave roof of the abdominal cavity and the convex floor of the thoracic cavity (Fig. 4-6). The diaphragm rises higher on the right side than on the left due to the larger right lobe of the liver. The central portion is aponeurotic and forms the strong **central tendon,** which is indistinctly divided into three leaflets. The muscle fibers of the diaphragm originate at the margins of the thoracic outlet and insert into the central tendon. The short and narrow **sternal portion** arises as small slips from the back of the xiphoid process; the extensive **costal portion,**

from the inner surface of the lower six costal cartilages; and the **vertebral portion,** from the arcuate ligaments and the upper lumbar vertebrae as a pair of muscular crura. Each crus is a thick, fleshy bundle that tapers inferiorly and becomes tendinous. The **left crus** attaches to the upper two lumbar vertebrae and the intervening vertebral disc, the **right crus** to the upper three lumbar vertebrae and intervening discs. Fibers of each crus spread out and ascend to attach to the central tendon, with the right crus encircling the esophagus. The lateral, medial, and median arcuate ligaments give partial origin to the diaphragm. The unpaired **median arcuate ligament,** opposite

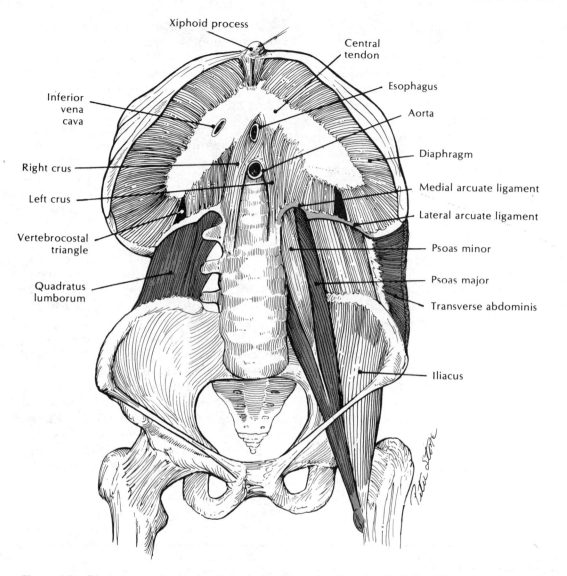

Xiphoid process

Central tendon

Esophagus

Aorta

Inferior vena cava

Diaphragm

Right crus

Medial arcuate ligament

Left crus

Lateral arcuate ligament

Vertebrocostal triangle

Psoas minor

Psoas major

Quadratus lumborum

Transverse abdominis

Iliacus

Figure 4-6. *Diaphragm and posterior abdominal wall muscles. (Langley LL, Telford IR, Christensen JB: Dynamic Anatomy and Physiology, 5th ed. New York, McGraw-Hill, 1980)*

the twelfth thoracic vertebra, arches between the right and left crura and crosses the aorta as it enters the abdomen. The paired **medial arcuate ligaments** are highly arched and pass laterally from the tendinous part of each crus to curve across the psoas major muscle and attach to the tip of the transverse process of the first or second lumbar vertebra. The paired **lateral arcuate liga-**

ments stretch across the quadratus lumborum from the tip of the transverse process of the first or second lumbar vertebra to the tip and lower margin of the twelfth rib. That portion of the diaphragm attaching to the lateral arcuate ligament is thin and sometimes devoid of muscular fibers, with only connective tissue separating the pleura from the renal fat. Through this weakened area, the **vertebrocostal triangle,** abdominal contents may herniate into the thorax.

The continuity of the diaphragm is interrupted by three large and several small apertures. Between the crura, at the level of the twelfth thoracic vertebra, the **aortic hiatus** is bridged anteriorly by the median arcuate ligament. The aortic hiatus transmits the aorta, azygos and hemiazygos veins, and thoracic duct. At the level of the tenth thoracic vertebra the oval **esophageal hiatus** is situated obliquely behind the central tendon, 2 cm to 3 cm to the left of the midline, and is surrounded by the right crus of the diaphragm. It transmits the esophagus, vagal trunks, and esophageal branches of the left gastric vessels.

Esophageal (Hiatal) Hernia

The most common site of diaphragmatic (hiatal) herniae is through an enlarged esophageal hiatus. An hiatal hernia is usually acquired and occurs most often in middle-aged individuals. The principal complaints of a patient with an hiatal hernia are "heartburn" and upper-abdominal fullness. This is often accompanied by regurgitation of gastric contents into the esophagus and mouth. These symptoms are usually relieved by sitting up or elevating the head and shoulders. The acquired hiatal hernia is caused by the cardia of the stomach protruding into the mediastinum through the weakened, enlarged esophageal hiatus in the diaphragm. Since the stomach may move back and forth through the hiatus, it is also called a sliding hernia.

At the level of the eighth thoracic vertebra the wide **caval opening** is located within the central tendon of the diaphragm about 3 cm to the right of the median plane. It transmits the inferior vena cava, terminal branches of the right phrenic nerve, and lymph vessels. Additional structures passing between the thorax and the abdomen include the superior epigastric vessels, the lower five throacoabdominal nerves, and the subcostal vessels and nerves. The sympathetic trunk lies behind the medial arcuate ligament, and the three splanchnic nerves pierce the crus on their side of origin to enter the abdomen.

Innervation to the diaphragm is derived from the **phrenic nerve** and from twigs of the **lower intercostal nerves.** The **inferior phrenic artery,** a branch of the abdominal aorta, is the chief arterial supply to the diaphragm. Additional blood is supplied from the superior phrenic, the **pericardiophrenic branch** of the internal thoracic artery, from irregular twigs of the thoracic aorta, and peripherally, through branches from the **musculophrenic artery** and intercostals.

Abdominal Cavity

Objectives

After the completion of the study of the peritoneal cavity and gastrointestinal tract the student should be able to

▶ *Describe the extent of the peritoneal cavity and its subdivisions*

▶ *List the boundaries of the lesser sac (omental bursa)*

▶ *Accurately define mesentery and visceral ligament and describe their functional importance*

▶ *Relate each visceral ligament (mesentery) to its developmental derivation*

▶ *Define the term retroperitoneal; list those organs that are retroperitoneal in position*

▶ *Define the subdivisions of the stomach*

and small intestine and describe their anatomic characteristics

▶ Discuss the three gross features that differentiate the colon from the small intestine

▶ Identify the subdivisions of the large intestine and describe their unique anatomic characteristics

▶ Give the origin, course, and distribution of arteries supplying the GI tract

The **abdominal cavity,** the largest cavity in the body, is bounded anteriorly by the rectus abdominis, laterally by the external, internal, and transverse abdominal, and more inferiorly by the iliacus muscles. The abdominal cavity is bounded posteriorly by the vertebral column and the psoas major and minor, and quadratus lumborum muscles. Superficially, it is subdivided, for descriptive purposes, into nine regions by two horizontal and two vertical arbitrary lines (or planes) (Fig. 4-7A and B). The horizontal lines are the **transpyloric,** at the level of the pylorus of the stomach, and the **intertubercular,** at the level of the iliac tubercles. The vertical lines are two lateral lines that bisect the clavicles. The resulting **subdivisions of the abdomen** are from superior to inferior, **right** and **left hypochondriac** and **epigastric, right** and **left lumbar** and **umbilical,** and **right** and **left inguinal** (iliac) and **pubic** (hypogastric).

Peritoneal Cavity (Fig. 4-8)

The **peritoneal cavity** is the largest serous cavity in the body. It is located in the abdominal and pelvic cavities, extending from the thoracic diaphragm to the pelvic diaphragm. In the fetus it communicates via the **processus vaginalis** with the serous cavity of the scrotum. Developmentally, the gastrointestinal tract is suspended into the abdominal cavity by specializations of peritoneum. These specializations, the **dorsal** and **ventral mesogastria,** consist of two layers of serous membrane. They are formed as parietal perito-

neum reflected from the body wall to attach to the gut and continue as visceral peritoneum.

As a result of the rotation of the gut the disappearance, fusion, shifting, shortening, or redundant growth of these peritoneal folds divide the pertioneal cavity into two distinct parts, the greater and lesser sac. In the adult the **lesser sac** (omental bursa) is situated posterior to the lesser omentum, stomach, and gastrocolic ligament. It is limited inferiorly by the transverse colon with its mesocolon and bounded on the left by the gastrolienal and lienorenal ligaments. To its right, the omental bursa communicates through the **epiploic foramen (of Winslow)** with the **greater sac** of the peritoneal cavity.

In the adult, mesenteries or visceral ligaments that extend from the anterior abdominal wall to suspend organs into the abdominal cavity are derivatives of the ventral mesogastrium. These include the falciform, coronary, right and left triangular ligaments of the liver, and the lesser omentum. The latter may be subdivided into gastrohepatic and hepatoduodenal ligaments. Dorsal mesogastrium derivatives include the greater omentum, "the" mesentery, gastrosplenic and lienorenal ligaments, and transverse and sigmoid mesocolons. Structures within the abdominal cavity not suspended from the body wall by a mesentery, or from other viscera by a visceral ligament, are retroperitoneal in position (Fig. 4-9). In studying the blood supply and innervation of abdominal viscera, their course should be related to visceral ligaments. All vessels and their autonomic plexuses begin in a retropertioneal position. To reach an organ suspended into the abdominal cavity, they must pass between the two serous membrane layers of its mesentery or visceral ligament. In the male, the peritoneum forms a closed cavity, while in the female, it communicates with the exterior through the openings of the uterine tubes.

Peritonitis
Peritonitis is an inflammation of the peritoneum. It may be general or localized. It is characterized by an accumulation of a

Text continues on page 162

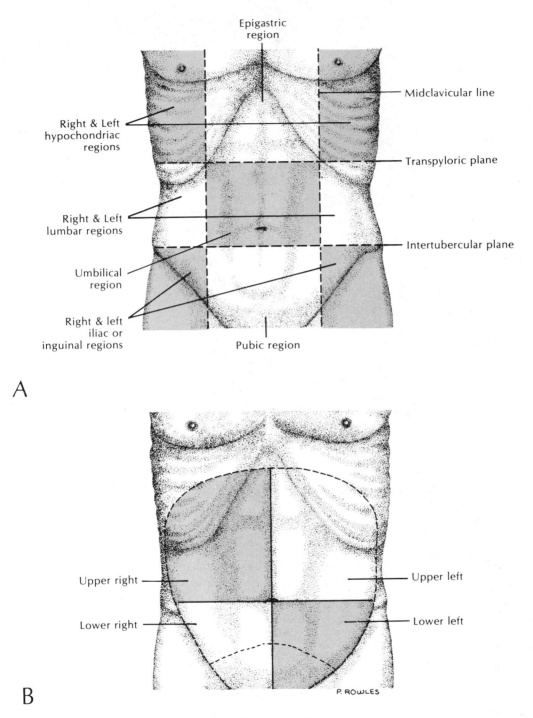

Figure 4-7. **A.** *Regions of the abdomen.* **B.** *Quadrants of the abdomen.*

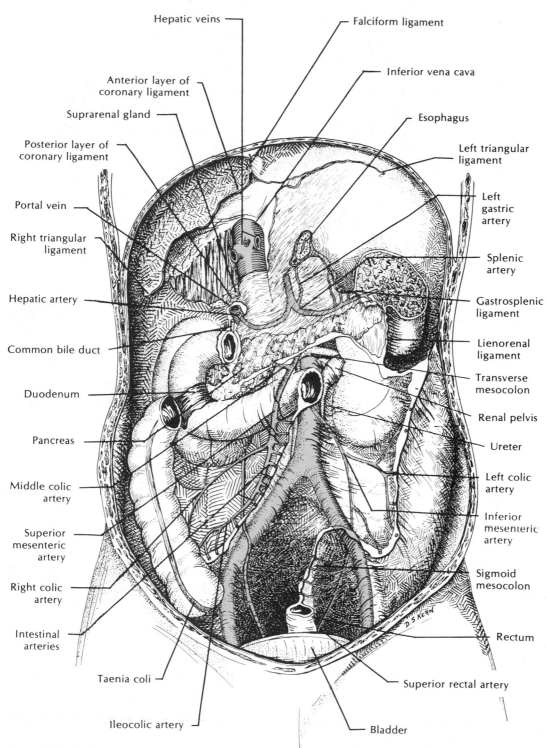

Hepatic veins

Falciform ligament

Anterior layer of coronary ligament

Inferior vena cava

Suprarenal gland

Esophagus

Posterior layer of coronary ligament

Left triangular ligament

Portal vein

Left gastric artery

Right triangular ligament

Splenic artery

Hepatic artery

Gastrosplenic ligament

Common bile duct

Lienorenal ligament

Duodenum

Transverse mesocolon

Pancreas

Renal pelvis

Middle colic artery

Ureter

Superior mesenteric artery

Left colic artery

Right colic artery

Inferior mesenteric artery

Intestinal arteries

Sigmoid mesocolon

Taenia coli

Rectum

Ileocolic artery

Superior rectal artery

Bladder

D.S. KERN

Figure 4-8. *Peritoneal cavity.*

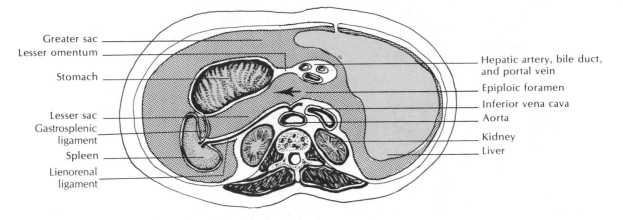

Figure 4-9. *Epiploic foramen and related structures.*

large amount of pertioneal fluid (ascites) containing fibrin and many leucocytes (pus). In the supine patient, the infected fluid tends to collect at two sites: in the pelvic cavity, and in the right posterior subphrenic space. Tapping or draining of excess fluid from the abdomen is accomplished by inserting, under local anesthesia, a trocar and cannula or a needle and plastic tubing through the anterior abdominal wall, and aspirating the peritoneal fluid (paracentesis). When a patient with peritonitis is examined, stretching of the infected parietal peritoneum is very painful. The pain is especially severe when the digital pressure over the inflamed area is suddenly released, because this causes the abdominal wall to rebound suddenly. Clinically, this is called rebound tenderness, which localizes the site of inflammation of the parietal peritoneum and often occurs over the infected organ.

Nerves

Parasympathetic innervation of the abdominal viscera is derived from vagal and sacral nerves. The **left** and **right vagus nerves** form a plexus around the esophagus and, upon passing through the diaphragm, are reconstituted as the anterior and posterior vagal trunks, respectively. The **anterior vagal trunk** innervates the liver, gallbladder, bile ducts, pylorus of the stomach, duodenum, and pancreas. The **posterior vagal trunk** supplies the rest of the stomach and then joins the superior mesenteric plexus. Here it intermingles with sympathetic filaments and is distributed to the kidney, small intestine, and ascending and transverse colon by forming plexuses around arteries to these structures. The distal portion of the large intestine beyond the splenic flexure is innervated by **sacral parasympathetic nerves.** Postganglionic parasympathetic cell bodies are usually located within the substance of the organ they supply.

The principal sources of sympathetic innervation to abdominal viscera are the **greater** (T_5 to T_9), **lesser** (T_{10} and T_{11}) and **least** (T_{12}), **splanchnic nerves.** Additional sympathetic fibers, the **lumbar splanchnic** and **sacral splanchnic nerves,** arise directly from the lumbar or sacral segments of the sympathetic chain. The sympathetic nerves form plexuses around the main branches of the abdominal aorta. The large **celiac ganglia** and **plexus** lie adjacent to the celiac artery; smaller plexuses surround the phrenic, hepatic, gastric, splenic, renal, and inferior mesenteric, suprarenal, and gonadal arteries. Ganglia associated with these plexuses contain postganglionic sympathetic cell bodies. Fibers dis-tributed by the plexuses consist of postganglionic sympathetic, preganglionic parasympathetic, and visceral afferent fibers.

Gastrointestinal Tract

Stomach

The **stomach** is the first abdominal subdivision of the alimentary canal. Its size, position, and config-uration are determined by its physiologic state, the impingement of other abdominal viscera, and general body build. Classically, it is described as pear-shaped, with its blunt upper end related to the left dome of the diaphragm (Fig. 4-10). The upper and lower ends of the stomach are rela-tively fixed. Its midportion moves as the position of the other viscera or its contents may require.

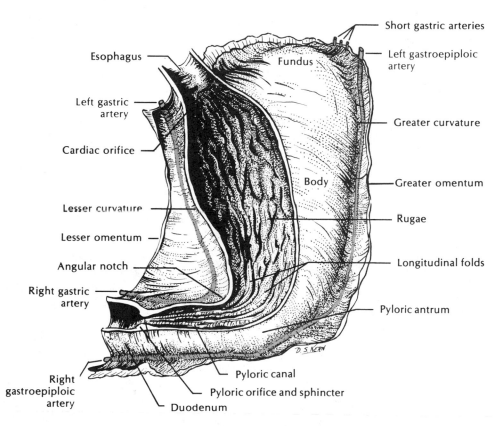

Figure 4-10. *Stomach.*

Gastric Ulcer
In a gastric ulcer, one type of peptic ulcer, the mucosa is eroded or lost and a "craterlike" depression penetrates to various depths of the gastric wall. Although the cause of gastric ulcers is unknown, one prominent theory is the loss of the mucous protection of the mucosa and back diffusion of HCL acid. Common complications of gastric ulcers are bleeding, recurrent pain, and gastric outlet obstruction. Empirically, surgeons have discovered that removal of 50% of the stomach (hemigastrectomy) will cure most patients with gastric ulcer.

The stomach is subdivided into the cardiac portion, fundus, body, and pyloric portion, with anterior and posterior surfaces, and greater and lesser curvatures. The esophagus joins the stomach at the **cardiac orifice,** with the limited **cardiac portion** located adjacent to this opening. The **fundus** is the full-rounded uppermost part above the level of the esophageal junction, while the main part of the stomach, the **body,** lies below the fundus. The **pyloric region** includes the dilated antrum, the pyloric canal, and the pyloric orifice, which is surrounded by a thickened muscular coat, the **pyloric sphincter.** The anterior surface of the stomach is closely related to the diaphragm, the left lobe of the liver, and the left rectus abdominis muscle. The posterior surface is related to many structures, which collectively make up the "bed of the stomach." These include the body of the pancreas, splenic artery, medial border of the left kidney, left suprarenal gland, spleen, diaphragm, transverse colon, and the transverse mesocolon.

Pyloric Stenosis
Pyloric stenosis is due either to true hypertrophy, or a continuous spasm of the pyloric sphincter of the stomach. It appears in infants usually between the second and twelfth weeks of life and is much more frequent in the male than female. Clinical signs include projectile vomiting of food, no substantial bowel movements, and a loss of weight. Such a condition is life-threatening and requires surgical relief by longitudinal division of the pyloric sphincter, which allows the stomach to empty food easily into the duodenum.

The **lesser curvature** of the stomach is concave and affords attachment for the **lesser omentum** (gastrohepatic and hepatoduodenal ligaments), a double layer of peritoneum extending between the stomach and the liver. At the lesser curvature the peritoneum separates to cover the anterior and posterior surfaces of the stomach. From the convex **greater curvature** of the stomach, the serosal coverings fuse to form the **greater omentum,** which extends inferiorly in an apronlike fashion to cover most of the abdominal contents. Inferiorly it reflects back on itself and ascends to pass over the transverse colon and attach to the posterior body wall. That portion of the greater omentum between the transverse colon and the greater curvature of the stomach is the **gastrocolic ligament.** Between the transverse colon and the body wall, the greater omentum fuses with the transverse mesocolon. Below the transverse colon, the layers (two descending and two ascending) fuse to form a four-layered segment of the greater omentum.

Blood Vessels and Nerves (Table 4-4)

The stomach receives a rich blood supply from all branches of the **celiac artery** (Fig. 4-11). This vessel originates from the anterior aspect of the uppermost part of the abdominal aorta as a short trunk and gives rise to splenic, common hepatic, and left gastric arteries. The **left gastric,** the smallest of the three branches, ascends retroperitoneally behind the lesser sac toward the esophagus to pass forward and then descend between the layers of the lesser omentum. It parallels and supplies the lesser curvature of the stomach, gives branches to the lower esophagus, and terminates

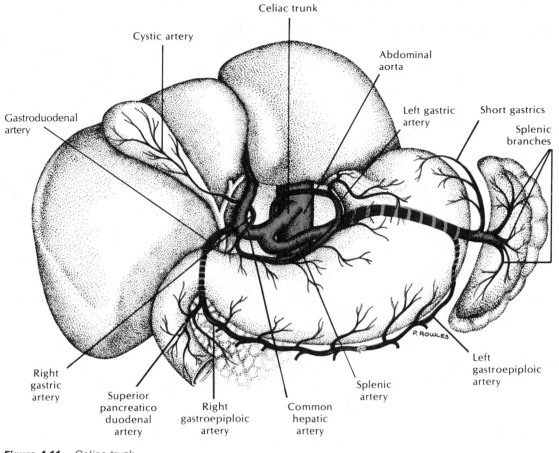

Figure 4-11. Celiac trunk.

by anastomosing with the **right gastric artery.** The latter vessel arises as a branch of the common hepatic artery. It enters the lesser omentum from its retroperitoneal position and passes to the left, between the layers of the lesser omentum along the lesser curvature, to supply the stomach. The splenic artery gives branches to the body of the pancreas, sends **short gastric branches** and the **left gastroepiploic branch** to the stomach. The former pass through the gastrolienal ligament to the upper part of the greater curvature of the stomach. The **left gastroepiploic,** a long branch passing first through the gastrolienal ligament and continuing between the layers of the greater omentum, supplies the greater curvature of the stomach and the greater omentum before anastomosing with the **right gastroepiploic artery.** The latter vessel is a branch of the gastroduodenal from the common hepatic artery. It passes between the layers of the greater omentum to supply the right portion of the greater curvature of the stomach. Veins draining the stomach correspond in position to the arteries and are named similarly, except for the **pyloric vein,** which is the companion of the right gastric artery, and the **coronary vein,** the companion of the left gastric artery.

As the esophagus enters the abdomen the plex-

Table 4-4
Arterial Supply to the Stomach, Liver, Pancreas, and Spleen.

Artery	Origin	Course	Distribution	Anastomoses
Celiac trunk	Abdominal aorta	Short trunk divides into left gastric, splenic, and common hepatic arteries	Supplies portion of gastrointestinal tract derived from foregut	No direct anastomoses
Left gastric	Celiac trunk	Ascends retroperitoneally to esophageal hiatus where it passes between layers of gastrohepatic ligament	Distal portion of esophagus, and lesser curvature of stomach	Esophageals, short gastrics, inferior phrenics
Splenic	Celiac trunk	Passes retroperitoneally along superior border of pancreas; thence between the layers of lienorenal ligament to reach hilum of spleen	Body of pancreas, spleen, and greater curvature of the stomach	Lumbars
Left gastroepiploic	Splenic artery at hilum of spleen	Passes between layers of gastrosplenic ligament to reach greater curvature of stomach	Left portion of greater curvature	Right gastroepiploic
Short gastric	Splenic artery at hilum of spleen	Passes between layers of gastrosplenic ligament to reach fundus of stomach	Fundic aspect of stomach	Left gastroepiploic, left gastric, esophageals
Common hepatic	Celiac trunk	Passes retroperitoneally to reach the hepatoduodenal ligament	Liver, gallbladder, stomach, pancreas, duodenum	No direct anastomoses
Proper hepatic	Continuation from common hepatic	Passes between layers of hepatoduodenal ligament to porta hepatis; divides into right and left hepatics	Respective physiological lobes of liver	No direct anastomoses
Cystic	Proper hepatic or right hepatic	Arises as artery is in the hepatoduodenal ligament	Gallbladder and cystic duct	No direct anastomoses
Right gastric	Common hepatic	Interdigitates between layers of gastrohepatic ligament	Right portion of lesser curvature of stomach	Left gastric
Gastroduodenal	Common hepatic	Descends retroperitoneally behind the gastroduodenal junction	Stomach, pancreas, and duodenum	No direct anastomoses

(Continued)

Table 4-4 (Continued)

Artery	Origin	Course	Distribution	Anastomoses
Right gastroepiploic	Gastroduodenal	Passes between layers of greater omentum	Right portion of greater curvature of stomach	Superior pancreaticoduodenal, left gastroepiploic
Superior pancreaticoduodenal	Gastroduodenal	Descends on superficial aspect of head of pancreas	Proximal portion of duodenum and head of pancreas	Right gastroepiploic, inferior pancreaticoduodenal
Inferior pancreaticoduodenal	Superior mesenteric	Ascends retroperitoneally on superficial aspect of head of pancreas	Distal portion of duodenum and head of pancreas	Superior pancreaticoduodenal, right gastroepiploic, middle colic

iform arrangement of the left and right vagi around the esophagus is reconstituted as the anterior and posterior vagal trunks. They are distributed with the celiac and superior mesenteric plexuses and provide parasympathetic innervation to the abdominal viscera. Parasympathetic innervation through the vagal components extends distally to the level of the descending colon. Sympathetic innervation is also distributed through the celiac and superior mesenteric plexuses.

Small Intestine

The **small intestine,** a convoluted tube extending from the pylorus to the ileocecal valve, is subdivided into the duodenum, the jejunum, and the ileum. It is situated centrally in the abdominal cavity and is flanked laterally and superiorly by the large intestine. As the small intestine proceeds distally it gradually diminishes in diameter. Anteriorly it is related to the greater omentum and the anterior abdominal wall, and posteriorly to the posterior abdominal wall, the pancreas, the kidney, and occasionally the rectum.

Duodenal Ulcer

A duodenal ulcer, another type of peptic ulcer, results from excessive HCl secretion by the parietal cells of the stomach.

Duodenal ulcers often occur in individuals who lead a stressful, highly competitive life. The acid chyme pouring from the stomach into the duodenum causes an ulcer in the first (superior) portion of the duodenum. Duodenal ulcers occur most frequently in the first 2 cm, which is called the duodenal bulb. If the ulcer occurs on the anterior wall, it may perforate into the peritoneal cavity. If it occurs on the posterior wall, it may erode the head of the pancreas or the gastroduodenal artery, and cause massive hemorrhage. Patients with duodenal ulcers who have such serious complications are usually treated surgically. The bleeding vessel is ligated; the perforation is closed; and the duodenal ulcer is most commonly treated with a vagotomy and pyloroplasty.

The **duodenum,** the first segment of the small intestine (about 25 cm), has the widest lumen and the thickest wall of any region of the small intestine (Fig. 4-12). It follows a C-shaped course and is divided into four portions. The first (superior) portion, arising from the pylorus of the stomach, passes posteriorly and superiorly, and at the neck of the gallbladder makes a sharp inferior bend to become the second (descending) part. It is connected by the **hepatoduodenal ligament** to the porta hepatis. The bile and the pancreatic ducts penetrate the posteromedial surface of the second

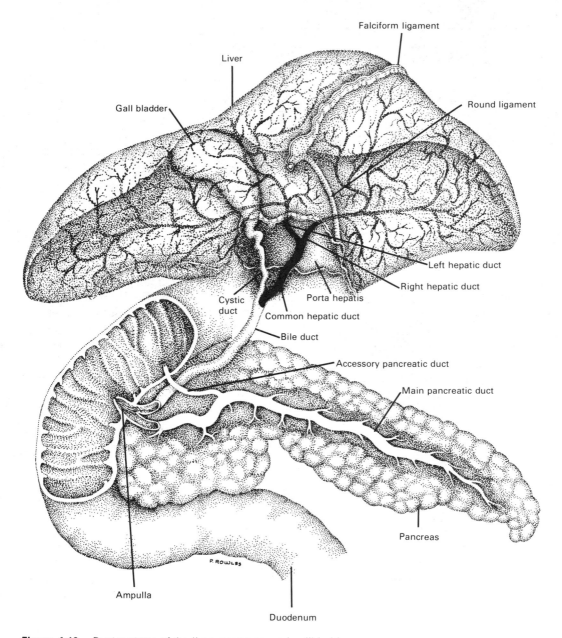

Figure 4-12. *Duct systems of the liver, pancreas, and gallbladder.*

part to form the **hepatopancreatic ampulla (of Vater).**

The second segment passes inferiorly and opposite the third and fourth lumbar vertebrae; it bends to the left to become the third (horizontal) part.

This third portion passes horizontally across the vertebral column, crura of the diaphragm, inferior vena cava, and aorta to ascend slightly and become the fourth (ascending) part. The fourth and terminal part of the duodenum passes upward to the left side of the aorta and head of the pancreas to bend sharply at the duodenojejunal flexure, where it becomes the jejunum. Except for about 2 cm to 3 cm at its origin and termination, the duodenum is entirely retroperitoneal. A fibromuscular band, the **ligament of Trietz,** attaches the small intestine to the posterior abdominal wall at the duodenojejunal flexure. The internal structure of the duodenum presents the **duodenal papilla** at the entrances of the common bile duct and the pancreatic duct.

Meckel's Diverticulum

Meckel's diverticulum is the most common anomaly of the small intestine. It is the result of failure of the yolk sac duct to atrophy. If present, it consists of a blind pouch of varying lengths and extends from the ileum about a meter proximal to the ileocecal valve. It may end freely or be attached to the anterior abdominal wall. When inflamed, the diverticulum may produce symptoms similar to appendicitis. Because it is prone to infections, it is generally removed incidentally if discovered during an abdominal operation. It is more common in men than in women.

The junctional area between the jejunum and the ileum has no gross morphologic points of distinction. The gross division is arbitrary; however, histologic characteristics change so that sections taken from the two areas can be easily distinguished. The **jejunum** is wider and has thicker walls than the ileum and prominent circular folds (plicae circulares). The **ileum** has fewer circular folds, has aggregates of lymph nodules (Peyer's patches), and terminates by joining the medial aspect of the cecum at the **ileocecal valve.** The jejunum and ileum are attached to the posterior abdominal wall by an extensive fold of peritoneum, the fan-shaped **mesentery,** which projects 20 cm to 25 cm into the abdominal cavity. The **root of the mesentery,** 15 cm to 18 cm long, is attached obliquely along the posterior abdominal wall and successively crosses the horizontal part of the duodenum, the aorta, the inferior vena cava, the right ureter, and the right psoas major and minor muscles.

Intussusception

Intussusception is a term that denotes the invagination or telescoping of a part of the intestine into the lumen of another part of the intestine. It is a common cause of bowel obstruction in infants. It most frequently involves the prolapse of the ileum into the lumen of the cecum.

Arteries (Table 4-5)

The duodenum receives its blood supply from the **superior pancreaticoduodenal branch** of the gastroduodenal artery, which passes between the head of the pancreas and the duodenum to its distribution. The **inferior pancreaticoduodenal branch** of the superior mesenteric artery follows a similar course to give twigs to the pancreas and duodenum. Twelve to fifteen **intestinal branches** arise from the superior mesenteric artery. They lie parallel to each other between the layers of the mesentery. Distally they form a series of two to five loops or arcades from which small branches arise that encircle and supply the intestine. The latter are referred to as the **arteria recti** (straight arteries) of the intestine.

Volvulus

Volvulus is a twisting of the intestine around its mesenteric axis, an adjacent intestinal coil, or a postoperative adhe-

Table 4-5
Arterial Supply to Small and Large Intestines

Artery	Origin	Course	Distribution	Anastomoses
Superior mesenteric	Abdominal aorta	Passes along root of the mesentery to ileocecal junction	Supplies that portion of the gastrointestinal system derived from midgut	No direct anastomoses
Intestinals (12–15)	Superior mesenteric	Interdigitated between the layers of the mesentery	Jejunum and ileum	With each other as arcades; ileal branches with ileocolic
Middle colic	Superior mesenteric	Ascends retroperitoneally to interdigitate between the layers of the transverse mesocolon	Transverse colon	Ascending branches of right and left colics
Right colic	Superior mesenteric	Passes retroperitoneally to ascending colon	Ascending colon	Middle colic and ileocolic
Ileocolic	Terminal branch of superior mesenteric	Passes along root of mesentery to divide into ileal and colic branches	Ileum and cecum	Intestinals and right colic
Appendicular	Ileocolic	Passes between layers of mesoappendix	Vermiform appendix	Right colic
Inferior mesenteric	Abdominal aorta	Descends retroperitoneally toward the pelvic cavity	Supplies that portion of the gastrointestinal system derived from hindgut	No direct anastomoses
Left colic	Inferior mesenteric	Passes retroperitoneally to reach the descending colon	Descending colon	Middle colic and sigmoidals
Sigmoidals	Inferior mesenteric	Interdigitate between layers of sigmoid mesocolon to reach the sigmoid colon	Sigmoid colon	Left colic and superior rectal
Superior rectal	Terminal branch of inferior mesenteric	Descends retroperitoneally to rectum	Proximal portion of rectum	Sigmoids and middle rectal
Middle rectal	Internal iliac	Passes retroperitoneally to rectum	Midportion of the rectum	Superior and inferior rectals
Inferior rectal	Internal pudendal	Crosses ischiorectal fossa to rectum	Supplies distal rectum and anal canal	Middle rectal

* The anastomoses between ileocolic, right colic, middle colic, left colic, sigmoidal, and rectal arteries form the marginal artery (of Drummond). This provides an anastomotic pathway from the ileocecal junction to the anal canal.

sive band. The sigmoid colon and the cecum are predisposed to this condition, especially if their mesenteries are unusually long, which affords these gut segments greater mobility. The twisting may return to normal or reduce spontaneously, but if it persists, the blood supply to the bowel is impaired and gangrene may set in. If volvulus blocks the superior mesenteric artery, virtually all of the blood supply to the small intestine and half of the large intestine is lost.

Large Intestine

The **large intestine** extends from the ileum to the anus (Fig. 4-13). It is about 1.5 meters in length and diminishes in diameter from its origin toward its termination. It differs from the small intestine in its greater luminal size, more fixed position, the presence of **sacculations** or haustra, the presence of fat tabs on its external coat, the **appendices epiploicae,** and the disposition of its external longitudinal muscular coat into three longitudinal bands, the **taeniae coli.**

The **cecum** is a large blind pouch 5 cm to 8 cm long at the beginning of the large intestine. The ileum opens into it medially through a longitudinal slit, the **ileocecal orifice,** guarded by the **ileocecal valve.** Below the orifice, the vermiform appendix opens into the cecum, which usually lies in the right iliac fossa immediately above the left half of the inguinal ligament.

The **vermiform appendix,** a blind tube about 5 mm thick and 10 cm long, is suspended by a mesentery, the mesoappendix, which is continuous with the mesentery of the small intestine. The appendix has no fixed position, but commonly moves with the cecum and is most usually lateral or anterior to it.

Appendicitis
The vermiform appendix is quite mobile and its location variable. It most frequently lies either lateral or anterior to the cecum. Other positions include: medially towards the ileum, inferiorly towards the middle of the inguinal ligament, or posteriorly behind the cecum (retrocecal). The surgeon must keep these various positions in mind when searching for the appendix through a small muscle-splitting incision. Because an inflamed appendix may perforate and spew pathogenic organisms into the peritoneal cavity that may cause generalized peritonitis and death, it must be located and removed to prevent these infectious complications.

The **ascending colon** passes superiorly on the posterior abdominal wall from the right iliac fossa to become the transverse colon as it bends to the left at the **hepatic flexure.** It is normally retroperitoneal, but it may possess a limited mesentery. The **transverse colon** extends to the left from the hepatic flexure. It passes transversely across the superior region of the abdominal cavity to the spleen, where it makes a sharp inferior bend called the **splenic flexure** to become the descending colon. It is always suspended from the posterior abdominal wall by the **transverse mesocolon,** which varies in length and thereby permits the location of the transverse colon to vary in position from just inferior to the liver, to the level of the iliac fossae. From the splenic flexure the **descending colon** passes inferiorly along the lateral border of the left kidney in the angle between the psoas major muscle and the quadratus lumborum. It becomes the **sigmoid colon** at the level of the iliac crest. Being fixed in position only at its junctions with the descending colon and the rectum, the sigmoid colon curves on itself, resulting in an S-shaped configuration from which it receives its name. From the sigmoid colon, the **rectum** continues inferiorly to follow the sacral curvature to the pelvic diaphragm, where it makes a ninety-degree bend to become the **anal canal.** Externally the rectum has no sacculations. The taenia coli spread out to form a uniform outer muscular coat, and the peritoneum covers only its anterior and lateral aspects. From

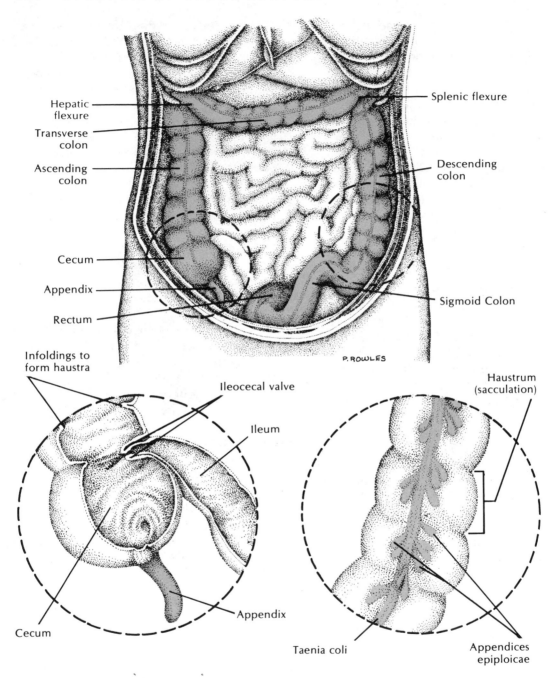

P. ROWLES

Figure 4-13. *Large intestine.*

the rectum, the peritoneum reflects onto the bladder in the male to form the **rectovesical pouch,** and onto the uterus in the female as the **rectouterine pouch.**

Megacolon or Hirschsprung's Disease

Megacolon, or Hirschsprung's disease, is a congenital condition. It is usually diagnosed in infants and young children. In megacolon the large intestine, usually the sigmoid colon, is greatly enlarged, hypertrophied, and often elongated, causing tremendous abdominal distension. Megacolon is caused by a failure of ganglia to develop in Auerbach's and Meissner's plexuses. Absence of these ganglia results in loss of muscular tone and peristalsis in the involved gut wall and causes pronounced constipation and distension. Successful surgical treatment of the disease consists of resecting the affected colon and bringing well-innervated colon to the distal rectum for an anastomosis. If successful, normal bowel function is restored.

The **anal canal** is surrounded by the levator ani and external anal sphincter muscles as it passes through the pelvic diaphragm posterior to the perineal body (Fig. 4-14). The **external anal sphincter** is disposed into a **subcutaneous portion** just deep to the skin and a **superficial part** attaching to the tip of the coccyx and perineal body. A **deep portion** associated with the puborectalis muscle (a portion of the levator ani) also attaches anteriorly to the perineal body. Internally the mucosa of the upper half of the anal canal forms five to ten vertical folds designated as the **anal** or **rectal columns.** Each fold contains small veins that result in **internal hemorrhoids** if they become overdistended. The lower ends of these columns are joined by small cresentic folds of mucosa, the **anal valves** (anal sinuses), with the scalloped **pectinate line** present at their lower limit. In this region, the anal mucosa merges with the skin of the anus as the **mucocutaneous line.** Venous varicosities in the area below the pectinate line result in **external hemorrhoids.**

Arteries (See Table 4-5)

The blood supply to the large intestine is from branches of the **superior** and **inferior mesenteric** and internal iliac arteries. The **middle colic branch** of the superior mesenteric artery passes between the layers of the transverse mesocolon to divide into right and left branches supplying the transverse colon. Arising at about the middle of the superior mesenteric artery, the **right colic branch** courses to the right behind the peritoneum toward the middle of the ascending colon where it divides into ascending and descending branches. They supply the ascending colon and then anastomose, respectively, with the middle colic and ileocolic arteries. The **ileocolic artery,** the terminal branch of the superior mesenteric, sends a colic branch to supply the cecum and the appendix, and small ileal branches to the distal portion of the ileum. The **inferior mesenteric artery** arises from the anterior aspect of the aorta, 3 cm to 5 cm above its bifurcation. Coursing inferiorly in a retroperitoneal position, the inferior mesenteric artery gives origin to the **left colic artery.** This vessel passes to the left and divides into ascending and descending branches to supply the left half of the transverse colon and the descending colon, respectively. Three or four **sigmoid branches** of the inferior mesenteric artery pass between the layers of the sigmoid mesocolon to supply the lower part of the descending and sigmoid colon. The inferior mesenteric then terminates as the **superior rectal artery,** which continues inferiorly in a retroperitoneal position to ramify on the upper part of the rectum. Additional blood to the rectum is derived from the **middle rectal branch** of the internal iliac and the **inferior rectal branch** of the internal pudendal artery.

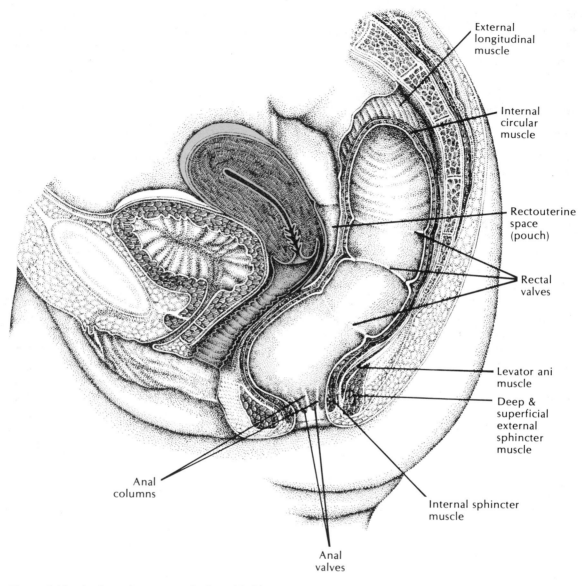

External
longitudinal
muscle

Internal
circular
muscle

Rectouterine
space
(pouch)

Rectal
valves

Levator ani
muscle

Deep &
superficial
external
sphincter
muscle

Internal sphincter
muscle

Anal
valves

Anal
columns

Figure 4-14. *Anal canal, uterus, and urinary bladder.*

Liver

Objectives

After completion of the study of the liver, pancreas, and spleen the student should be able to

▶ Describe the surfaces, borders, lobes, impressions, and peritoneal coverings of the liver

▶ Describe the mesenteric attachments of the liver to the body wall, diaphragm, and stomach; relate these to their embryonic origin

▶ Relate structures present at the porta hepatis to the lesser omentum and the epiploic foramen

▶ Follow the duct system of the liver and pancreas to the duodenum

▶ Draw and label the portal system of veins, and locate sites of possible portocaval anastomoses

▶ Delineate the relationships of the pancreas to other abdominal viscera

▶ List and locate the contributions of the sympathetic and parasympathetic components to the gut innervation

The **liver,** the largest gland in the body, is roughly wedge-shaped, with the base of the wedge directed to the right (Fig. 4-15). The **superior (convex) surface** continues undemarcated into the posteroinferior surface. The **anterior surface** of the liver is triangular, slightly convex, and related to the anterior body wall, diaphragm, and ribs. The **posterior surface** is roughly triangular and markedly concave from left to right as it passes in front of the vertebral column. The oblong **inferior (visceral) surface** is very uneven and, when hardened *in situ,* bears the impressions of structures in contact with it. Anteriorly the

inferior surface is separated from the anterior surface by a sharp border.

The **porta hepatis** is a relatively deep and wide area, approximately 5 cm long, through which the portal vein, hepatic artery, bile ducts, nerves, and lymphatics pass. The porta hepatis separates the **quadrate lobe** inferiorly from the **caudate lobe** superiorly, and its boundaries serve as the site of attachment of the **hepatoduodenal ligament.** The **fissure for the ligamentum venosum** extends superiorly from the left end of the porta hepatis. It holds the ligamentum venosum, which is the remnant of the fetal ductus venosus. It affords attachment for the **gastrohepatic ligament,** which, with the hepatoduodenal ligament, forms the **lesser omentum.** The liver is partially divided by the fissures for the ligamentum venosum and ligamentum teres into a large **right** and a small **left lobe.** Two circumscribed areas on the medial aspect of the right lobe are further demarcated by the **fossae for the gallbladder** and the **inferior vena cava.** Between the above fissures and fossae the porta hepatis divides a superior segment, the **caudate lobe,** from an inferior segment, the **quadrate lobe.** The latter two subdivisions are part of the right lobe of the liver. The larger right lobe rises higher than the left, pushing the right dome of the diaphragm upward and thus partially displacing the right lung.

Structures on the posterior surface of the liver include the fissure for the ligamentum venosum, the fossa for the inferior vena cava, and the caudate lobe, which projects into the lesser peritoneal sac. Between the layers of the coronary ligament, the posterior surface is not covered by peritoneum and is in direct contact with the endoabdominal fascia of the diaphragm as the **"bare area"** of the liver. On the inferior surface the left lobe presents the gastric impression for the stomach and the **tuber omentale,** a bulging prominence above the lesser curvature of the

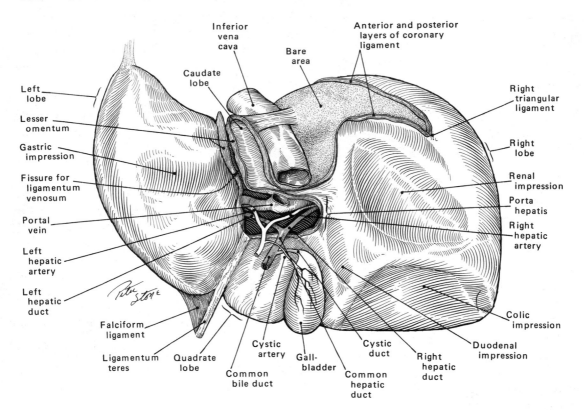

Figure 4-15. *The liver. Posteroinferior aspect.*

stomach. The fossa for the gallbladder, impressions of the duodenum, right colic flexure, and right kidney are associated with the inferior surface of the right lobe.

Ligaments

The **falciform ligament** is a thin, sickle-shaped anteroposterior fold consisting of two apposing layers of peritoneum. One of its three borders is attached to, and reflected over, the anterior surface of the liver; a second is attached to, and reflected over, the diaphragm and the anterior abdominal wall to the level of the umbilicus, while the third inferior border is free and encloses the **ligamentum teres.** The latter represents the obliterated left umbilical vein. At the upper extent of the border of the falciform ligament attaching to

the anterior surface of the liver, the peritoneal layers diverge laterally and reflect onto the diaphragm. The right reflection forms the **anterior layer of the coronary ligament,** which passes laterally to bend sharply at the **right triangular ligament,** where it becomes the **posterior layer of the coronary ligament.** The peritoneum forming the coronary ligament reflects from the liver onto the diaphragm to enclose an area devoid of peritoneum. Here the liver is in direct contact with the diaphragm and is designated as the "bare area." The left divergence of the falciform ligament, the **left triangular ligament,** reflects onto the left lobe corresponding to, and continuous posteriorly with, the posterior layer of the coronary ligament. The two folds of peritoneum composing this left divergence are not widely separated.

Gallbladder and Biliary Tree

The **gallbladder** is a small piriform sac that serves as the reservoir for bile (see Fig. 4-12). It holds 60 ml to 150 ml of fluid. It lies in a small fossa on the visceral (inferior) surface of the liver, and is divided into the fundus, body, and neck. The inferior extremity of the sac, the **fundus,** is wide, usually protrudes beyond the inferior margin of the liver, and is covered by peritoneum. It is in contact with the transverse colon, the anterior body wall, and the ninth costal cartilage at the point the latter is crossed by the linea semilunaris. The anterior surface of the **body** of the gallbladder is in direct contact with the liver, while the posterior surface and sides of the body are covered by peritoneum. Posteriorly the body is related to the transverse colon and the second part of the duodenum. The narrow **neck** of the gallbladder continues as the cystic duct. Both structures are closely applied to the liver, and related inferiorly to the first part of the duodenum. The 2 cm or 3 cm long **cystic duct** enters the hepatoduodenal ligament at the right end of the porta hepatis and runs a short distance with, and then joins, the common hepatic duct to form the common bile duct.

Bile, secreted by the liver cells, is carried away from the liver lobules by way of **bile canaliculi** to **intralobular ductules.** These unite to become **interlobar ducts,** which in turn join to form **right** and **left hepatic ducts,** whose junction forms the **common hepatic duct** at the porta hepatis. This latter duct, joined by the cystic duct, forms the **common bile duct.** It descends in the free margin of the lesser omentum anterior to the portal vein and lateral to the hepatic artery. The common bile duct continues inferiorly, between the duodenum and the head of the pancreas, and terminates by uniting with the main pancreatic duct before penetrating the second part of the duodenum. The musculature in the wall of the common bile duct thickens to form a **sphincter** at its junction with the pancreatic duct. Within the wall of the duodenum the combined ducts expand slightly as the **hepatopancreatic** **ampulla (of Vater).** This bulges the mucous membrane inward to form the major **duodenal papilla** with an opening at its summit (Fig. 4-16).

Gallstones

Gallstones result from precipitations, especially cholesterol, of chemical compounds of the bile. These precipitations may become lodged in the gallbladder or the biliary ducts. Their development is usually accompanied by repeated inflammations of the gallbladder. Stones lodging along the excretory duct system may cause an obstruction or stasis of bile flow. Acute, severe pain (biliary colic) is caused by this obstruction with stretching of the duct system and smooth muscle spasm. Biliary obstruction distends the ducts and results in a cessation of flow of the bile. The individual becomes jaundiced (yellow) due to the impaired clearance of bilirubin from the blood.

Arteries (See Table 4-4)

Blood is supplied to the liver by the common hepatic artery, a branch of the celiac trunk. The **hepatic artery** runs inferiorly and to the right. It courses retroperitoneally behind the lesser sac to reach the first part of the duodenum. Here it continues below the epiploic forearm to gain a position between the layers of the hepatoduodenal ligament and passes to the porta hepatis. Within the hepatoduodenal ligament it divides into the **left** and **right hepatic branches,** which supply their respective hepatic lobes. It gives a third branch, the **cystic artery,** to supply the cystic duct and the gallbladder. The cystic artery may arise from the right hepatic artery.

Hepatic Portal Circulation (Table 4-6)

The **portal vein** originates behind the head of the pancreas by the union of the **splenic** and **superior mesenteric veins.** It ascends behind the first part of the duodenum to pass in the free

Text continues on page 180

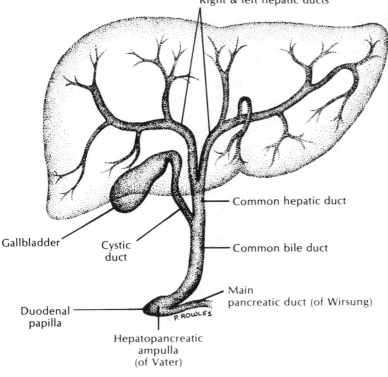

Right & left hepatic ducts

Common hepatic duct

Common bile duct

Gallbladder

Cystic duct

Main pancreatic duct (of Wirsung)

Duodenal papilla

Hepatopancreatic ampulla (of Vater)

P. ROWLES

Figure 4-16. *Biliary tree.*

Table 4-6
Veins of the Hepatic Circulation

Vein	Origin	Course	Distribution	Anastomoses
Portal	Coalescence of superior mesenteric with splenic	Passes between layers of the hepatoduodenal ligament to reach the porta hepatis	Supplies nutrient-rich blood to liver	No direct anastomoses
Right and left lobars	Portal vein	Pass to respective physiological lobes of liver	Physiological lobes of liver	No direct anastomoses
Liver sinusoids (capillary plexuses)	From terminal branches of right and left portal veins	Extend from periphery of liver lobule to central vein	Capillary plexuses carrying blood to liver cells	No direct anastomoses
Central	Originate at center of liver lobule from liver sinusoids	Coalescence to form → intralobular → interlobular → hepatic veins	Drain blood into sublobular veins	No direct anastomoses
Hepatic	Arise from lobar tributaries	Extend through liver to reach inferior vena cava	Carry blood to inferior vena cava	No direct anastomoses

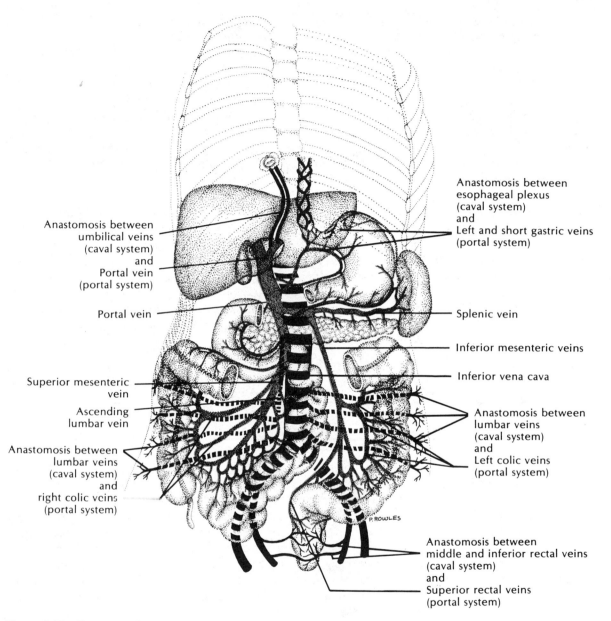

Anastomosis between
umbilical veins
(caval system)
and
Portal vein
(portal system)

Portal vein

Superior mesenteric
vein

Ascending
lumbar vein

Anastomosis between
lumbar veins
(caval system)
and
right colic veins
(portal system)

Anastomosis between
esophageal plexus
(caval system)
and
Left and short gastric veins
(portal system)

Splenic vein

Inferior mesenteric veins

Inferior vena cava

Anastomosis between
lumbar veins
(caval system)
and
Left colic veins
(portal system)

Anastomosis between
middle and inferior rectal veins
(caval system)
and
Superior rectal veins
(portal system)

P. ROWLES

Figure 4-17. *Portosystemic anastomosis.*

margin of the lesser omentum and divides into right and left branches to enter their respective lobes of the liver. These branches eventually subdivide to gain a position at the margin of the liver lobule, where they course with the branches of the hepatic artery and the bile duct. The portal vein drains the spleen, pancreas, gallbladder, and all of the alimentary canal distal to the esophagus, except for the lower rectum and anal canal. Tributaries of the portal system, in addition to the splenic and superior mesenteric veins, include the **short gastric, right** and **left gastroepiploic,** and **coronary veins** draining the stomach, small **pancreatic veins** from the pancreas, and the **inferior mesenteric vein** draining blood from the upper rectum and the distal half of the colon. The **hepatic veins** draining the liver begin as **central veins** of the liver lobules. These small vessels drain into **sublobular veins** as tributaries of the hepatic veins, which empty into the inferior vena cava before it passes through the diaphragm.

Portosystemic (Portocaval) Venous Anastomosis

Tributaries of the portal vein communicate with systemic veins in several locations, particularly the lower ends of the esophagus and rectum, and the umbilicus. Ordinarily these channels are collapsed with little blood flow because blood drains from the abdominal viscera through the portal vein to the liver. If venous flow through the liver is impeded by a blood clot or scarring in the liver (cirrhosis), the portal vein pressure rises markedly (Fig. 4-17). Portal hypertension dilates the veins of the portal system, including some or all of these anastomotic channels. This can cause varicosities of the veins around the umbilicus (caput medusae), the esophagus (esophageal varices), and the rectum (hemorrhoids).

Pancreas

The **pancreas** is an elongated endocrine and exocrine gland that lies obliquely on the upper part of the posterior abdominal wall. It extends from the concavity of the duodenum to the spleen (see Fig. 14-12). It is soft and pliable, contains a minimum of connective tissue, and is subdivided into a head, body, and tail. The flattened, expanded **head** nestles in the concavity of the duodenum. Posteriorly the head is related to the aorta and inferior vena cava, with the common bile duct enbedded in its lateral margin. It is related posteriorly to the superior mesenteric vessels and is in contact with the transverse colon. The **uncinate process** is an extension from the head, to the isthmus of the pancreas. Laterally the anterior surface of the **body** forms part of the stomach bed. It bulges slightly in the median plane as the **tuber**

omentale immediately inferior to the celiac artery. The splenic artery courses along the superior border of the pancreas, and the transverse mesocolon reflects from the posterior abdominal wall along its anterior surface. Posteriorly the body of the pancreas rests on the aorta, the superior mesenteric artery, the left crus of the diaphragm, the left psoas muscles, and the left kidney. The **tail** is thick and blunt and related to the hilus of the spleen.

Diabetes Mellitus

Diabetes mellitus is a deficiency of insulin production from the islet cells of the pancreas. Insulin is essential in carbohydrate metabolism. It facilitates cellular uptake of glucose and the conversion of glucose to glycogen. Loss of these functions result in increased blood sugar levels (hyperglycemia), increased sugar in the urine (glycosuria), increased urine output (polyuria), thirst, hunger, emaciation, and weakness. Diabetes mellitus is controlled by strict dieting, by properly spaced injections of insulin, or by the oral administration of antidiabetic drugs.

Two ducts drain the pancreas. The **main pancreatic duct** courses along the entire length of the gland, emerging to join the common bile duct and pierce the duodenal wall 7 cm to 10 cm beyond the pylorus, as the duodenal papilla. The much smaller **accessory pancreatic duct** commonly appears in the head of the pancreas as an offshoot from the main pancreatic duct. It usually opens independently into the duodenum 2 cm to 3 cm above the major duodenal papilla.

Arteries and Nerves (See Table 4-4)

The pancreas receives blood from small twigs of the **splenic artery** as it courses along the superior border of the pancreas, from the **superior pancreaticoduodenal branch** of the gastroduodenal artery, and from the **inferior pancreaticoduodenal branch** of the superior mesenteric artery. Innervation to the gland is derived from an extension of the **celiac plexus** surrounding its arterial supply.

Spleen

The **spleen** (lien) is an oblong, flattened, highly vascular organ. It is located in the left hypochondriac region of the abdomen, behind the stomach and inferior to the diaphragm. It has diaphragmatic and visceral surfaces, superior and inferior extremities, and anterior, posterior, and inferior borders. The convex **diaphragmatic surface** is molded to fit the diaphragm. The **visceral surface** is divided by a ridge into an anterior (gastric) portion and an inferior (renal) portion. A fissure for the passage of vessels and nerves, the **hilus,** separates the gastric from the renal portions. The gastric portion is concave and in contact with the posterior wall of the stomach, while the somewhat flattened renal portion is related to the left kidney and left suprarenal gland. The **superior extremity** is directed toward the vertebral column at the level of the eleventh thoracic vertebra. The flat, triangular **inferior extremity** (colic surface) rests on the left colic flexure in contact with the tail of the pancreas. The notched **anterior border** is free and sharp and separates diaphragmatic and gastric surfaces. The rounded blunt **posterior border** demarcates diaphragmatic from renal surfaces, while the **inferior border** divides diaphragmatic from colic surfaces.

The spleen is almost entirely surrounded by peritoneum and held in position by two peritoneal ligaments. The short **lienorenal ligament** extends from the upper half of the left kidney to the hilus and contains the splenic artery and vein, sympathetic nerves, and lymphatics. The **gastrolienal ligament** transmits the gastroepiploic and the short gastric artery and vein, nerve filaments, and lymphatics to the stomach as it passes forward from the hilus to become continuous with the greater omentum.

Arteries (See Table 4-4)

The **splenic artery,** the largest branch of the celiac trunk, follows a tortuous course along the superior border of the pancreas behind the posterior wall of the lesser sac. It terminates in five or six branches that pass through the lienorenal ligament to supply the spleen. The artery gives **pan-**

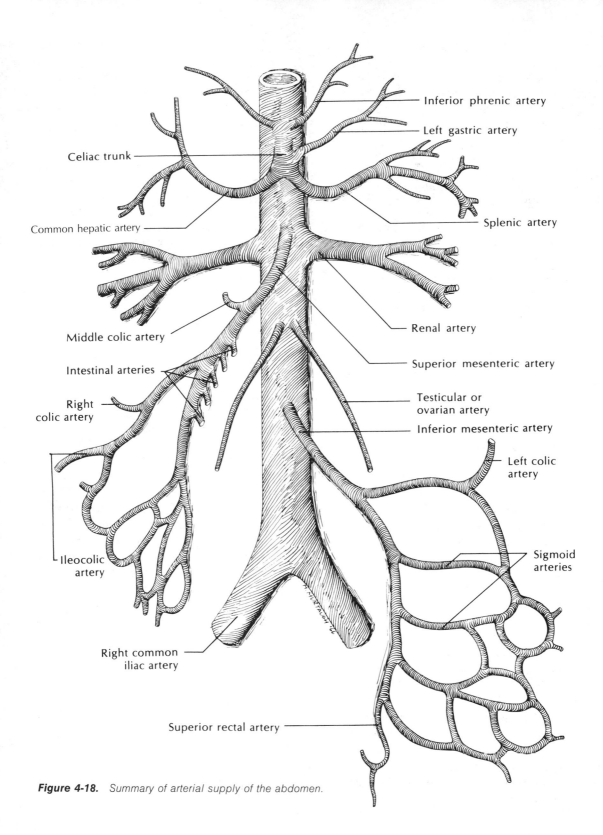

Figure 4-18. *Summary of arterial supply of the abdomen.*

Inferior phrenic artery

Left gastric artery

Celiac trunk

Splenic artery

Common hepatic artery

Renal artery

Middle colic artery

Superior mesenteric artery

Intestinal arteries

Testicular or ovarian artery

Right colic artery

Inferior mesenteric artery

Left colic artery

Ileocolic artery

Sigmoid arteries

Right common iliac artery

Superior rectal artery

creatic branches in its retroperitoneal position and sends the **left gastroepiploic** and **short gastric** branches through the gastrolienal ligament to supply the stomach. Figure 4-18 demonstrates arterial distribution of the abdomen.

Removal of the Spleen

Although the spleen is well protected from traumatic injuries in general, it is the most frequently damaged organ from blunt abdominal trauma; particularly from severe blows over the lower left chest or upper abdomen that fracture the protecting ribs. Such a crushing injury may rupture the spleen, which causes severe intraperitoneal hemorrhage and shock. This requires a prompt splenectomy to prevent the patient from bleeding to death.

Kidney

Objectives

After the completion of the study of the kidney and suprarenal gland the student should be able to

▶ *Describe the relationships and fascial support of the kidney*

▶ *Delineate the component parts of the nephron and follow the blood supply to the kidney and the suprarenal gland*

▶ *Draw and label extensions of the renal pelvis*

▶ *Give the origin, course, termination, and relationships of the ureter*

The **kidney,** a retroperitoneal structure embedded in fascia and fat, is ovoid in outline, with the medial border markedly concave. It lies obliquely in the upper part of the posterior abdominal wall. The left kidney is opposite the twelfth thoracic and upper three lumbar vertebrae. The right kidney is slightly lower. The middle of the renal sinus is approximately 5 cm from the median plane, with the upper pole of the kidney closer to the midline than the lower. The kidney presents anterior and posterior surfaces, medial and lateral borders, and upper and lower poles.

Relations of the **anterior surface** vary on each side. On the right the anterior surface is related to the liver, duodenum, right colic flexure, and small intestine. On the left side the anterior surface is related to the spleen, splenic vessels, left colic flexure, and small intestine. The portions of the kidney in contact with the suprarenal gland, pancreas, colon, and duodenum are devoid of peritoneum, while the remainder of the anterior surface is covered by peritoneum. The **posterior surface** of each kidney has similar relations, being embedded in areolar and fatty tissue, and is entirely devoid of peritoneum. The kidney rests on the lower part of the diaphragm, arcuate ligaments, psoas major, quadratus lumborum, and transversus abdominis muscles, one or two lumbar arteries, and subcostal, ilioinguinal, and iliohypogastric nerves.

The adipose tissue that completely surrounds the kidney is separated by a thin membranous sheet into **perirenal fat,** adjacent to the kidney, and **pararenal fat,** external to this membrane. From the lateral border of the kidney this membrane splits to pass anterior and posterior to the kidney as the **renal fascia.** Traced medially, the anterior layer of renal fascia passes anterior to the renal vessels, over the aorta, and becomes continuous with the same layer of the opposite side. Superiorly it passes over the suprarenal gland to become continuous with the posterior layer, which is the thicker layer of the two, and passes medially behind the aorta and the vena cava to unite with connective tissue over the vertebral column. Inferiorly the renal fascia fuses with the extraperitoneal connective tissue and loses its identity. Thus, in emaciation a loss of the fat surrounding the kidney may be accompanied by a **downward displacement (ptosis)** of the kidney because of the nonclosure of this fascia inferiorly.

Nephroptosis

The fat surrounding the kidney is a major factor in its stabilization. In emaciation loss of this fat may permit the kidney to displace inferiorly. If this occurs it may cause a "kink" in the ureter because the ureter is firmly bound by parietal peritoneum to the posterior abdominal wall. Urine will then accumulate in the renal pelvis and result in an hydronephrosis. A similar mobility of the kidney may occur from a blow to the lumbar region of the back ("kidney punch" in boxing), or jarring of the body while riding trail bikes. The latter is the rationale for racing motorcyclists to wear wide supportive belts.

Internal Anatomy

On sagittal section the internal anatomy of the kidney presents, adjacent to the cortex, a fibrous tunic **(capsule)** and an inner medullary region (Fig. 4-19). The **cortex** consists of granular appearing tissue containing glomeruli and elements of the nephron. At intervals, extensions of cortical tissue project centrally between the pyramids of the medulla as the **renal columns.** The **medulla** is composed of a series of eight to sixteen conical masses, the renal or medullary **pyramids,** containing the collecting tubules. The bases of the pyramids are directed toward the cortex and their apices converge at the **renal sinus,** forming within the sinus the prominent **renal papillae.** Four to thirteen cup-shaped, sleevelike projec-

tions surround one or more renal papillae as the **minor calyces.** They join to form two or three **major calyces,** which empty into, and are continuous with, the funnel-shaped **renal pelvis,** the proximal dilatation of the ureter (Fig. 4-19).

Hydronephrosis

Hydronephrosis is the distention of the pelvis and calyces of the kidney. It is caused by an obstruction of the lower urinary tract, and a backup of urine into these structures. If the condition persists, permanent damage to the parenchyma of the kidney results. One common cause of ureteral obstruction is a kidney stone. Hydronephrosis may also be caused by an abnormal position of the kidney (nephroptosis) that allows the ureter to kink. Hydronephrosis is also an infrequent complication of pregnancy. The developing fetus exerts pressure on the ureter as the latter crosses the brim of the pelvis. This may obstruct urine flow.

The **nephron,** the functional unit of the kidney, is located in cortical tissue. It begins as a blind-ending tube, **Bowman's capsule,** that is invaginated by the **glomerulus** (Fig. 4-20). From Bowman's capsule, the **proximal convoluted tubule, Henle's loop, distal convoluted tubule** and **collecting ducts** complete the nephron. An ultrafiltrate of plasma passes from the capillary tuft comprising the glomerulus across the wall of Bowman's capsule. Nutritious products and fluid are resorbed into the bloodstream at the capillary plexus between the efferent arteries and the renal venules. For every 100 ml of fluid crossing Bowman's capsule, 1 ml arrives at the minor calyx as urine.

Blood Vessels and Nerves (Table 4-7)

Arising from the aorta about 2 cm to 3 cm below the origin of the superior mesenteric artery, the **renal artery** passes transversely to reach the si-

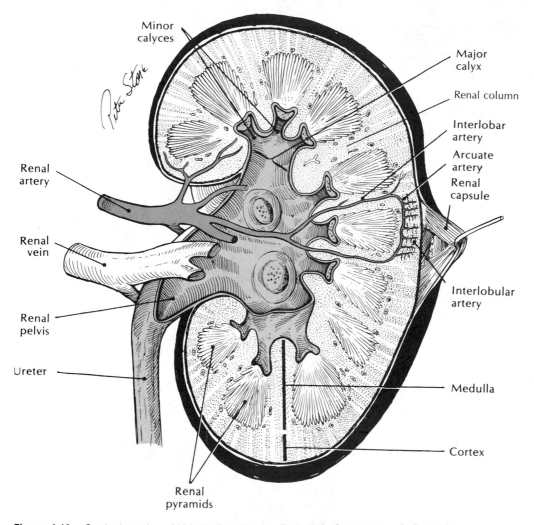

Minor calyces

Major calyx

Renal column

Interlobar artery

Arcuate artery

Renal capsule

Renal artery

Renal vein

Renal pelvis

Ureter

Interlobular artery

Medulla

Cortex

Renal pyramids

Figure 4-19. *Sagittal section of kidney. (Langley LL, Telford IR, Christensen JB: Dynamic Anatomy and Physiology, 5th ed. New York, McGraw-Hill, 1980)*

nus of the kidney. The right renal artery, longer and lower than the left, passes behind the inferior vena cava, the head of the pancreas, and the second part of the duodenum to reach the kidney. The left vessel passes behind the renal vein, the pancreas, and the splenic vein to enter the kidney. Each gives an **inferior suprarenal branch** as it enters the sinus, then divides into interlobar branches that traverse the renal sinus. The **in-terlobar branches** course in the renal columns to reach the cortex of the kidney, where they bifurcate at right angles to give rise to **arcuate arteries.** The latter course parallel to the cortex, to anastomose and form arcades. From the arcades **interlobular arteries** pass peripherally into the cortex, giving a series of **afferent arterioles** to the glomeruli.

The large renal veins are anterior to the renal

Table 4-7
Arterial Supply to the Kidney and Suprarenal Glands

Artery	Origin	Course	Distribution	Anastomoses
Renal	Aorta	Passes retroperitoneally to renal sinus	At renal sinus divides into lobar branches	No direct anastomoses
Lobars	Renal	Pass through renal sinus to kidney substance	Divide into interlobars	No direct anastomoses
Interlobars	Lobar arteries	Pass between renal pyramids to corticomedullary junction; divide in a T-shape manner	Medullary substance of the kidney	Adjacent interlobars
Arcuates	T-shaped branches of interlobars	Lie in junctional area between the renal cortex and renal medulla	Cortical tissue	Adjacent interlobars
Interlobulars	Arcuate	Pass through cortex toward kidney surfaces	Cortical tissue	Adjacent interlobulars
Afferent arterioles	Interlobulars	Pass to capillary plexus of glomerulus	Cortical tissue and glomerulus	Adjacent afferent arterioles
Glomerulus	Afferent arterioles	Small knot of capillaries surrounded by Bowman's capsule	An ultrafiltrate of plasma passes from glomerulus into Bowman's capsule	No direct anastomoses
Efferent arterioles	Glomerulus	Extend from glomerulus to second capillary plexus that will form renal venules	Elements of renal nephron; nutritive materials from ultrafiltrate at Bowman's capsule pass back into bloodstream	Adjacent efferent arterioles
Superior suprarenal	Inferior phrenic	Descends retroperitoneally to suprarenal gland	Suprarenal gland	Middle and inferior suprarenals
Middle suprarenal	Abdominal aorta	Passes retroperitoneally to suprarenal gland	Suprarenal gland	Inferior and superior suprarenals
Inferior suprarenal	Renal artery	Ascends retroperitoneally to suprarenal gland	Suprarenal gland	Middle and superior suprarenal arteries

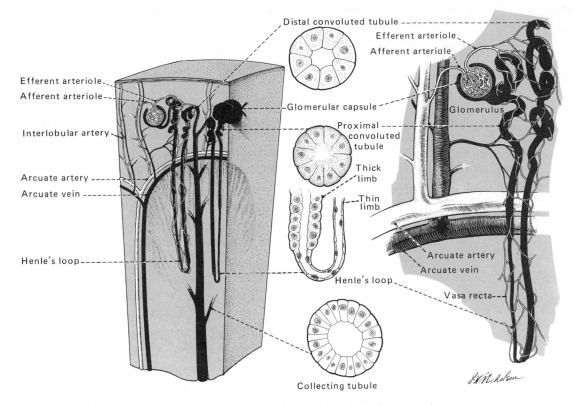

Distal convoluted tubule
Efferent arteriole
Afferent arteriole
Efferent arteriole
Afferent arteriole
Interlobular artery
Glomerular capsule
Glomerulus
Arcuate artery
Proximal convoluted tubule
Arcuate vein
Thick limb
Thin limb
Henle's loop
Arcuate artery
Arcuate vein
Henle's loop
Vasa recta
Collecting tubule

Figure 4-20. *Schematic diagram of a nephron and its blood supply. In the center the tubules are shown in cross section or longitudinal section.*

arteries. The right vein empties directly into the inferior vena cava. After receiving suprarenal, renal, left gonadal, and left phrenic tributaries, the left renal vein also joins the inferior vena cava. Innervation to the kidneys is from the renal autonomic plexus, whose fibers accompany the renal artery into the kidney.

Renal Hypertension
In many kidney diseases blood flow through the organ is compromised. The ischemic kidney reacts by secreting the enzyme, renin, which activates a plasma protein fraction, angiotensin I. The latter is converted by another enzyme into angiotensin II, which is a powerful vasoconstrictor. This causes a narrowing of arterioles over the entire body. The resulting high blood pressure (renal hypertension) ensures a continuance of a sufficiently high blood pressure for producing an adequate volume of urine. In this way, the needs of the kidney are met, but at the price of chronic high blood pressure.

Ureter

The **ureter** carries urine from the renal pelvis to the urinary bladder. It is approximately 25 cm long, has thick muscular walls, and a narrow lumen. It is situated half in the abdominal and half in the pelvic cavities. Descending retroperitoneally on the psoas major muscle, it passes anterior to the bifurcation of the iliac artery and posterior to the gonadal and right or left colic arteries. The ureter then turns anteriorly at the level of the ischial spine to reach the posterior aspect of the urinary bladder (Fig. 4-21).

Normal Constrictions of Ureter
Normal constrictions of the ureter are located: (1) at the lower end of the pelvis of the kidney, or the ureteropelvic junction, (2) at the pelvic brim where it crosses the iliac vessels, and (3) at its oblique entry through the bladder wall. Kidney stones tend to become lodged at these narrow points in the ureter.

In the male, the ureter is crossed on its medial side by the ductus deferens and, as it approaches

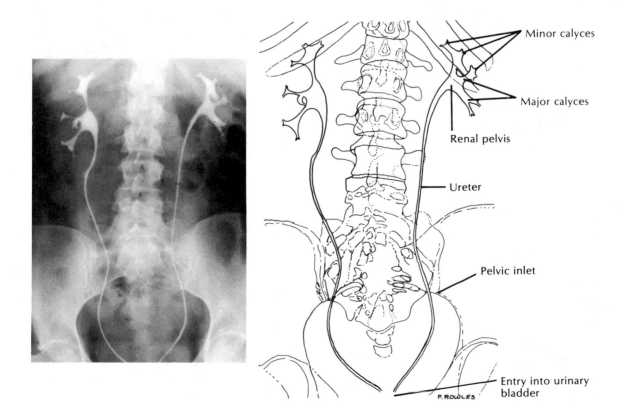

Minor calyces

Major calyces

Renal pelvis

Ureter

Pelvic inlet

Entry into urinary bladder

P. ROWLES

Figure 4-21. *Anteroposterior pyelogram.*

the bladder, lies anterior to the upper end of the seminal vesicle. In the female, it is related to the posterior border of the ovary. Along the lateral pelvic wall it turns anteromedially to pass close to the vagina and cervix in its course toward the bladder. The ureter receives its **blood** and **nerve supply** regionally from the renal, gonadal, and vesicular arteries and nerves.

Surgical Trauma to Ureters

The pelvic ureters are the most vulnerable pelvic organs in surgery performed on other pelvic structures, such as the uterus. During an abdominal hysterectomy, the ureter may be accidentally ligated with the ovarian vessels. It may be crushed by the clamp applied to the serous layers on the broad ligament, or ligated as a tie is placed on the uterine artery and the transverse cervical ligament. While the possibility of ureteral injury is greatest in gynecologic procedures, the risk of injury is always present in any operation that involves the pelvis or lower abdomen.

Suprarenal Gland

The **suprarenal (adrenal) gland** is an endocrine organ adjacent to the upper pole of the kidney. On the right side the gland is related to the diaphragm posteriorly, the inferior vena cava medially, and the liver laterally. On the left side it rests on the left crus of the diaphragm anteriorly, the stomach superiorly, and the pancreas inferiorly. The two glands are separated from each other by the celiac axis and plexus and are enclosed in the renal fascia. The suprarenal gland is supplied by six to eight small **superior suprarenal** twigs from the inferior phrenic artery; the **middle suprarenal,** a branch of the aorta, and the **inferior suprarenal,** a branch of the renal artery (See Table 4-7). Veins draining the gland correspond to the arteries; however, the principal drainage is by way of a prominent single **central vein** that emerges from the hilus to empty on the left side into the renal vein, and on the right side into the inferior vena cava.

MAJOR ANATOMIC AND CLINICAL POINTS

Abdominal Wall

☐ The abdominal contents can be localized by arbitrary divisions of the abdomen into four or nine regions.

☐ Three layers of muscle are in the anterolateral abdominal wall. Vessels and nerves course in the fascial cleft between the second and third layer.

☐ Innervation to the anterolateral abdominal wall is by the lower intercostal (thoracoabdominal) nerves.

☐ The inguinal ligament and its extensions are formed by the inferior free border of the aponeurosis of the external abdominal oblique.

☐ Coverings of the spermatic cord and scrotal wall have counterparts in the abdominal wall related to the embryonic evagination of the processus vaginalis.

☐ An indirect inguinal hernia follows the course of the descent of the testis. This condition is most commonly caused in young patients by a persistence of a patent processus vaginalis.

□ A direct inguinal hernia passes through Hesselbach's triangle.

□ The descent of the testes is not complete until the eighth or ninth month of fetal life.

□ The rugosal appearance of the skin of the scrotum is caused by the smooth muscle in the dartos tunic.

□ The cremaster reflex tests the integrity of the genitofemoral (L_1 and L_2) nerve.

Peritoneal Cavity

□ The peritoneal cavity is a closed sac in the male; it communicates to the exterior by way of the uterine tubes in the female. It contains only peritoneal fluid.

□ The omental bursa is formed by the rotation of the gut.

□ Mesenteries and visceral ligaments provide pathways for vessels and nerves to reach the organs that are suspended into the abdominal cavity.

□ Retroperitoneal organs have one surface contiguous to the body wall.

□ The peritoneal cavity extends into the pelvic cavity.

□ The lesser omentum, falciform ligament, coronary ligament, and left and right triangular ligaments are derived from the ventral mesogastrium; all others are from the dorsal mesogastrium.

□ Whenever it is possible, surgical procedures approach abdominal organs retroperitoneally to maintain the integrity of the peritoneum.

□ A diffuse infection in the peritoneal cavity can be life threatening; the greater omentum has a protective role in that it can adhere to an inflamed viscus and localize the infection.

□ In ovulation the ovum ruptures into the peritoneal cavity and is drawn by peristalsis into the uterine tube. If fertilization occurs in the cavity the zygote can implant on the peritoneum and result in an abdominal pregnancy.

Gastrointestinal Tract

□ The blood supply of the gut should be studied in concert with visceral ligaments, i.e., the vessels to the stomach must traverse the greater or the lesser omentum.

□ The hepatic portal system has an extra set of venous capillaries (sinusoids) interposed between arterial supply to the gut and venous drainage to the heart.

□ The venous flow from the GI tract drains principally to the liver.

□ Embryonic derivation of the GI tract specifically relates to its blood supply; the celiac trunk supplies the foregut derivatives; the superior mesenteric, midgut derivatives; and the inferior mesenteric, hindgut derivatives.

□ Positive differentation can be made between the small bowel and the colon by identifying the taeniae coli, the epiploic appendages, or the haustra.

□ A "hidden" appendix can be located by following taeniae coli proximally on the ascending colon or cecum; they lead directly to the appendix.

□ Occlusion of venous flow through the liver leads to enlargement of portosystemic anastomoses; clinical manifestations include esophageal varicosities, varicosities of the rectal veins (hemorrhoids), and dilatation of cutaneous veins of abdominal wall (caput medusae).

□ In resection of a portion of the gut (e.g., the stomach), vessels in the visceral ligaments must be identified and preserved to retain blood supply to portions not removed.

Liver, Pancreas and Spleen

□ The liver receives venous blood (portal vein) draining from the GI tract and arterial blood from the hepatic artery; the liver drains by way of the hepatic veins into the inferior vena cava.

□ Vestigial remnants of embryonic circulation are present in the adult. The left umbilical vein becomes the ligamentum teres hepatis; in the ductus venosum, the ligamentum venosum.

□ Both biliary secretions and exocrine pancreatic secretions empty into the duodenum.

□ Gallstones in the biliary ducts can block the passage of bile and result in jaundice.

□ Because of the proximity of the duodenum and the pancreas, perforation of a duodenal ulcer can result in a pancreatitis.

□ A tumor in the head of the pancreas can block passage of exocrine secretion from both the pancreas and liver; this will result in an obstructive jaundice.

☐ Traumatic injury to the abdomen can rupture the spleen. If this occurs, an emergency splenectomy may be necessary because of the extensive blood supply to the organ.

Posterior Abdominal Wall and Diaphragm

☐ The origin of the diaphragm is peripheral, the insertion, central. As the muscle contracts, the domes flatten out and increase the vertical dimension of the thoracic/pleural cavity.

☐ The kidney is embedded in fat and is surrounded by renal fascia, which is open inferiorly.

☐ Renal circulation has two sets of capillaries between the renal artery and the renal vein.

☐ The renal pelvis and major and minor calyces are extensions of the ureter.

☐ The medulla of the suprarenal gland is directly innervated by preganglionic sympathetics; that is, no postganglionic neurons are involved.

☐ The location of the inferior vena caval opening in the central tendon and the aortic hiatus behind the median arcuate ligament preclude constriction of these vessels when the diaphragm contracts.

☐ In debilitating disease with excessive weight loss the decrease of fat around the kidney can result in a nephroptosis. This can cause a "kink" in the ureter, thereby blocking urinary flow and causing a hydronephrosis.

☐ Retrograde movement of a urinary bladder infection can cause an inflammation of the kidney.

☐ In pregnancy the ureter can be blocked by the enlarged uterus as the ureter crosses the brim of the pelvis.

QUESTIONS FOR REVIEW

1. Laterally the inguinal ligament attaches to the

   A. Anterior superior iliac spine
   B. Anterior inferior iliac spine
   C. Crest of the ilium
   D. Ischial spine

2. Concerning the rectus sheath, all of the following are correct except

   A. The anterior wall of the rectus sheath completely covers the rectus abdominis muscle from the xiphoid process to the pubic bone
   B. The rectus abdominis muscle is contained in the rectus sheath, whereas the pyramidalis muscle lies posterior to the rectus sheath
   C. Both superior and inferior epigastric arteries and veins are contained within the rectus sheath
   D. The aponeurosis of the external oblique and the superficial lamina of the aponeurosis of the internal oblique form the anterior wall of the rectus sheath in the umbilical region

   E. The posterior wall of the rectus sheath at the pubis is formed by the endoabdominal (transversalis) fascia

3. The cremaster muscle and its fascia are

   A. Smooth muscle
   B. Derived from the external abdominal oblique muscle
   C. Derived from the internal abdominal oblique muscle
   D. Derived from the transversus abdominis muscle

4. The linea alba is

   A. The white line around the iris of elderly people
   B. The light reflex of the tympanic membrane
   C. Formed by the fusion of the eighth, ninth, and tenth costal cartilages
   D. A reflection of the inguinal ligament
   E. Formed by the midline fusion of the aponeuroses of the anterolateral abdominal wall muscles

5. The omental bursa communicates with the greater sac of of the peritoneal cavity through an opening called the

   A. Pancreatic foramen
   B. Gastric cecum
   C. Foramen rotundum
   D. Iliac sinus
   E. Epiploic foramen (of Winslow)

6. In the embryo all of the gut tube is supported by a mesentery, some of which is obliterated in the adult. All of the following adult structures are supported by a peritoneal mesentery except the

   A. Transverse colon
   B. Stomach
   C. Ileum
   D. Jejunum
   E. Descending colon

7. Which of the following statements about the peritoneum are false?

   A. It is a serous membrane
   B. It is a closed sac in both the male and female
   C. It has parietal and visceral portions that are continuous
   D. It is present in both the abdominal and pelvic cavities

8. The umbilical arteries undergo atrophy and obliteration within 3 to 4 days after birth, to become the umbilical ligaments. The intra-abdominal remnant of the left umbilical vein forms the

   A. Hypogastric ligament
   B. Ligamentum teres hepatis
   C. Ligamentum arteriosum
   D. Ductus venosus

9. The gastrointestinal tract is supplied with arterial blood by all of the following arteries except the

   A. Celiac
   B. Superior mesenteric
   C. Inferior mesenteric
   D. External iliac
   E. Internal iliac

10. The ascending colon

    A. Forms the left colic flexure
    B. Begins at the cecum

   C. Is usually suspended into the abdominal cavity by a mesentery
   D. Is supplied by a branch of the inferior mesenteric artery

11. The vermiform appendix is the wormlike appendage of the

    A. Transverse colon
    B. Duodenum
    C. Jejunum
    D. Ileum
    E. Cecum

12. The first section of the duodenum

    A. Is in contact with the common bile duct
    B. Is entirely retroperitoneal in position
    C. Is suspended by the ligament of Treitz
    D. Receives only secretions produced by the liver

13. All of the following are lobes of the liver except the

    A. Caudate
    B. Left
    C. Right
    D. Quadrate
    E. Uncinate

14. The common bile duct and the pancreatic duct join to form the hepatopancreatic ampulla (of Vater), which perforates the medial side of which portion of the duodenum?

    A. The first
    B. The second
    C. The third
    D. The fourth
    E. The fifth

15. The bare area of the liver refers to that part in contact with the

    A. Duodenum
    B. Kidney
    C. Anterior body wall
    D. Stomach
    E. Diaphragm

16. The myenteric plexus (of Auerbach) is located

    A. External to the longitudinal layers of the muscularis externa
    B. In the submucosa of the bowel

C. Between the circular and longitudinal layers of the tunica muscularis
D. In the epithelial layer of the bowel
E. In the mucous layer of the bowel

17. The muscles of the posterior abdominal wall consist of all of the following except the

A. Rectus abdominis
B. Psoas major
C. Quadratus lumborum
D. Iliacus

18. In the formation of the lumbar plexus, branches of the ventral rami from the 2nd, 3rd, and 4th lumbar nerves form the

A. Iliohypogastric and ilioinguinal nerves
B. Lateral femoral cutaneous and genitofemoral nerves
C. Obturator and genitofemoral nerves
D. Genitofemoral, obturator, and femoral nerves
E. Obturator and femoral nerves

19. Renal pyramids consist primarily of the

A. Bowman's capsule
B. Proximal convoluted tubules
C. Henle's loop
D. Distal convoluted tubule
E. Collection tubules

20. Concerning the kidney, all of the following statements are true except

A. Most glomeruli are located in the cortex
B. Interlobar arteries traverse each renal pyramid
C. Renal papillae are the apices of renal pyramids
D. The minor calyces surround renal papillae

21. The _____ nerve supplies motor innervation to the rectus abdominis muscle at the umbilical region.

A. Iliohypogastric
B. Genitofemoral
C. Vagus
D. Tenth intercostal (T_{10})
E. Hypogastric

22. The inferior free border of the aponeurosis of the external abdominal oblique muscle

A. Forms the floor (inferior wall) of the inguinal canal

B. Inserts only into the linea alba
C. Forms the posterior wall of the inguinal canal
D. Contributes to the internal spermatic layer of fascia around the spermatic cord
E. Forms the internal (deep) inguinal ring

23. The lacunar ligament is a subdivision or continuation of the

A. Inguinal ligament
B. Internal abdominal oblique aponeurosis
C. Transversus abdominis aponeurosis
D. Conjoint tendon
E. Rectus abdominis aponeurosis

24. The inguinal canal is described as having a roof, floor, anterior wall, and posterior wall (*i.e.,* four sides). The internal abdominal oblique or its aponeurosis contributes to the formation of how many of these?

A. One
B. Two
C. Three
D. Four

25. Which of the following structures is not found in the inguinal canal?

A. Spermatic cord
B. Round ligament of the uterus
C. Ilioinguinal nerve
D. Genital branch of the genitofemoral nerve
E. Iliohypogastric nerve

26. The endoabdominal (transversalis) fascia is represented in the coverings of the spermatic cord and testis as the

A. Cremasteric muscle and fascia
B. Internal spermatic fascia
C. External spermatic fascia
D. Tunica vaginalis
E. Dartos tunic

27. The adult structures derived from the ventral mesogastrium include all of the following except the

A. Gastrohepatic ligament
B. Hepatoduodenal ligament
C. Falciform ligament
D. Left triangular ligament of the liver
E. Gastrolienal ligament

28. Which vessel is found in the gastrosplenic ligament?

A. Left gastric
B. Splenic
C. Left gastroepiploic
D. Right gastric
E. Right gastroepiploic

29. Structures traversing the hepatoduodenal ligament include all of the following except the

A. Common bile duct
B. Hepatic artery
C. Hepatic vein
D. Lymphatic drainage of the liver
E. "Hepatic" (autonomic) plexus

30. The falciform fold of the peritoneum connected to the vermiform appendix is known as the

A. Teniae omentalis
B. Teniae libra
C. Teniae mesocolica
D. Mesoappendix
E. Greater omentum

31. As the peritoneum covering the stomach continues from the greater curvature it forms a visceral ligament called the

A. Gastrohepatic
B. Greater omentum
C. Lesser omentum
D. Transverse mesocolon
E. Ligament of Treitz

32. Which one of the following statements concerning the stomach is false?

A. It receives blood from the artery of the embryonic foregut.
B. It has a pyloric orifice located at the level of L_1.
C. Its cardiac orifice is located at the level of T_{10}.
D. It has a lesser omentum containing the gastroepiploic blood vessels.
E. It forms a portion of the anterior wall of the omental bursa (lesser sac).

33. The stomach is contiguous to and forms impressions on all of the following organs except the:

A. Liver
B. Pancreas

C. Spleen
D. Right kidney

34. The left colic and right colic arteries normally arise as direct branches of the _____ arteries, respectively.

A. Inferior mesenteric and superior mesenteric
B. Left gastroepiploic and right gastroepiploic
C. Superior rectal and ileocolic
D. Inferior mesenteric and sigmoid
E. Superior mesenteric and inferior mesenteric

35. Classically, the _____ and _____ veins join to form the hepatic portal vein.

A. Hepatic, splenic
B. Splenic, superior mesenteric
C. Hepatic, superior mesenteric
D. Splenic, inferior mesenteric
E. Hepatic, inferior mesenteric

36. The peritoneum surrounding the developing liver is directly continuous with the

A. Lesser omentum
B. Greater omentum
C. Gastrocolic ligament
D. Transverse mesocolon
E. Mesentery of the small bowel

37. The lesser omentum connects the _____ curvature of the stomach, with the _____

A. Greater, liver
B. Lesser, transverse colon
C. Greater, transverse colon
D. Lesser, liver
E. Lateral, spleen

38. Which one of the following is not a branch of the superior mesenteric artery?

A. Intestinal branches
B. Superior pancreaticoduodenal
C. Ileocolic
D. Right colic

39. Which of the following is not found in association with the greater curvature of the stomach?

A. Greater omentum
B. Hepatogastric ligament
C. Right gastroepiploic artery

D. Left gastroepiploic artery
E. Gastrocolic ligament

40. All of the following are peritoneal ligaments or folds attached to the liver except the

A. Coronary
B. Falciform
C. Greater omentum
D. Lesser omentum
E. Left triangular

41. All of the following structures pass through the porta hepatis except the

A. Left and right hepatic ducts
B. Hepatic portal vein
C. Left and right hepatic arteries
D. Hepatic vein
E. Vagal preganglionic parasympathetic nerve fibers

42. The left boundary of the caudate lobe of the liver is the

A. Ligamentum venosum and its fissure
B. Porta hepatis
C. Fossa for the inferior vena cava
D. Anterior lamina of the coronary ligament
E. Gallbladder and its fossa

43. Two vessels that anastomose in the substance of the pancreas are the superior and inferior pancreaticoduodenal arteries. Respectively, the arteries from which these usually branch are the

A. Left gastric and celiac
B. Left gastroepiploic and celiac
C. Gastroduodenal and superior mesenteric
D. Common hepatic and inferior mesenteric

44. In traumatic injuries the spleen may be ruptued and result in massive internal bleeding. In an emergency splenectomy, clamping which of the following structures would stop the bleeding?

A. Gastrohepatic ligament
B. Hepatoduodenal ligament
C. Transverse mesocolon
D. Gastrolienal ligament
E. Lienorenal ligament

45. Which of the following nerves is not a branch of the lumbar plexus?

A. Femoral
B. Obturator
C. Subcostal
D. Thoracoabdominals
E. Ilioinguinal

46. What nerve innervates the cremaster muscle?

A. Femoral
B. Obturator
C. Iliohypogastric
D. Ilioinguinal
E. Genitofemoral

47. All of the following veins drain directly into the inferior vena cava except the

A. Hepatic veins
B. Right suprarenal vein
C. Left gonadal vein
D. Left inferior phrenic vein
E. Median sacral vein

48. The ascending lumbar veins normally drain directly into the

A. Superior vena cava
B. Inferior epigastric vein
C. Inferior mesenteric vein
D. Azygos system
E. Gonadal vein

49. What structure is not in direct relationship to the right kidney?

A. The diaphragm
B. The liver
C. The duodenum
D. The pancreas

50. The medial arcuate ligament that gives partial origin to the thoracic diaphragm spans over the

A. Right crus
B. Psoas major
C. Quadratus lumborum
D. Aorta
E. Esophagus

FIVE

Perineum and Pelvis

Perineum

At the completion of the study of the perineum the student should be able to

▶ *Define the following: perineum, anal triangle, urogenital triangle and diaphragm, and the ischiorectal fossa and its extensions*

▶ *Give the boundaries of and label on a diagram the contents of the deep and superficial perineal compartments*

▶ *List the structures that would be pierced by an icepick if it penetrated the skin just lateral to the midline of the bulb of the penis, and extended into the pelvic cavity*

▶ *List the coverings of the spermatic cord; give their counterparts in the abdomen*

▶ *List the component parts of the penis, its associated nerves and vessels, and its homologues in the female*

Surface Anatomy

The **perineum,** conforming to the deep-lying outlet of the pelvis, is a diamond-shaped area at the lower end of the trunk between the thighs and buttocks (Fig. 5-1). It is bounded anteriorly by the pubic symphysis, laterally by the ischial tuberosities, and posteriorly by the coccyx. A line passing transversely between the ischial tuberosities through the central point of the perineum, divides the perineum into an anterior **urogenital triangle** and a posterior **anal triangle.** The **central point of the perineum,** located between the anus and the urethral bulb in the male, and between the anus and the vestibule in the female, overlies the deeply placed **perineal body.** In the male, a slight median ridge, the **perineal raphe,** passes forward from the anus to become continuous with the **median raphe** of the scrotum and the **ventral raphe** of the penis.

Fasciae and Superficial Perineal Compartment

The **superficial fascia** of the perineum, consisting of a superficial fatty layer and a deeper membranous layer, is continuous with the superficial fascia of the abdominal wall, thigh, and buttock (Fig. 5-2). In the male the fat is lost as the superficial fascia is extended over the scrotum, where both superficial and membranous layers fuse and gain smooth muscle fibers to form the **dartos tunic** of the scrotum. In the female the fat in the superficial layer increases as it passes over the labia majora and the mons pubis. The **membranous layer** (superficial perineal or Colles' fascia) is attached posteriorly to the posterior margin of the urogenital diaphragm and laterally to the is-

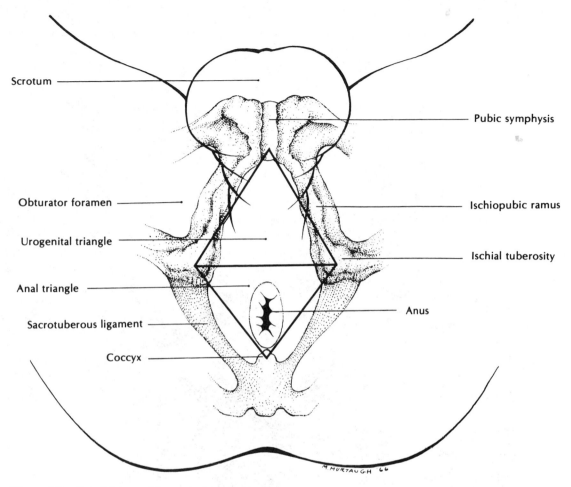

Scrotum

Obturator foramen

Urogenital triangle

Anal triangle

Sacrotuberous ligament

Coccyx

Pubic symphysis

Ischiopubic ramus

Ischial tuberosity

Anus

M. MURTAUGH 66

Figure 5-1. *Subdivisions of the male perineum.*

chiopubic rami. Anteriorly in the male it is continuous with the dartos tunic and, at the superior extent of the scrotum, with the membranous layer of the superficial fascia of the abdomen. In urethral rupture fascial attachments of the membranous layer prevent urine from passing beyond the posterior margin of the urogenital diaphragm or into the thigh, but may allow it to spread a considerable distance into the subcutaneous tissue over the abdomen.

The **superficial perineal compartment,** between the external perineal fascia and the superfi-

cial layer of the urogenital diaphragm, contains the crura of the penis or clitoris, the bulb of the penis or vestibule, the superficial transverse perineal muscles, and perineal vessels and nerves. The **external perineal (Gallaudet's) fascia** covers the ischiocavernosus, bulbocavernosus, and superficial transverse perineal muscles and is firmly adherent to the inferior layer of the urogenital diaphragm. Within the superficial perineal compartment, the crura of the penis are attached to the inferior aspect of the ischiopubic rami and the inferior layer of the urogenital diaphragm.

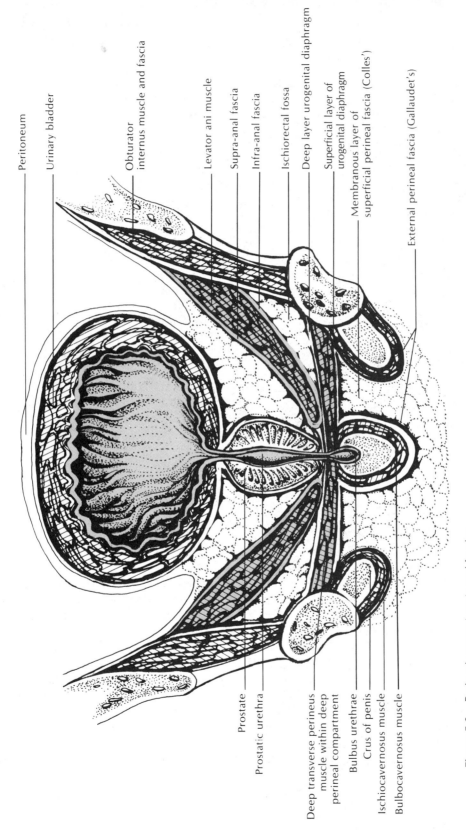

Peritoneum

Urinary bladder

Obturator internus muscle and fascia

Levator ani muscle

Supra-anal fascia

Infra-anal fascia

Ischiorectal fossa

Deep layer urogenital diaphragm

Superficial layer of urogenital diaphragm

Membranous layer of superficial perineal fascia (Colles')

External perineal fascia (Gallaudet's)

Prostate

Prostatic urethra

Deep transverse perineus muscle within deep perineal compartment

Bulbus urethrae

Crus of penis

Ischiocavernosus muscle

Bulbocavernosus muscle

Figure 5-2. *Perineal compartments and fasciae.*

They are covered by the ischiocavernosus muscles. Anteriorly the crura converge toward each other to form the two corpora cavernosa of the penis. The bulb of the penis (urethral bulb) is attached to the inferior layer of the urogenital diaphragm and surrounds the urethra. It is covered by the bulbocavernosus muscle and passes forward to lie inferior to the corpora cavernosa of the penis as the corpus cavernosus urethra (spongiosum) of the penis.

The **perineal body** (tendinous center of the perineum) is a fibromuscular mass located in the median plane between the anal canal and the urogenital diaphragm, with which it is fused. Several muscles attach, at least in part, to it. These include the superficial and deep transverse perinei, the bulbocavernosus, the central portion of the levator ani, and the external anal sphincter. The perineal body is of special importance in the female, as it may be torn or damaged during parturition.

Urogenital Diaphragm and Deep Perineal Compartment

The **urogenital diaphragm** is located in the infrapubic angle between the ischiopubic rami (see Fig. 5-2). It is formed by connective tissue membranes spanning this area as the superficial and deep layers of the urogenital diaphragm and by the structures located between them. The **superficial layer of the urogenital diaphragm** (inferior perineal membrane) is composed of strong bands of fibrous connective tissue that pass between ischiopubic rami and separate the deep perineal compartment from the external genitalia. A similar structure, the **deep layer of the urogenital diaphragm** (superior perineal membrane), also passes between the ischiopubic rami and separates the deep perineal compartment from the anterior recess of the ischiorectal fossa. Anteriorly these membranes fuse to form the **transverse ligament of the pelvis,** and poste-

riorly they fuse at the posterior extent of the deep transverse perineus muscle.

The **deep perineal compartment** is the area between the superficial and deep layers of the urogenital diaphragm. It contains the deep transverse perineus and external urethral sphincter muscles, the internal pudendal vessels and pudendal nerve, and in the male the bulbourethral (Cowper's) glands.

Muscles (Table 5-1)

Muscles of the perineum include the external anal sphincter, the structures within the deep perineal compartment, the superficial transverse perineal muscle, and muscles covering the crura and bulb of the penis or clitoris and vestibule (Fig. 5-3).

The **external anal sphincter** is a ring of voluntary muscle dispersed around the distal part of the anal canal. It is arranged into subcutaneous, superficial, and deep portions. The **subcutaneous part** surrounds the anal orifice, attaching to the skin of the anus. The **superficial segment** extends from the tip of the coccyx to the perineal body. Encircling the anus, it holds the anus in the median plane. The larger **deep portion** surrounds and attaches to the anal canal like a heavy collar. Its muscle fibers are closely associated superiorly with the levator ani muscle (puborectalis), and decussate anteriorly to interlace with the superficial transverse perineal muscle, while other fibers insert into the perineal body.

Within the superficial perineal compartment, thin sheetlike muscles, the median bulbocavernosus and two lateral ischiocavernosi, cover, respectively, the bulb and the two crura of the penis. At the posterior extent of the urogenital diaphragm, superficial transverse perinei muscles arise from the ischial tuberosities and insert into the central tendon of the diaphragm.

The **bulbocavernosus** originates at the central tendon of the perineum and the median raphe

Table 5-1
Muscles of the Perineum

Muscle	Origin	Insertion	Action	Nerve
External anal sphincter	Skin and fascia surrounding anal orifice and tip of coccyx	Central tendon of perineum	Closes anal orifice	Inferior rectal
Bulbocavernosus	Male—median raphe, ventral surface of bulb, and central tendon of perineum	Male—corpus cavernosum urethra, subpubic triangle, and root of penis	Male–compresses bulb and bulbous portion of urethra; anterior fibers believed to act in erection	Perineal branch of pudendal
	Female—central tendon of perineum	Female—dorsum of clitoris and superficial layer of urogenital diaphragm between crura of clitoris	Female—compresses bulb and vaginal orifice	
Ischiocavernosus	Pelvic surface of inferior ramus of ischium surrounding crus	Crus near pubic symphysis	Maintains erection of penis or clitoris by compression of crus	Perineal branch of pudendal
Superficial transverse perineus	Ramus of ischium near tuberosity	Central body of perineum	Supports central body of perineum	Perineal branch of pudendal
Deep transverse perineus	Internal aspect of ischiopubic ramus	Median raphe, central tendon of perineum, and external anal sphincter	Fixes central body of perineum	Perineal branch of pudendal
External urethral sphincter	Ischiopubic ramus	Fibers interdigitate around urethra	Closes and compresses urethra	Perineal branch of pudendal

In the female, an additional muscle of the deep perineal space is specialized as the constrictor vaginae and acts to compress the vagina and greater vestibular glands.

on the superficial (ventral) aspect of the bulb. Its fibers spread over the bulb of the penis to insert into the superficial perineal membrane. The muscle fibers interdigitate with those of the opposite side, and the most distal fibers spread over and attach to the corpus cavernosum penis. In the female the counterparts of the bulbocavernosus covering the bulb of the penis in the male are separate muscles that cover the bulb of the vestibule on either side of the pudendal cleft. They may act as a weak sphincter of the vagina. Originating from the central tendon, their fibers pass anteriorly to insert into the body of the clitoris.

The **ischiocavernosus** originates from the tuberosity of the ischium and the ischiopubic ramus. It covers and inserts into the crus of the penis/clitoris.

The **superficial transverse perineus,** a narrow muscular slip, passes from its origin on the inner surface of the ischial tuberosity to insert medially into the central tendon of the perineum.

Two muscles are located within the deep perineal space. Centrally the **external urethral sphincter** surrounds the membranous portion of the urethra. Peripherally, in the same plane, the **deep transverse perineus** passes from the ischiopubic rami to interdigitate with its opposite member and insert into a fibrous raphe.

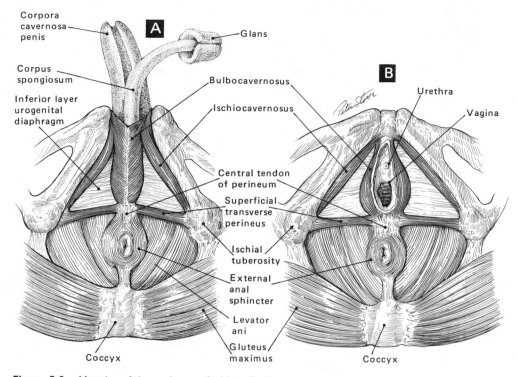

Figure 5-3. *Muscles of the perineum.* **A.** *Male.* **B.** *Female.*

Ischiorectal Fossa

Each of the paired lateral **ischiorectal fossae** (the area between the perineum and the pelvis) is bounded anterolaterally by the fascia of the obturator internus, the ischial tuberosity, and the ischiopubic ramus; medially by the inferior layer of fascia covering the levator ani, and posterolaterally by the fasia of the gluteus maximus. Each continues anteriorly toward the body of the pubis as an **anterior recess,** situated between the ischiopubic rami, the levator ani, and the deep layer of the urogenital diaphragm. Anterosuperiorly the anterior recess is limited at the point of origin of the levator ani. The **posterior recess** of each fossa extends posteriorly between the coccygeus

muscle medially, and the gluteus maximus muscle and sacrotuberous ligament posterolaterally.

Each fossa is filled by an **ischiorectal fat pad,** composed of adipose tissue traversed by irregular connective tissue septa, which are continuous with the subcutaneous fatty layer of the perineum. The **pudendal (Alcock's) canal,** a fascial tunnel located on the medial aspect of the ischial tuberosity internal to the obturator internus muscle, is formed by a splitting of the obturator fascia. The pudendal canal transmits the pudendal nerves and internal pudendal vessels that supply structures in the urogenital triangle. The **inferior rectal vessels and nerves,** arising from nerves and vessels noted above, pass from the canal across the ischiorectal fossa to supply the external anal sphincter, lower rectum, and anal canal (Table 5-2).

Table 5-2
Arterial Supply to Perineum

Artery	Origin	Course	Distribution	Anastomoses
Internal pudendal	Internal iliac	Leaves pelvic cavity through greater sciatic foramen; hooks around spine of ischium; through lesser sciatic foramen to pudendal canal	Perineum and external genitalia	No direct anastomoses
Inferior rectal	Internal pudendal	From pudendal canal crosses ischiorectal fossa	Distal portion of rectum and anal canal	With middle rectal
Perineal	Continuation of internal pudendal	At termination of pudenal canal, superficial branch passes to superficial compartment; deep branch to deep compartment	Superficial branch to structures in superficial compartment; deep branch to structures in deep compartment	Deep external pudendal
Posterior scrotal (labial)	Superficial branch of perineal	Superficial fascia over scrotum/labium majus	Skin of posterior scrotum/labium majus	External pudendal and obturator
Artery to the bulb (penis, vestibuli)	Deep branch of perineal	Pierces inferior layer of urogenital diaphragm to reach bulb	Supplies cavernous tissue of erectile body	No direct anastomoses
Central artery of penis/clitoris	Deep branch of perineal	Pierces inferior layer of urogenital diaphragm to reach crus of penis/clitoris	Supplies cavernous tissue of erectile bodies	No direct anastomoses
Dorsal artery of penis/clitoris	Termination of deep branch of perineal/internal pudendal	Passes from urogenital diaphragm over transverse perineal ligament to dorsum of penis/clitoris	Skin of penis/clitoris	External pudendal

External Male Genitalia

Penis

The **penis,** formed by two corpora cavernosa and the corpus spongiosum, is covered by skin and fascia (Figs. 5-3 to 5-5). The thin elastic skin is loosely connected to the deeper parts and extends over the distal end of the penis for a variable distance, as the **prepuce.** A narrow median fold extends from the inferior aspect of the glans to the prepuce, as the **frenulum.**

Balanitis
Balanitis is an inflammation of the glans penis. This is significant (especially in the very young) in the male who has not been circumcised and has a narrow preputial opening. If the foreskin is retracted over a

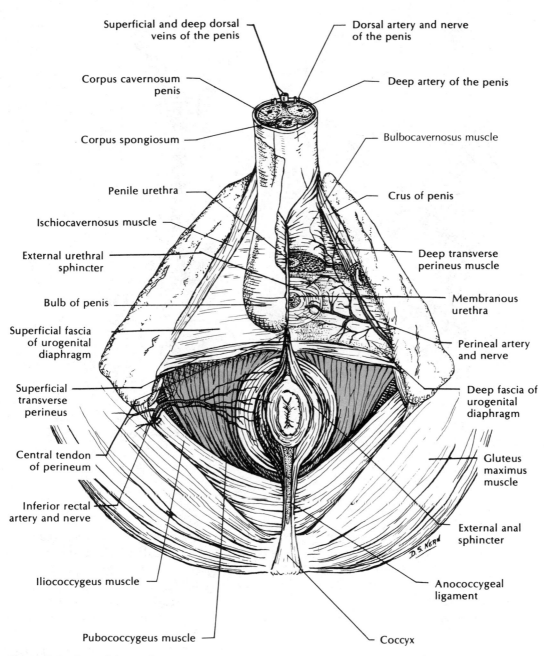

Superficial and deep dorsal veins of the penis

Dorsal artery and nerve of the penis

Corpus cavernosum penis

Deep artery of the penis

Corpus spongiosum

Bulbocavernosus muscle

Penile urethra

Crus of penis

Ischiocavernosus muscle

External urethral sphincter

Deep transverse perineus muscle

Bulb of penis

Membranous urethra

Superficial fascia of urogenital diaphragm

Perineal artery and nerve

Superficial transverse perineus

Deep fascia of urogenital diaphragm

Central tendon of perineum

Gluteus maximus muscle

Inferior rectal artery and nerve

External anal sphincter

Iliococcygeus muscle

Anococcygeal ligament

Pubococcygeus muscle

Coccyx

Figure 5-4. *Root of the penis.*

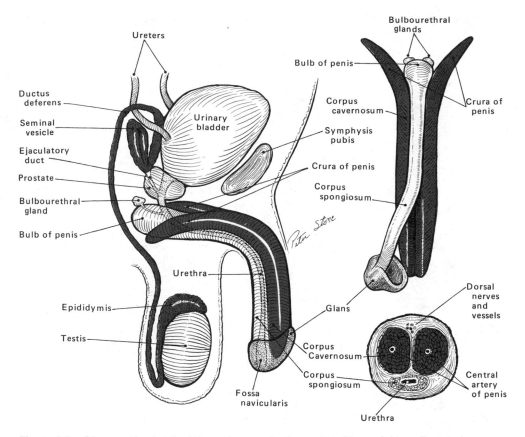

Figure 5-5. *Diagramatic sketch of the male reproductive system.* Upper right: *a dissected penis.* Lower right: *cross section of a penis.*

glans that is edematous and swollen from inflammation, it may act as a tourniquet, compress the veins, decreasing the venous return and thereby cause further enlargement to the glans. Surgical intervention may be necessary to release the constriction by the prepuce.

The **superficial fascia** is directly continuous with dartos tunic of the scrotum. The **deep (Buck's) fascia** forms a tubular investment of the shaft of the penis to the corona, thus surrounding the corpora with a strong capsule. The **suspen-**

sory ligament, a strong, fibroelastic triangular band of deep fascia, extends from the anterior border of the pubic symphysis to the penis, where it divides into right and left lamellae. The latter fuse to the deep fascia at the sides of the shaft of the penis. Vessels and nerves to the dorsum of the penis pass between the two lamellae.

The **corpora cavernosa penis** are a pair of elongated bodies extending from the perineum to the corona of the penis. They consist of **erectile tissue** (dilated vascular spaces) filled with blood, and each is surrounded by a dense white fibrous capsule, the **tunica albuginea.** The apposing sides of this capsule are imperfectly fused to form the **septum.** The **deep arteries of the penis** are terminal branches of the internal pudendal arteries and are located in the center of each corpus cavernosum penis. The **corpus cavernosum urethra (spongiosum),** a structure similar to the corpus cavernosum penis, has a thinner, less dense capsule, lies in a groove on the under surface of the corpora cavernosa penis, and contains centrally the penile portion of the **urethra.** The anterior extremity of the corpus cavernosum urethra expands distally to form the **glans.** The junctional area between the glans and the shaft of the penis is the **corona.**

The **superficial dorsal vein** of the penis lies in the superficial fascia in the median plane and drains into either the right or the left external pudendal vein. The **deep dorsal vein,** also in the median plane but deep to the deep fascia, passes between the lamellae of the suspensory ligament below the infrapubic ligament to drain into the pudendal and prostatic plexuses. Two **dorsal arteries,** terminal branches of the internal pudendal vessels, course lateral to the deep dorsal vein and supply the glans and skin. The dorsal arteries are accompanied by two **dorsal nerves,** the terminal branches of the pudendal nerve, which course lateral to the dorsal arteries and supply twigs to the skin and glans.

Erection

Enlargement of the penis or clitoris is under the influence of the parasympathetics (nervi erigentes) and results from stimulation of the genitalia or erotic thoughts. The physiological mechanism involves a relaxation of the smooth muscle in the walls of arteries supplying the cavernous spaces, thereby enlarging their caliber and increasing the volume of blood flowing into the organ. The inelasticity of the ensheathing tunica albuginea results in increased pressure on the thin walled veins draining the cavernous tissue and tumescence ensues. Following orgasm the sympathetics take over, the arteries become constricted, arterial flow is reduced, and the penis or clitoris gradually returns to its flaccid condition.

Scrotum

The **scrotum,** a pendulous purselike sac of skin and fascia, contains the testis, the epididymis, and the spermatic cord. The skin is rugose, contains smooth muscle fibers and sebaceous glands, and is covered with sparse hair. **Dartos tunic,** forming part of the wall of the scrotum, is a continuation of the two layers of the superficial fascia of the abdomen (Camper's and Scarpa's layers), which fuse into a single layer. It is devoid of fat, highly vascular, and interspersed with smooth muscle fibers. The **septum** (a midline extension of deep fascia) is incomplete superiorly. It passes into the interior of the scrotum to divide it into two chambers.

Hydrocele

A hydrocele is a collection of fluid in the serous cavity of the scrotum. It may be congenital due to the persistence of a communication between the processus vaginalis and the peritoneal cavity. By gravity, peritoneal fluid drains slowly into the processus with no pain or distress. The hydrocele may attain huge size, greatly distending the scrotum. An acquired hydrocele often arises from inflammation of the serous covering of the testis.

Spermatic Cord

The **spermatic cord** is formed at the deep inguinal ring and passes through the inguinal canal to exit at the superficial ring and enter the scrotum, where it is attached to the testis (Fig. 5-6). It contains the vas deferens, a hard cordlike tube that transmits spermatozoa from the epididymis to the urethra, the testicular artery from the aorta, and the artery of the vas deferens. Other structures

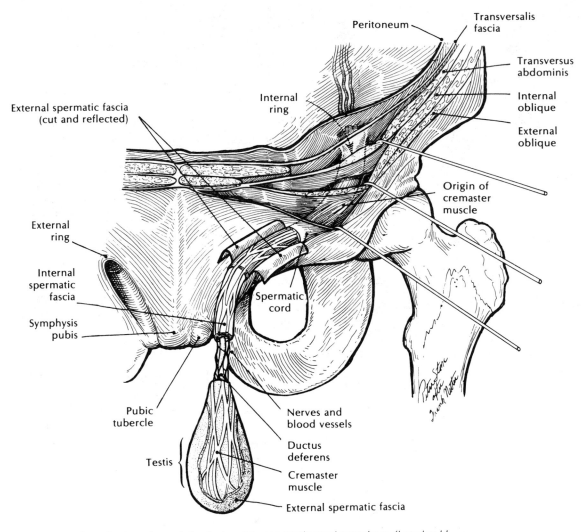

External spermatic fascia
(cut and reflected)

Peritoneum

Transversalis
fascia

Internal
ring

Transversus
abdominis

Internal
oblique

External
oblique

Origin of
cremaster
muscle

External
ring

Internal
spermatic
fascia

Symphysis
pubis

Spermatic
cord

Pubic
tubercle

Testis

Nerves and
blood vessels

Ductus
deferens

Cremaster
muscle

External spermatic fascia

Figure 5-6. *Contributions from abdominal wall to spermatic cord coverings. (Langley LL, Telford IR, Christensen JB: Dynamic Anatomy and Physiology, 5th ed. New York, McGraw-Hill, 1980)*

within the spermatic cord include the pampini-form plexus of veins, which join to form the testicular vein and drain on the right side into the inferior vena cava and on the left side to the renal vein (through internal spermatic veins); lymph vessels draining to nodes at the aortic bifurcation; sympathetic nerves from the renal and aortic plexuses; and the remnant of the processus vaginalis.

Vasectomy

Vasectomy is usually performed by sectioning or ligating the ductus (vas) deferens. A bilateral procedure blocks the passage of sperm from the testis to the urethra. Although the vasectomized male is sterile, he is able to engage in normal sexual activities and does not show signs of male hormone deficiency since the hormonal activity of the testis is not impaired in this procedure.

The **coverings of the spermatic cord** are continuous with the fasciae of the abdominal wall. The **external spermatic fascia,** attached to the crura of the inguinal ligament, is continuous above with the fascia of the external oblique muscle. The **cremasteric fascia,** with interspersed skeletal muscle fibers, is an extension of the internal oblique and its fascia. The **internal spermatic fascia** is continuous with the endoabdominal (transversalis) fascia of the abdominal cavity, and the **subserous fascia** is continuous with the extraperitoneal connective tissue.

Cremasteric Reflex
The cremasteric reflex is elicited by stroking the skin over the medial side of the thigh, which causes contraction of the cremaster muscle. In a positive cremasteric reflex, the testis is drawn upwards towards the superficial inguinal ring. This muscle forms the middle component of the coverings of the spermatic cord, and is an extension from the internal abdominal oblique. As the processus vaginalis evaginates the abdominal wall, some of the muscle fibers are drawn into the scrotum, where they become incorporated into the scrotal wall as a looplike muscular investment, the cremaster muscle.

Testis

The **testes** are two flattened, oval glands suspended in the scrotum by the spermatic cords that produce sperm and certain male sex hormones. During early development they are attached to the upper posterior body wall behind the peritoneum. In late fetal life they descend retroperitoneally to pass through the inguinal canal, emerging at the superficial ring to descend into the scrotum. They are preceded by an evagination of peritoneum, the **processus vaginalis.** External to the processus vaginalis, the wall of the developing scrotum is derived from several layers of the abdominal wall. These are described above as the coverings of the spermatic cord.

Torsion of the Testis
Torsion of the testis is a term synonymous with torsion of the spermatic cord. The testis undergoes rotation about its vertical axis. This occludes the spermatic veins, but usually not the arteries. The tissues distal to the twist become edematous, and develop an infiltrating hematoma from the distended veins. If arterial circulation is impaired, gangrene of the testis may occur.

As the testis enters the scrotum it invaginates into the processus vaginalis to form a parietal layer lining the inner surface of the internal spermatic fascia, and a visceral layer firmly attached to the front and sides of the testis and epididymis. The posterolateral aspect of the visceral layer is tucked between the body of the epididymis and the testis to form a slitlike recess, the **sinus of the epididymis.**

In longitudinal section the testis presents the **tunica albuginea,** a dense, tough, fibrous, outer coat, and the **mediastinum,** a longitudinal thickened ridge along the posterior edge of the testis. The mediastinum is traversed by nerves, blood and lymph vessels, and a network of channels, the **rete testis.** Radiating fibrous septa, passing from the mediastinum to the tunica albuginea, separate the testis into about 250 **lobules,** which contain the **seminiferous tubules.** These unite to form **straight tubules,** passing to the rete testis within the mediastinum. The rete testis continues as the **efferent ducts** to the head of the epididymis.

Descent of the Testis
The testis develops high on the posterior abdominal wall. In late fetal life it descends retroperitoneally, traverses the inguinal canal, and shortly before birth enters the scrotum. During descent the

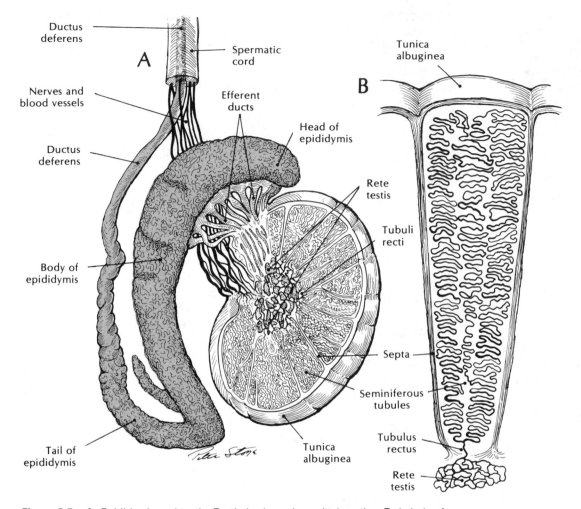

Figure 5-7. **A.** *Epididymis and testis. Testis is shown in sagittal section.* **B.** *Lobule of testis. (Langley LL, Telford IR, Christensen JB: Dynamic Anatomy and Physiology, 5th ed. New York, McGraw-Hill, 1980)*

testis carries with it the nerves, blood vessels, and lymphatics it acquired during development. Hence, pain from testicular disease may be referred to the renal region, and conversely, kidney disease may cause scrotal pain. Furthermore, testicular cancer will spread initially through lymphatics to the upper lumbar and para-aortic lymph nodes, and only much later to the inguinal nodes.

Anomalies arise from imperfect descent of the testis. It may become arrested in descent anywhere along its path, for example, within the abdomen, along the inguinal canal, especially at the superficial ring, or high in the scrotum. Failure of the testis to descend fully into the scrotum (cryptorchidism) will result in an inhibition of spermatogenesis. Normal sperm development occurs only in the scrotum where the temperature is about 5°F lower than in the abdomen. Indirect inguinal hernias are frequently associated with an undescended testis.

Epididymis

The comma-shaped **epididymis** is composed of a **head,** which receives fifteen to twenty efferent ducts from the rete testis; a **body,** separated from the posterior aspect of the testis by the sinus of the epididymis; and a **tail,** which is continuous with the ductus deferens (Fig. 5-7). The spirally coiled efferent ducts form a series of small masses, the **lobules** of the epididymis.

Epididymitis
Epididymitis (inflammation of the epididymis) is usually caused by tuberculosis, gonorrhea, or high fever. If the inflammatory process is bilateral, sterility may result. Since the seminiferous tubules of the testis are terminal extensions of the duct of the epididymis (through the efferent ducts and rete testis), inflammation involving one of these structures often spreads to the other, particularly from epididymitis to orchitis.

External Female Genitalia (Fig. 5-8)

Subcutaneous fat anterior to the pubic symphysis forms a rounded median eminence, the **mons pubis** (mons veneris). The **pudendal cleft,** a midline fissure in the urogenital triangle, is flanked by the **labia majora,** two elongated swellings that are the lateral boundaries of the vulva. The labia majora converge anteriorly at the mons pubis to unite at the lower border of the symphysis pubis as the **anterior commissure.** Posteriorly the labia majora do not unite; however, the forward projection of the perineal body gives the appearance of a posterior commissure, which lies between the vagina and the anus.

The **labia minora** are two thin folds of skin devoid of hair and subcutaneous fat but richly supplied with blood vessels and nerve endings. They flank the vaginal orifice and diverge posteriorly to blend with the labia majora. A transverse fold of skin, the **fourchette,** passes between the posterior terminations of the labia minora. Anteriorly each labium minus divides into two small folds that extend above and below the distal extremity of the clitoris. These folds unite with similar folds of the opposite side to form the **prepuce** dorsally, and the **frenulum of the clitoris** ventrally.

The cleft between the labia minora, the **vestibule,** receives the openings of the vagina, the urethra, and the ducts of the greater vestibular glands. The **external urethral orifice,** a median slitlike aperture, opens posterior to the glans clitoris. The urethral margins are slightly everted. The minute **paraurethral ducts (of Skene)** open into the pudendal cleft.

The vaginal opening is located posterior to the urethral orifice and is narrowed in the virgin by a crescent-shaped fibrovascular membrane, the **hymen.** The **fossa navicularis** is that portion of the floor of the vestibule between the vaginal orifice and the fourchette. **Greater vestibular glands (of Bartholin)** are located bilaterally between the labia minora and the vaginal opening. During coitus the glands are compressed to release a mucuslike secretion to lubricate the lower end of the vagina.

Bartholin's Gland Cyst
Bartholin's gland cyst develops when the duct of the gland becomes blocked, creating a cyst or an abscess. Cysts are best treated by complete excision. An abscess is usually caused by gonorrhea and often is started during sexual excitement when the duct is patent and the gonoccoceal organism present. The cyst may form a relatively large mass that protrudes into

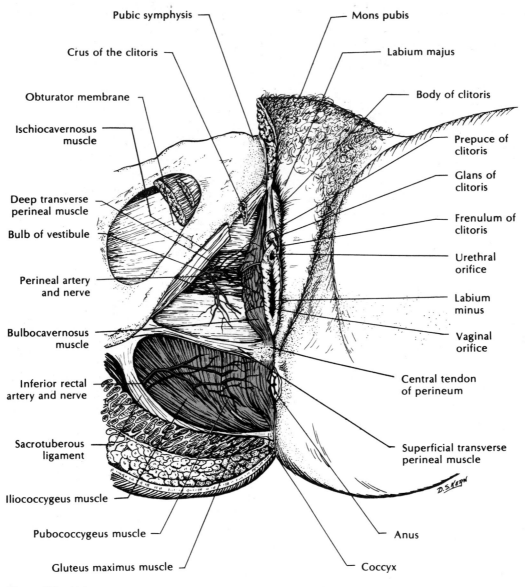

Figure 5-8. *Vulva.*

the vestibule because the gland is within the superficial perineal compartment, and the superficial fascia limiting the compartment is distensible.

Paired elongated masses of erectile tissue forming the **bulbs of the vestibule** are located at the sides of the vaginal orifice and are attached to the superficial layer of the urogenital diaphragm. They are covered by the bulbocavernosus muscle.

The **clitoris,** an erectile organ corresponding anatomically to the male penis, is composed of a body, two crura, and a glans. It differs from the penis in that it is smaller and is not traversed by the urethra. The **body,** formed by the union of the crura, is entirely embedded in the tissues of the vulva and suspended from the pubic symphysis by the **suspensory ligament.** The **crura** of the clitoris are attached to the perineal surface of the ischiopubic rami and to the inferior layer of the urogenital diaphragm. They are covered by the ischiocavernosus muscles. The glans is a small, rounded elevation at the free end of the body. Like the crura, it is composed of erectile tissue and contains abundant sensory nerve endings.

Pelvis

Objectives

At the completion of the study of the pelvic cavity the student should be able to

▶ *Describe the anatomic position of the pelvic cavity; define the birth canal, and give the boundaries of the true and false pelvic cavities*

▶ *Identify the muscles forming the wall of the pelvic cavity and describe the component parts of the pelvic diaphragm*

▶ *Discuss the extension of the peritoneal cavity into the pelvis*

▶ *Draw and label the formation of the sacral plexus and list those branches that supply structures of the pelvis and the perineum*

▶ *Give the origin, course, and distribution of branches of the iliac arteries*

The funnel-shaped **pelvis,** the portion of the trunk inferoposterior to the abdominal cavity, is bounded by the innominate, sacral, and coccygeal bones. The **pelvis minor,** or true pelvic cavity, is below the brim of the pelvis, while the **pelvis major,** or false pelvic cavity, is located between the iliac fossae and is a part of the abdominal cavity. The pelvis contains the lower part of the alimentary canal, the distal ends of the ureters, the urinary bladder, and in the female, most of the internal reproductive organs. The peritoneal cavity extends from the abdominal cavity into the pelvic cavity.

Pelvic Mensuration
Pelvic mensuration is a procedure that measures the size of the inlet and outlet of the pelvis. The dimensions and shape of the bony pelvis can be most accurately determined by x-ray studies. Because of the danger of radiation to the fetus such procedures may be contraindicated even though such information would be helpful to the obstetrician in determining if the baby can be delivered vaginally or if a cesarean section is indicated.

For descriptive purposes, the pelvic boundaries can be divided into lateral and posterior walls, and the floor (Fig. 5-9). The **bony frame-**

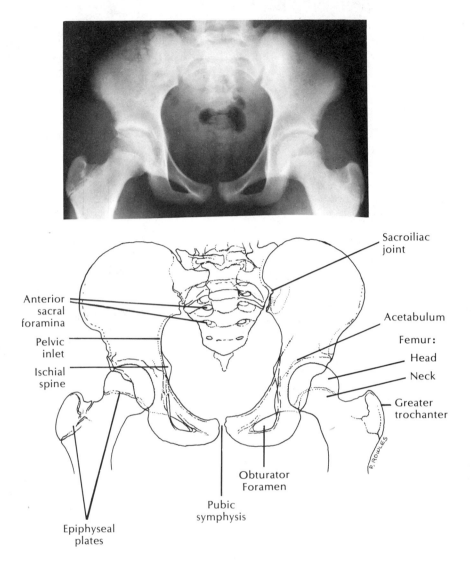

Figure 5-9. *Radiograph and schematic of an adolescent pelvis.*

work of the true pelvis is formed by the portion of the paired innominate bones below the arcuate line, the sacrum, and the coccyx. Most of the pelvic surface of the innominate bone is covered by the obturator internus muscle and its fascia. The gap between each innominate bone and the sacrum is partially filled by the **sacrotuberous** and **sacrospinous ligaments.** The former extends between the posterior iliac spine, lower sacrum, coccyx, and the ischial tuberosity. An extension of the sacrotuberous ligament, the **falciform process,** passes to the lower margin of the ramus of the ischium. The triangular **sacrospinous ligament** lies internal to the sacrotuberous ligament. Its base is attached to the lower sacrum and the upper coccyx. Its apex is attached to the ischial spine. The sacrotuberous and sacrospinous ligaments convert the lesser and greater sciatic notches into foramina, with the sacrospinous ligament demarcating the **lesser** from the **greater sciatic foramen.** The latter transmits the piriformis muscle, the superior and inferior gluteal vessels and nerves, the internal pudendal vessels, and the pudendal, sciatic, and posterior femoral cutaneous nerves. The lesser sciatic foramen is traversed by the tendon of the obturator internus, and internal pudendal vessels and pudendal nerve, and nerves to the obturator internus.

The curved **posterior wall** of the pelvis, which faces anteroinferiorly, is composed of the sacrum and coccyx. It is covered internally by the piriformis and coccygeus muscles and their fasciae. The **floor** of the pelvis includes all structures giving support to pelvic viscera, that is, the peritoneum, the pelvic and urogenital diaphragms, and structures associated with them. Posteriorly the rectum passes through the pelvic floor. The urethra in the male, and the urethra and vagina in the female, penetrate the floor anteriorly. The **peritoneum** covers pelvic viscera to a variable extent, reflecting from the rectum onto the urinary bladder to form the **rectovesical pouch** in the male, and onto the uterus in the female to form the **rectouterine pouch (of Douglas).** Anterior to the rectouterine pouch, the peritoneum reflects from the anterior surface of the uterus onto the bladder and forms the **uterovesical pouch.** Subserous connective tissue between the peritoneum and the pelvic diaphragm varies in thickness and contains blood vessels and nerve plexuses to pelvic viscera, the lower part of the ureter, and the proximal part of the ductus deferens. Localized connective tissue thickenings form sacrogenital and uterosacral ligaments that aid in the support of pelvic organs.

Paracentesis in the Rectouterine Space
The rectouterine space (pouch of Douglas) is the inferiormost portion of the peritoneal cavity. Therefore, blood and pus in the peritoneal cavity tend to settle in this region, causing pelvic abscesses that must be drained. Drainage may be ac-

complished by opening the abscess through the wall of the vagina (paracentesis). The needle is passed through the posterior vaginal fornix into the peritoneal cavity.

Muscles

The levator ani and coccygei muscles, and fasciae covering the upper and lower surfaces of these muscles form the **pelvic diaphragm** (Fig. 5-10). The **levator ani** is a thin, wide, curved sheet of muscle, variable in thickness, which forms the muscular floor of the true pelvis and separates the pelvic cavity from the ischiorectal fossae. A narrow midline gap (genital hiatus) permits passage of the vagina in the female and of the urethra in both sexes. The levator ani is subdivided into three parts, the pubococcygeus, puborectalis, and iliococcygeus muscles, which may be differentiated by the position, direction, and attachment of their fibers.

The **pubococcygeus** forms the main part of the levator ani and originates from the posterior aspect of the body of the pubis and tendinous arch of the levator ani. Its more lateral fibers insert into the perineal body, the wall of the anal canal, and the anococcygeal body. In the male, its most medial fibers insert into the prostate as the **levator prostatae muscle,** while in the female these fibers insert into the urethra and vagina as the **pubovaginalis muscle.** Fasciculi of the latter encircle the urethra and vagina to form the **sphincter vaginea.**

The most conspicuous portion of the pubococcygeus, the **puborectalis muscle,** passes posteriorly from the pubis and unites with fibers from the opposite side to form a muscular sling behind the rectum near its anorectal junction.

Defecation
In order to pass a solid stool the axis of the rectum and anal canal must assume a more or less straight line. Therefore, defe-

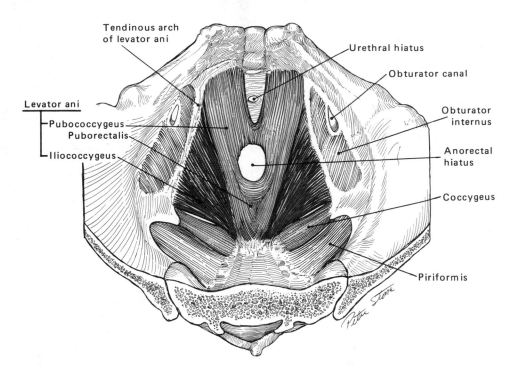

Figure 5-10. *Pelvic diaphragm. View from the interior of the pelvic cavity. Muscles form-ing the floor of the pelvic cavity of a male are depicted.*

cation necessitates a relaxation of the pu-borectalis (rectal sling), which allows the anal canal to become aligned with the rectum.

The **iliococcygeus,** although extending over a relatively large area, is often the most poorly de-veloped portion of the levator ani. It orginates from the tendinous arch and the ischial spine and passes obliquely inferiorly to insert into the sides of the coccyx and the anococcygeal raphe.

Located posterior to the levator ani, the **coccy-geus** (ischiococcygeus) may be present only as tendinous strands. It arises from the pelvic aspect of the ischial spine, and its fibers spread out to insert into the lateral and lower margins of the sacrum and the upper part of the coccyx. It lies on the internal aspect of the sacrospinous ligament.

Fasciae (Fig. 5-11)

The **endopelvic fascia** is a continuation of the endoabdominal fascia, which passes onto the lat-eral pelvic wall as the iliac fascia to become the **obturator fascia** at the brim of the pelvis. The latter, the most definite layer of fascia in the pelvis, lines the internal surface of the obturator internus and, at the margins of this muscle, fuses with the periosteum. Inferiorly, it joins the falciform pro-cess of the sacrotuberous ligament and, at the an-terior margin of the obturator foramen, fuses with the obturator membrane to form the floor of the obturator canal. The obturator fascia gives origin to most of the levator ani and, below the origin of this muscle, forms the lateral walls of the ischio-rectal fossae. Internal to the ischial tuberosity, the

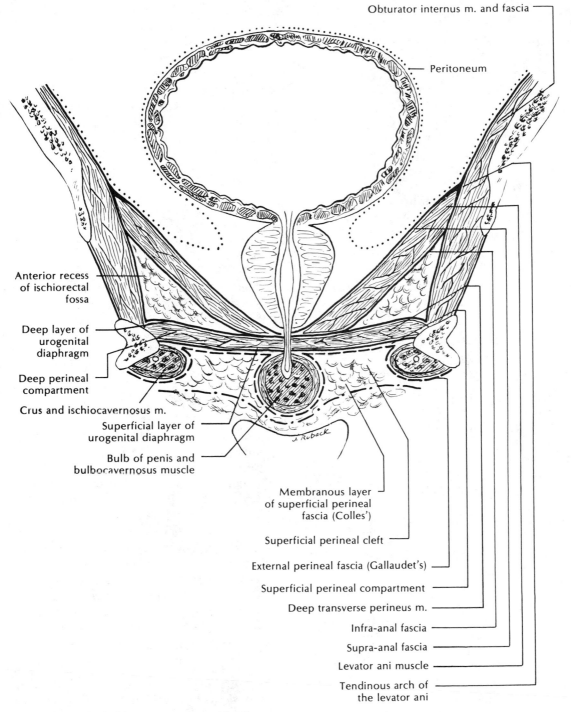

Obturator internus m. and fascia

Peritoneum

Anterior recess
of ischiorectal
fossa

Deep layer of
urogenital
diaphragm

Deep perineal
compartment

Crus and ischiocavernosus m.

Superficial layer of
urogenital diaphragm

Bulb of penis and
bulbocavernosus muscle

Membranous layer
of superficial perineal
fascia (Colles')

Superficial perineal cleft

External perineal fascia (Gallaudet's)

Superficial perineal compartment

Deep transverse perineus m.

Infra-anal fascia

Supra-anal fascia

Levator ani muscle

Tendinous arch of
the levator ani

Figure 5-11. Perineal compartments and fasciae.

obturator fascia splits to form the pudendal canal.

The **supra-anal fascia,** attaching at the tendinous arch of the levator ani, covers the pelvic surface of the levator ani and coccygeus muscle. **Infra-anal fascia,** thinner than the above, covers the extra-pelvic surface of the levator ani and coccygeus muscles. Anteriorly a midline thickening of these fasciae forms the **puboprostatic ligament** (**pubovesical** in the female); posteriorly the fasciae thin out to cover the coccygeus and fuse with the sacrospinous ligament.

The thin **obturator membrane** closes the obturator foramen except anterosuperiorly, where a gap, the **obturator canal,** transmits the obturator nerve and vessels to the adductor region of the thigh. The obturator membrane also gives partial origin to the obturator internus and externus muscles.

Nerves

Innervation to pelvic structures is from the **sacral** and **coccygeal spinal nerves** and the **sacral portion of the autonomic nervous system.** Each of the five sacral and the coccygeal nerves divides into dorsal and ventral rami within the sacral canal. Dorsal rami of the first through fourth sacral nerves pass through the posterior sacral foramina and divide into a medial muscular branch to the erector spinae muscles and a lateral cutaneous branch that forms a series of loops to give perforating branches (cluneal nerves) to the skin over the buttock. Dorsal rami of the fifth sacral and coccygeal nerves pass through the sacral hiatus to supply the skin over the coccyx and around the anus. The ventral rami of the first four sacral nerves pass through sacral foramina into the pelvis, while the ramus of the fifth sacral nerve passes between the coccyx and the sacrum. The ventral rami of the first and second sacral nerves are the largest components of the sacral plexus, and thereafter the ventral rami decrease in size.

The ventral ramus of the coccygeal nerve passes below the rudimentary transverse process of the coccyx to form, with the fourth and fifth sacral nerves, the small coccygeal plexus. The second, third, and fourth sacral nerves give off **pelvic splanchnic nerves,** (nervi erigentes), which transmit parasympathetic preganglionic fibers to the pelvic autonomic plexuses supplying the large intestine below the level of the splenic flexure, as well as the pelvic viscera.

Sacral Plexus

Situated largely anterior to the piriformis muscle, the **sacral plexus** is formed by the ventral rami of the fourth and fifth lumbar nerves, and the first four sacral nerves (Fig. 5-12). The ventral ramus of the fourth lumbar nerve divides into an upper and lower segment. The latter joins with the ventral ramus of the fifth lumbar to form the **lumbosacral trunk,** which descends to join the sacral plexus. Twelve named nerves are described as arising from the sacral plexus. Seven are directed to the buttock and lower limb, and five supply pelvic structures.

Gluteal Injections
Since the gluteal region is a common site for injections, the sciatic nerve may be injured by a poorly placed intramuscular injection. Injury is usually avoided if the injection is made in the upper outer quadrant of the buttock, which is far removed from the sciatic nerve and large blood vessels. Most nerve damage from injections affect the peroneal division of the sciatic nerve, which causes loss of power to dorsiflex the ankle or to extend the toes, and is called foot drop. Other complications of parenteral injections include embolism, hematoma, abscess, intravascular injection of drugs, and sloughing of skin.

The **superior gluteal nerve** (L_4, L_5, and S_1) passes posteriorly through the greater sciatic foramen above the level of the piriformis muscle. It is

Roots

Anterior divisions

Posterior divisions

Nerves

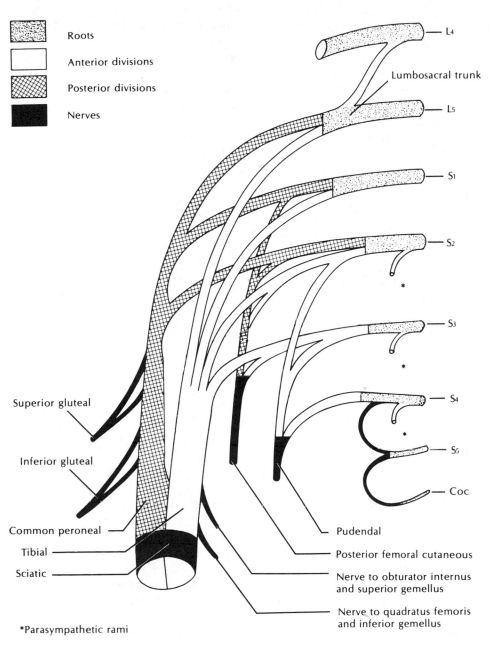

L4

Lumbosacral trunk

L5

S1

S2

*

S3

*

S4

*

S5

Coc

Superior gluteal

Inferior gluteal

Common peroneal

Tibial

Sciatic

Pudendal

Posterior femoral cutaneous

Nerve to obturator internus
and superior gemellus

Nerve to quadratus femoris
and inferior gemellus

*Parasympathetic rami

Figure 5-12. *Sacral plexus.*

accompanied in its course by the superior gluteal artery and vein. It innervates the gluteus medius and minimus muscles and terminates in the tensor fascia lata muscle. The **inferior gluteal nerve** (L_5, S_1, and S_2) passes posteriorly through the greater sciatic foramen below the level of the piriformis muscle to supply the gluteus maximus.

The **nerve** to the **quadratus femoris** and **inferior gemellus** muscles (L_4, L_5, and S_1) leaves the pelvis through the greater sciatic foramen inferior to the piriformis muscle. The **nerve** to the **obturator internus** (L_5, S_1, and S_3) exits by way of the same foramen and, after giving a branch to the superior gemellus, curves around the ischial spine to pass through the lesser sciatic foramen to supply the obturator internus.

The **posterior femoral cutaneous nerve** (S_1, S_2, and S_3) also leaves the pelvis through the greater sciatic foramen, passing inferior to the piriformis muscle. It accompanies the inferior gluteal vessels to the inferior border of the gluteus maximus. It courses down the thigh, superficial to the biceps femoris but deep to the fascia lata. It innervates skin over the posterior aspect of the thigh and leg.

Inferior cluneal (perforating cutaneous) **nerves** (S_2 and S_3) pierce the gluteus maximus and deep fascia midway between the coccyx and the ischial tuberosity to supply skin of the lower gluteal region.

The largest nerve in the body, the **sciatic** (L_4, L_5, S_1, S_2 and S_3), enters the gluteal region through the lower part of the greater sciatic foramen. It traverses the gluteal region in the interval between the greater trochanter of the femur and the ischial tuberosity and descends under cover of the gluteus maximus. It is related anteriorly, in sequence from superior to inferior, to the nerve to the quadratus femoris, the tendon of the obturator internus, the two gemelli, and the quadratus femoris muscles. As it enters the posterior compartment of the thigh, it descends on the posterior aspect of the adductor magnus muscle to divide into the **tibial (medial popliteal)** and **common peroneal (lateral popliteal) nerves.**

Its distribution will be considered with the inferior extremity.

The **nerve to the piriformis** (S_1 and S_2) enters the anterior aspect of this muscle directly, while the **nerve to the levator ani and coccygeus** (S_3 and S_4) descends on the deep aspect of these muscles to innervate them. The **nerve to the external anal sphincter** (S_4) passes either through the coccygeus or between this muscle and the levator ani to continue forward in the ischiorectal fossa. It supplies the external anal sphincter and the skin surrounding the anus.

The **pudendal nerve** (S_2, S_3, and S_4) supplies most of the perineum. It emerges from the pelvis through the greater sciatic foramen inferior to the piriformis muscle, crosses the posterior aspect of the ischial spine medial to the internal pudendal artery, then passes through the lesser sciatic foramen to enter the pudendal canal. Its branches include several inferior rectal twigs, which pass across the ischiorectal fossa to supply the external anal sphincter and anal canal. It also gives rise to perineal nerves, which supply cutaneous innervation to the perineum, and muscular branches to muscles in the urogenital triangle. Its terminal branch is the dorsal nerve of the penis or clitoris.

The **inferior rectal nerve** may arise anywhere along the course of the pudendal nerve, but usually pierces the fascia forming the pudendal canal to cross the ischiorectal fossa and to supply the external anal sphincter and the overlying skin. It also innervates the lining of the anal canal below the pectinate line. The **perineal nerve** divides into a **superficial** and a **deep branch** within the pudendal canal. The former supplies **posterior scrotal** or **labial nerves,** which are distributed to the posterior aspect of the scrotum or the labium majus and to the skin of the perineum. The **deep branch** pierces the medial wall of the pudendal canal to supply twigs to the levator ani and the external anal sphincter. It also sends branches to structures in the superficial perineal space, namely, the bulbocavernosus, ischiocavernosus, superficial transverse perineus muscles, and the bulb of the penis (vestibule).

Table 5-3
Nerve Distribution to Perineum and Pelvis

Nerve	Origin	Course	Distribution
Pudendal	Anterior divisions of sacral plexus	Passes through greater sciatic foramen below level of piriformis; into ischiorectal fossa via lesser sciatic foramen; in ischiorectal fossa traverses pudendal canal, to reach UG diaphragm	Supplies perineum and external genitalia
Inferior rectal	Pudendal in ischiorectal fossa	Passes transversely across ischiorectal fossa	External anal sphincter and skin around anus
Superficial perineal	Pudendal at posterior border of UG diaphragm	Traverses superficial perineal compartment to reach skin	Skin of perineum; posterior aspect of labium majus or scrotum
Deep perineal	Continuation of pudendal at posterior border of UG diaphragm	Traverses deep perineal compartment	Muscles of superficial compartment and deep compartment
Dorsal nerve of penis (clitoris)	Terminal branch of pudendal at anterior part of UG diaphragm	Dorsum of penis or clitoris	Skin of penis or clitoris
Autonomic plexuses	Form around branches of internal iliac artery; parasympathetic (preganglionic) from pelvic splanchnic (S_2, S_3, S_4) sympathetic (postganglionic) from sacral splanchnics	Follow arteriolar branchings to walls of organs; postganglionic parasympathetics are in walls of organs	Parasympathetics—nerve of erection (vasodilatation); contraction of detrusor muscle of bladder and muscular wall of rectum; relaxation of internal sphincters of rectum and bladder. Sympathetics have opposite action and aid in ejaculation.

The deep perineal subsequently pierces the posterior margin of the urogenital diaphragm to supply the deep transverse perineus muscle, the sphincter urethra, and the external urethral sphincter. It then continues anteriorly to pass between the lamellae of the suspensory ligament of the penis/clitoris. The terminal branch of the pudendal is the **dorsal nerve of penis/clitoris,** which runs forward on the dorsum of the penis/clitoris supplying the skin, prepuce, and glans of those organs (Table 5-3).

Pudendal Nerve Block

Pudendal nerve block is perhaps the safest anesthesia for childbirth and is also effective for procedures on the female external genitalia. The primary innervation to the skin and muscles of the perineum is the pudendal nerve. Anesthesia is possible through two routes. In the perineal approach, a needle is inserted through the skin of the perineum towards the ischial tuberosity. The anesthetic is infiltrated just medial to and behind the ischial spine, where the pudendal nerve lies in the pudendal (Alcock's) canal. In the transvaginal approach, the needle is passed through the lateral vaginal wall to a point just medial to the ischial spine. In both procedures, it is necessary to block the pudendal nerves bilaterally to obtain adequate anesthesia.

Coccygeal Plexus

The **coccygeal plexus** is formed by the lower division of the ventral ramus of the fourth sacral nerve, the ventral ramus of the fifth sacral, and the coccygeal nerve. It supplies twigs to the sacrococcygeal joint and the skin over the coccyx.

Autonomic Nerves

The **pelvic splanchnic nerves (nervi erigentes)** are slender filaments passing from the second, third, and fourth sacral nerves. They transmit **preganglionic parasympathetic fibers** to the pelvic plexuses that have their cell bodies at the above-named levels of the spinal cord. They also supply parasympathetic innervation to the large intestine distal to the splenic flexure.

Pelvic autonomic plexuses receive their **sympathetic contribution** from either the inferior extension of the sympathetic trunk, or the downward continuation of the preaortic plexuses. The sacral portion of the sympathetic trunk consists of three or four ganglia lying on the anterior aspect of the sacrum, just medial to the sacral foramina. It ends in a fusion of the two sympathetic trunks in the midline to form the **ganglion impar.** Gray rami communicantes pass to sacral and coccygeal spinal nerves and are distributed by way of those nerves to the inferior extremity, perineum, and pelvis.

The **superior hypogastric autonomic plexus** is located between the common iliac arteries. It receives its component fibers from the lower lumbar splanchnic nerves and the descending fibers of the inferior mesenteric plexus. It continues inferiorly, divides, and passes to either side of the rectum as the **inferior hypogastric plexus.** Further subdivisions are designated according to the vessels they follow or the organs they supply, as for example, the middle rectal, prostatic, vesical, uterovaginal, and cavernous plexuses of the penis or clitoris. These plexuses contain postganglionic sympathetic fibers, preganglionic parasympathetic fibers, and visceral afferent (sensory) fibers.

Arteries (Table 5-4)

The **internal iliac (hypogastric),** the smaller of the two terminal branches of the common iliac artery, arises at the level of the lumbosacral articulation (Fig. 5-13). It usually divides into an **anterior** and **posterior division** before giving origin to its several named branches. Branches from the trunks are not constant, but usually all visceral branches (the vesicular, uterine, vaginal, and rectal) arise from the anterior trunk, as do the obturator, inferior gluteal, and internal pudendal arteries. The iliolumbar, lateral sacral, and superior gluteal arise from the posterior trunk.

In the male, the **superior vesical (umbilical) artery** gives off a branch close to its origin, the **artery to the ductus deferens,** before it continues along the lateral pelvic wall to the apex of the bladder. This small branch supplies the lower end of the ureter, the ductus deferens, the seminal vesicle, and part of the bladder. The superior vesical artery gives one or more twigs to the bladder before becoming a solid cord, the obliterated umbilical artery, which is embedded in a fold of peritoneum on the internal aspect of the anterior abdominal wall as it extends toward the umbilicus.

The **obturator artery** passes along the lateral pelvic wall on the surface of the obturator internus to supply the muscle. With the obturator nerve and vein, it passes through the obturator canal to be distributed to the muscles in the adductor compartment of the thigh. The **uterine artery** passes inferiorly along the lateral pelvic wall and turns medially at the base of the broad ligament to pass to the cervicouterine junction. Here it gives rise to one or more **vaginal branches,** ascends on the uterus, which it supplies, and terminates as twigs to the uterine tube.

Ureter/Uterine Artery Relationship
In gynecological surgery, especially hysterectomy, the relationship of the ureter to the uterine artery is critically important.

Table 5-4
Arterial Supply to the Pelvic Organs

Artery	Origin	Course	Distribution	Anastomoses
Internal iliac	Common iliac	Crosses brim of pelvis to reach pelvic cavity	Primary blood supply to pelvic organs, gluteal muscles, perineum, and medial compartment of thigh	No direct anastomoses
Anterior division internal iliac	Internal iliac	Passes anteriorly to divide into visceral branches and the obturator	Pelvic viscera, medial compartment of thigh	No direct anastomoses
Posterior division internal iliac	Internal iliac	Passes on anterior aspect of piriformis, gives rise to superior gluteal, iliolumbar, lateral sacrals	Gluteal region posterior abdominal wall and vertebral canal	No direct anastomoses
Superior vesicular	Anterior division internal iliac	Passes towards superior aspect of urinary bladder; continues as the obliterated umbilical artery (medial umbilical ligament)	Superior aspect of the urinary bladder	Inferior vesicular
Artery to ductus deferens	Inferior vesicular	Passes retroperitoneally to ductus deferens	Ductus deferens	Inferior vesicular
Inferior vesicular	Anterior division of internal iliac	Passes retroperitoneally to inferior portion of urinary bladder	Urinary bladder, twigs to seminal vesicle and prostate	Superior vesicular
Uterine	Anterior division internal iliac	Passes retroperitoneally; crosses internal to ureter to reach base of broad ligament; interdigitates between layers of broad ligament	Uterus, uterine tube and vagina	Ovarian artery
Vaginal	Uterine artery	At junction of body with cervix of uterus descends to vagina	Vagina	Deep branch of perineal
Gonadal (testicular, ovarian)	Abdominal aorta	Descends retroperitoneally; testicular reaches the deep inguinal ring; ovarian crosses brim of pelvis to infundibulopelvic ligament of ovary	Respectively to testis and ovary	Ovarian anastomoses with uterine artery
Iliolumbar	Posterior division of internal iliac	Ascends into abdomen along psoas major	Iliac branch to iliacus, lumbar branch to psoas, quadratus lumborum and vertebral canal	Lumbars, lateral sacrals, deep iliac circumflex
Lateral sacrals	Posterior division of internal iliac	Course on superficial aspect of piriformis	Piriformis, vertebral canal	Iliolumbar, middle rectal

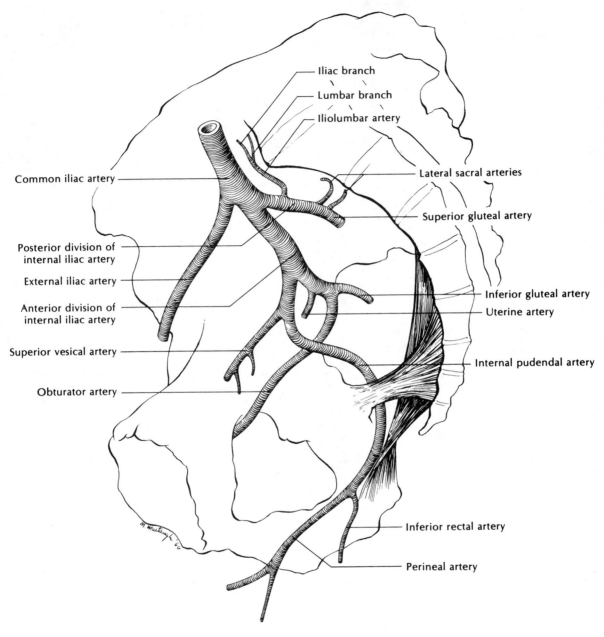

Figure 5-13. *Arterial supply of the pelvis.*

The ureter is contiguous to the artery as it courses towards the bladder. Thus, the surgeon must dissect or visualize the ureter to avoid clamping, ligating, or cutting it if it is necessary to transect the uterine artery.

The **middle rectal artery** is variable in both origin and size as it passes to the rectum, just superior to the pelvic diaphragm.

The **inferior gluteal artery** enters the gluteal region below the piriformis muscle at the lateral edge of the sacrotuberous ligament and supplies the gluteus maximus and other muscles in this area. It gives off the **artery to the sciatic nerve,** which passes inferiorly to enter the nerve. The **internal pudendal artery** emerges from the pelvic cavity, in company with the inferior gluteal artery and the pudendal nerve, then winds around the ischial spine to pass through the lesser sciatic foramen. With the nerve and a corresponding vein, it courses in the pudendal canal to be distributed to the perineum.

The posterior trunk of the internal iliac artery gives rise to three branches. One of these, the **iliolumbar artery,** divides near its origin into an iliac and a lumbar branch. The **iliac branch** crosses the pelvic brim and passes deep to the common iliac or external iliac artery and the psoas major muscle to supply structures in the iliac fossa. The **lumbar branch** ascends, coursing parallel to the lumbosacral trunk, to supply the psoas muscles, and then gives twigs to the structures within the vertebral canal. Usually, two **lateral sacral arteries** pass into the vertebral canal to supply the cord, meninges, and spinal nerves.

The **superior gluteal artery** is the largest branch of the internal iliac. It leaves the pelvic cavity through the greater sciatic foramen above the level of the piriformis muscle to enter the gluteal region. It usually courses between the lumbosacral trunk and the first sacral nerve. Passing between the gluteus maximus and medius muscles, it divides into a superficial and a deep branch. The former supplies the gluteus maximus,

and the latter further divides into superior and inferior branches to supply the gluteus medius, minimus, and tensor fascia lata.

Pelvic Organs

Objectives

At the completion of the study of the pelvic organs the student should be able to

▶ *Give the location, subdivisions, and relationships of the urinary bladder*

▶ *Label on a diagram the structures on the posterior wall of the prostatic portion of the urethra*

▶ *List the structures through which a sperm cell would pass on its way from the seminiferous tubule to the uterine tube*

▶ *Describe the component parts of the uterus and their relationship to the uterine tubes, ovary, and vagina*

Urinary Bladder

The size, shape, and position of the **urinary bladder** varies with its contents (Fig. 5-14). The distended bladder is spherical, rises into the abdomen, and has an average capacity in the adult of about 500 ml. The empty bladder lies on the pubis and the adjacent pelvic floor at the level of the pelvic inlet. It is usually slightly lower in position in the female than in the male. The flattened posterior surface forms the **base (fundus)** of the bladder. The apical superior surface is covered with peritoneum and is convex when filled, but concave and resting on the other bladder surfaces when empty. The inferolateral surfaces rest on the pelvic diaphragm and are continuous with the superior surface at the **apex,** from which the **urachus** (median umbilical ligament) extends to the umbilicus. The **body** of the bladder lies between the apex and the fundus, while the neck surrounds the internal urethral orifice.

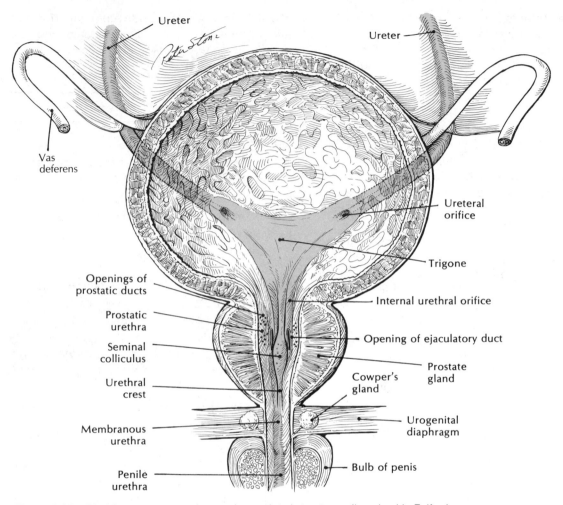

Figure 5-14. Bladder, prostate, urethra, and associated structures. (Langley LL, Telford IR, Christensen JB: Dynamic Anatomy and Physiology, 5th ed. New York, McGraw-Hill, 1980)

The **anterior surface** of the bladder has no peritoneal covering and faces anteroinferiorly toward the pubic symphysis, from which it is separated by the **prevesical space (of Retzius).** The **inferolateral surface** is separated from the levator ani and the obturator internus muscles by extraperitoneal tissue enclosing the vesicular vessels. In the male, it is related posteriorly to the ductus deferens, and its **posterior surface** is in direct contact with the anterior wall of the rectum, the ampulla of the ductus deferens, and the seminal vesicles. The neck is related inferiorly to the prostate and, in the female, adherent to the cervix of the uterus and the anterior wall of the vagina.

Prolapse of the Bladder
Prolapse of the bladder is usually a result of repeated obstetric trauma. If the pelvic floor is damaged from laceration of the perineum and separation of the levator

ani muscle, the pelvic diaphragm (floor) sags downward and becomes funnel shaped. Thus, the most inferior portion of the bladder will be below the opening of the urethra. In such conditions, recurrent infections may occur due to residual retention of urine. Retrograde infection may involve the kidney, resulting in nephritis. Surgical intervention may be necessary to realign the bladder to its normal anatomic position.

The **neck** is the least movable portion of the bladder. It is firmly anchored to the pelvic diaphragm and is continuous in the male with the prostate, where an external groove demarcates the separation of the two organs. The **medial puboprostatic (pubovesicular) ligament** passes from the body of the pubis to the anterior aspect of the prostate in the male, or the neck of the bladder in the female. The **lateral puboprostatic (pubovesicular) ligament** extends from the tendinous arch of the levator ani to the capsule of the prostate or the neck of the bladder. Lateral ligaments (condensations of subserous fascia) from the base of the bladder pass posterolaterally and posteriorly to continue as the **rectovesical folds** in the male, and the **rectouterine folds** in the female.

In the empty bladder the internal surface is thrown into folds and modified at the posterior aspect to form a smooth triangular area, the **trigone.** The angles of the trigone are marked by the orifices of the two ureters and the urethra. Between the ureteric orifices, a transverse ridge, the **interureteric fold** (plica interureterica), is formed by the underlying musculature. Lateral extensions of this fold, **plicae uretericae,** are formed by the passage of the ureters through the wall of the bladder. The **uvula,** a median longitudinal ridge above and behind the internal urethral orifice, is formed by the underlying median lobe of the prostate. The **internal urethral orifice,** situated at the lowest point of the bladder, is encircled by a thickening of the smooth muscle of

the wall of the bladder, the **internal urethral sphincter.**

Male Pelvic Organs

Prostate

The **prostate,** situated behind the pubic symphysis and below the urinary bladder, is formed by smooth muscle and collagenous fibers, in which is embedded secretory glandular tissue (Fig. 5-14). The **apex** is the lowermost part of the prostate, and the **base,** penetrated by the internal urethral orifice, lies horizontally and fuses with the wall of the more superiorly located bladder. Peripherally a narrow groove separates the two organs. The convex **inferolateral surfaces** are surrounded by the prostatic plexus of veins, while the narrow **anterior surface** is separated from the pubis by the retropubic fat pad. The flattened triangular **posterior surface,** which may have a more or less prominent median groove, can be palpated rectally. The prostate consists of two lateral lobes and a median lobe. No superficial demarcation is present between the **lateral lobes,** which are connected by the isthmus anterior to the urethra. The **median lobe,** responsible for the formation of the uvula, is variable in size and projects inwardly from the upper part of the posterior surface between the ejaculatory ducts and the urethra. The enlargement of the median lobe may block urinary flow.

Benign Hypertrophy of the Prostate
Benign hypertrophy of the prostate is present in 20% of men over 50 years of age. The incidence increases about 10% with each decade of life, reaching about 60% in men over 80 years old. The clinical significance of this condition is not so much the size of the gland but whether it interferes with the passage of urine. Hypertrophy of the median lobe is most likely to cause urinary obstruction. Sur-

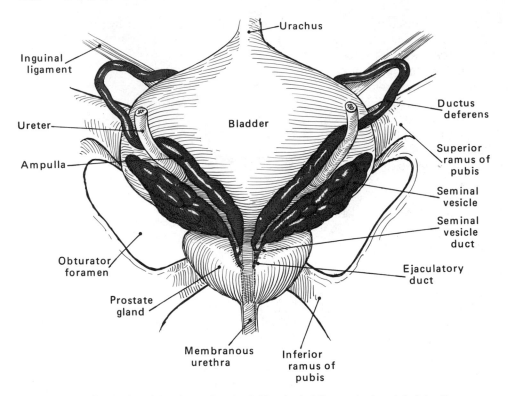

Figure 5-15. *Seminal vesicles (posterior view). The duct of the seminal vesicle joins the ductus deferens to form the ejaculatory duct.*

gery is indicated if urination becomes difficult or impossible, or if the bladder retains a large volume of urine after urination. Benign hypertrophy may be a predisposing factor in cancer of the prostate, which is the second most common malignancy in the male. It is detected by performing a careful rectal exam.

Ductus Deferens

The **ductus deferens** begins as a continuation of the ductus epididymis and passes as a component of the spermatic cord to the deep inguinal ring. From the deep ring it ascends anterior to the external iliac artery, then turns posteriorly to enter the pelvic cavity. It passes onto the lateral pelvic wall, medial to the umbilical artery and obturator

vessels, crosses the ureter and continues medially to reach the posterior aspect of the bladder. Near the base of the prostate it enlarges and becomes tortuous, as the **ampulla.** It is joined by the duct of the seminal vesicle to form the **ejaculatory duct.** The latter penetrates the base of the prostate and passes anteroinferiorly to enter the prostatic portion of the urethra just lateral to the colliculus seminalis.

Seminal Vesicle (Fig. 5-15)

The **seminal vesicles** are two large, sacculated pouches approximately 5 cm in length. They consist of blind-coiled tubes with several diverticula, which secrete the seminal fluid. When the bladder is empty, the seminal vesicles lie horizontally; when the bladder is distended, they are nearly vertical in position. The upper parts of the seminal vesicles are separated from the rectum by the rectovesical pouch. The terminal parts of the ureters and ampullae of the ductus deferens are located medially, and the prostatic and vesical venous plexuses laterally.

Urethra

The male **urethra** extends from the bladder to the external urethral orifice in the glans penis (see Fig. 5-14). It serves as a passage of both urine and semen and consists of prostatic, membraneous, and penile portions. The **prostatic portion** passes through the substance of the prostate where the posterior wall presents a longitudinal fold, the **urethral crest.** The latter is elevated at its midpoint into an enlargement, the **colliculus seminalis.** A small depression at the center, the **prostatic utricle,** corresponds developmentally to the vagina of the female. The ejaculatory ducts open as longitudinal slits at either side of the colliculus. More distally, small orifices of the ducts of the prostate are present.

The **membranous portion** penetrates the urogenital diaphragm, where it is surrounded by the **external urethral sphincter muscle** in the deep perineal compartment. Ducts of the two small **bulbourethral glands** lie at either side of the urethra in the deep pouch and pierce the inferior layer of the urogenital diaphragm to enter the penile urethra.

Extravasation of Urine

Extravasation of urine occurs most frequently following rupture of the penile portion of the urethra. Such damage is common in straddle injuries, for example, striking the frame of a bicycle, the rail of a fence, or the horn of a saddle. The bulb of the penis often tears and urine escapes through the superficial perineal pouch into the subcutaneous perineal cleft. It then passes anteriorly, in the fascial cleft, to reach the anterior abdominal wall deep to Scarpa's (deep membranous layer of superficial) fascia. Urine does not pass down the thighs because of the attachment of superficial fascia to the inguinal ligament and ischiopubic rami. It does not pass posteriorly because of a similar fusion of superficial fascia to the posterior border of the urogenital diaphragm.

The **penile portion** of the urethra is surrounded by the corpus spongiosum penis and, at the glans, flattens laterally to form the **fossa navicularis.** Numerous minute urethral glands (**of Littré**) open into this portion, with the largest of their orifices forming the **urethral lacunae (of Morgagni).**

The **female urethra,** approximately 3 cm to 4 cm long, extends inferiorly and slightly forward from the neck of the bladder to the external urethral orifice. It passes through the pelvic and urogenital diaphragms and opens between the labia minora, anterosuperior to the vaginal orifice and posteroinferior to the glans clitoris. The urethra is closed except during the passage of urine and is marked internally by longitudinal folds, the most prominent of which is located on the posterior aspect as the **urethral crest.** The female urethra is fused with the anterior wall of the vagina and fixed to the pubis by the **pubovesical ligament.**

Internal Female Reproductive Organs (Fig. 5-16)

Ovary

The **ovary** is located within a depression, the **ovarian fossa,** on the lateral pelvic wall at the level of the anterior superior iliac spine. It is about the size and shape of an almond and presents medial and lateral surfaces, anterior and posterior borders, and tubal and uterine poles. The **lateral surface** is in contact with parietal peritoneum; the **medial surface,** adjacent to the uterine tube, is in contact with the coils of the ileum.

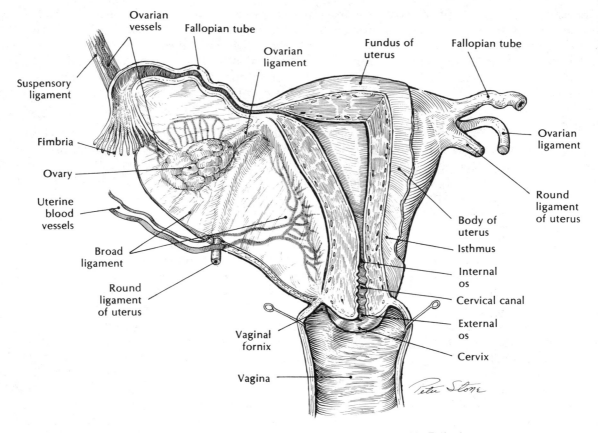

Figure 5-16. *Female reproductive organs with associated ligaments. (Langley LL, Telford IR, Christensen JB: Dynamic Anatomy and Physiology, 5th ed. New York, McGraw-Hill, 1980)*

A ligament, the **mesovarium,** extends between the anterior border at the hilus, and the broad ligament. The posterior free border is related anteriorly to the uterine tube and posteriorly to the ureter.

The suspensory (infundibulopelvic) ligament of the ovary is attached to the **upper (tubal) extremity** with the opening of the uterine tube in close proximity. This peritoneal fold provides the pathway for the major vessels to the ovary. The **lower (uterine) pole,** is directed toward the uterus and attached to the uterus by the ovarian ligament (false ligament of the ovary).

The **mesovarium** is a short, two-layered mes-enteric reflection of the posterior layer of the broad ligament of the uterus, which encloses the ovary. Between the broad ligament and the anterior border of the ovary the layers of the mesovarium are in apposition to each other. The **suspensory (infundibulopelvic) ligament** is a fold of peritoneum from the lateral pelvic wall to the superior pole of the ovary. It overlies the ovarian vessels and ovarian plexus of nerves. The (false) **ligament of the ovary** is a rounded cord containing some smooth muscle fibers that passes from the ovary to the uterus, where it attaches between the uterine tube and the round ligament of the uterus.

Uterine Tube

Uterine (fallopian) **tubes** convey ova from the ovaries to the uterine cavity and transmit spermatozoa in the opposite direction. Each tube is approximately 10 cm long and is located between the layers of the upper margin of the broad ligament. The uterine tube courses laterally from the uterus to the uterine end of the ovary. It arches over and terminates close to the upper pole of the ovary.

Uterine tube ligation

Uterine tube ligation is the tying off of the uterine tubes. This prevents the sperm from ascending and the ovum from descending in the tube, thus fertilization is prevented.

The uterine tube is divided into four segments: the infundibulum, ampulla, isthmus, and uterine portion. The **infundibulum** is somewhat funnel shaped, with the abdominal or pelvic opening located at the outlet of the funnel. The opening has a number of irregular processes, the **fimbriae** projecting from its margins. The slightly tortuous **ampulla** is the longest portion and has relatively thin walls. The narrow, thick-walled **isthmus** is adjacent to the uterus. The **uterine (intramural) portion** is embedded in the wall of the uterus and opens into the uterine cavity.

Tubal Pregnancy

Tubal pregnancy is the most common type of ectopic gestation. Implantation of the fertilized ovum generally occurs in the ampulla but may occur in any region of the uterine tube. The trophoblastic cells of the conceptus invade the mucosa and musculature of the tube. No decidua develops in the mucosa of the tube, as it does in the uterus. Moreover, the very thin tubal wall usually ruptures from the pressure of the developing embryo. The resulting hemorrhage is life threatening and demands immediate surgical intervention.

Uterus

The **uterus** is a somewhat pear-shaped organ lying entirely within the pelvic cavity, with its narrow end directed inferoposteriorly. It is composed of the fundus, body, isthmus, and cervix. The **fundus** is the rounded upper portion above the level of the openings of the uterine tubes. The **body** is the main portion of the uterus, which constricts at the **isthmus** to become the **cervix.** The latter pierces the anterior wall of the vagina at the deepest aspect of the vagina. The entire organ forms an angle slightly greater than 90° with the vagina and usually inclines toward the right and is slightly twisted (Fig. 5-17).

The anterior surface of the uterus is separated from the urinary bladder by the **uterovesical pouch;** the posterior surface from the rectum by the **rectouterine pouch.** The right and left margins of the uterus are related to the **broad ligament** with its enclosed structures. The uterine cavity is widest at the entry of the uterine tubes and narrowest at the isthmus. The **cervical canal,** wider above than below, is an extension of the uterine cavity below the **internal ostium (os)** and opens into the vagina at the **external ostium (os).** Vertical folds are present on the anterior and posterior walls of the cervical canal, with **palmate folds** passing obliquely from the vertical folds.

The uterus derives much of its support from direct connective tissue attachments to the surrounding organs, (*i.e.,* the vagina, rectum, and bladder). Additional support is derived from peritoneal reflections. The **peritoneum** passes from the posterior surface of the bladder onto the isthmus of the uterus, to continue over its anterior surface, fundus, and posterior surface as the uterovesical pouch. It then passes onto the rectum to form the **rectouterine pouch.** Laterally the peritoneum comes into apposition at the lateral border of the uterus and passes as the **broad ligament** to the pelvic wall, where it reflects to become parietal peritoneum. Superiorly the broad ligament encloses the uterine tubes and the ovaries.

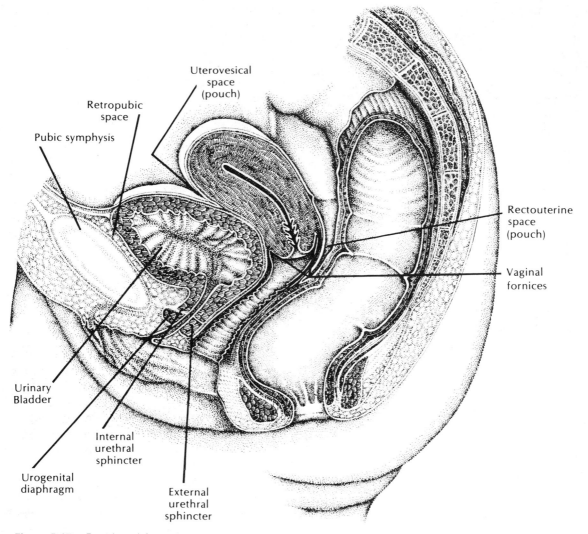

Figure 5-17. *Female pelvic organs.*

That portion of the broad ligament between the level of the mesovarium and the uterine tubes is designated the **mesosalpinx.** The **mesovarium** is an extension of the posterior layer of the broad ligament, which encloses the ovary. The **mesometrium** is the remainder of the broad ligament below the level of the mesovarium.

The **round ligament** of the uterus is a narrow band of fibrous connective tissue between the lay-

ers of the broad ligament. It attaches to the uterus just inferior and anterior to the entrance of the uterine tube. It passes laterally toward the anterior abdominal wall, hooks around the inferior epigastric vessels, traverses the inguinal canal, and blends with the subcutaneous tissue of the labium majus. The **transverse cervical (cardinal or Mackenrodt's) ligament** is a thickened band of fascia and connective tissue together with smooth muscle fibers at the junction of the cervix and the vagina, which gives major support to the uterus. It extends along the base of the broad ligament to attach to the lateral pelvic walls.

Prolapse of the Uterus
Prolapse of the uterus is usually a result of repeated obstetric trauma. If the pelvic floor is damaged from laceration of the perineum and separation of the levator ani muscle, the pelvic diaphragm (floor) sags downwards and becomes funnel shaped. The uterus becomes retroverted. This brings the longitudinal axis of the vagina and uterus in line. Thus, the uterus is able to descend along the vaginal canal, gradually inverting the vagina from above downward, causing prolapse of the vagina. If vaginal prolapse is not corrected, uterine prolapse may occur. In this condition the extruded uterus is often covered by the vagina.

Vagina

The **vagina** is a cylindrical musculomembranous tube, 7 cm to 10 cm in length. It extends from the vestibule to the uterus. It is constricted inferiorly, dilated at its middle, and narrowed superiorly,

where it surrounds and is attached to the periphery of the cervix of the uterus. Its posterior wall is longer than the anterior wall, and the space between the cervix and the vaginal wall is designated the **anterior** and **posterior fornices.** Anteriorly the vagina is related to the urethra and the fundus of the bladder; posteriorly it is separated superiorly from the rectum by the rectouterine pouch, and inferiorly from the anal canal by the perineal body; laterally it is attached to the levator ani and its fascia. The mucous membrane of the internal wall presents anterior and posterior longitudinal folds, the **vaginal columns,** and numerous transverse ridges, the **rugae vaginalis.** The entire structure forms an angle slightly greater than ninety degrees with the uterus.

Paracentesis
In peritonitis there is an accumulation of body fluid, blood, or pus in the peritoneal cavity. Drainage of the cavity is easily accessible in the female through the posterior fornix of the vagina. Only a few millimeters of tissue separate the vaginal vault from the rectouterine pouch of the peritoneal cavity. However, care must be taken to avoid penetrating the wall of the rectum with the needle.

MAJOR ANATOMIC AND CLINICAL POINTS

Pelvic Cavity

- The true pelvic cavity is coextensive with the birth canal.
- The muscular wall of the pelvic cavity is formed by the pelvic diaphragm (levator ani and coccygeus) inferiorly, obturator internis laterally, and piriformis posteriorly.
- Skeletal components of the pelvic cavity include the innominate bones (pubis, ilium, ischium), sacrum, and coccyx.

- The infrapubic angle is greater than ninety degrees in the female, and less than ninety degrees in the male.
- The peritoneal cavity extends into the pelvis.
- The major support of the uterus is provided by the transverse cervical (cardinal) ligament.
- The mesovarium, mesosalpinx, infundibulopelvic ligament, and mesometrium are all components of the broad ligament.
- The rectal sling (puborectalis) maintains the angulation of the anorectal junction.

Perineum

☐ The urogenital diaphragm is external to the pelvic diaphragm.

☐ The plane of the urogenital diaphragm is essentially parallel to the floor.

☐ Elements of the erectile bodies in both the male and the female are attached to the inferior layer of the urogenital diaphragm.

☐ The pudendal nerve is the primary innervation to the perineum.

☐ The deep perineal compartment is a confined space. It is limited laterally by the ischiopubic rami, and anteriorly and posteriorly by fusion of the superficial and deep layers of the urogenital diaphragm.

☐ In urethral rupture urine can release into the superficial perineal cleft and thence extravasate into the abdominal wall passing between the superficial (Scarpa's) fascia and the external investing layer of deep fascia.

☐ The urethra in the male passes through the prostate, pelvic diaphragm, urogenital diaphragm, and penis to reach the external urethral orifice.

☐ A cyst of the greater vestibular (Bartholin's) gland can project as a mass into the vestibule.

☐ Anesthesia of the perineum can be obtained by injection, by way of the vagina, into the pudendal canal to block the pudendal nerve.

Reproductive Organs

☐ Openings are present at each angle of the urinary trigone; for ureters (2), and urethra.

☐ The orientation of the seminal vesicle is dependent upon the physiological state of the urinary bladder. When the bladder is empty it lies horizontally, it lies vertically when bladder is distended.

☐ There is communication from the peritoneal cavity to the exterior by means of the uterine tubes, uterine cavity, and vagina.

☐ Both the urethra and anal canal have two sphincters; the internal sphincter consists of smooth muscle, the external sphincter is skeletal muscle.

☐ Hypertrophy of the median lobe of the prostate (uvula) can block urinary flow.

☐ Torsion of the spermatic cord can result in the obstruction of blood flow to the testis.

☐ Access to the peritoneal cavity in order to drain blood, fluid, or pus can be obtained by passing a trochar through the posterior fornix of the vagina.

Vasculature and Nerve Supply to the Pelvis and Perineum

☐ External hemorrhoids are painful because of cutaneous sensitivity; internal hemorrhoids are pain-free owing to the lack of sensitivity of the anal canal mucosa.

☐ The internal iliac artery supplies the pelvic structures, perineum, and medial compartment of the thigh.

☐ The sacral plexus is derived not only from ventral rami of sacral nerves, but also from L_5 and part of L_4.

☐ Pelvic splanchnic nerves transmit preganglionic parasympathetic fibers, and supply pelvic viscera and the large intestine distal to the splenic flexure.

☐ The anastomosis between the superior and inferior epigastric arteries is functional in blockage of the aorta.

☐ The anastomosis between rectal veins (superior, middle, and inferior) is functional in portocaval collateral circulation.

☐ Varicosities of the inferior rectal veins are external hemorrhoids; those of the middle and superior rectals are internal hemorrhoids.

☐ Blood-borne metastases can reach the cerebral circulation from the pelvic cavity by way of the vertebral plexus of veins.

QUESTIONS FOR REVIEW

1. All of the following are parts of the broad ligament except the

   A. Mesovarium
   B. Mesometrium
   C. Mesosalpinx
   D. Suspensory (infundibulopelvic) ligament of the the ovary
   E. Round ligament of the uterus

2. The uterosacral ligaments extend from the

   A. Fundus of the uterus to the lower third of the sacrum
   B. Uterine cervix to the lateral aspect of the sacrum
   C. Body of the uterus to the entire sacrum
   D. Cornua of the uterus to the internal inguinal ring
   E. Insertion of the round ligament to the sacral promontory

3. Major support for the cervix and upper vagina comes from the

   A. Uterosacral ligaments
   B. Round ligaments
   C. Transverse cervical (cardinal) ligament
   D. Ovarian ligaments
   E. Cul-de-sac of Douglas

4. The female pelvic skeleton is composed of the

   A. Sacrum, coccyx, ilium, and ischium
   B. Sacrum, coccyx, and two innominate bones
   C. Sacrum, ilium, ischium, and pubis
   D. Ilium, ischium, pubis, and coccyx
   E. Pubis, coccyx, and two innominate bones

5. Structures that are common to the penis and clitoris include all the following except the

   A. Urethra
   B. Crura
   C. Prepuce
   D. Ischiocavernosus muscle
   E. Glans

6. Which of the following statements concerning the testicular veins is true?

   A. The right vein drains into the inferior vena cava and the left vein drains into the left renal vein.
   B. The right vein drains into the inferior vena cava and the left vein drains into the portal vein
   C. The right vein drains into the right renal vein and the left vein drains into the inferior vena cava.
   D. Both veins drain into the inferior vena cava.
   E. Both veins drain into their respective renal veins.

7. The deep perineal compartment is located between the superficial (inferior) and deep (superior) layer of the urogenital diaphragm. It contains all of the following structures except the

   A. Deep transverse perineus muscle
   B. External urethral sphincter muscle
   C. Branches of the internal pudendal artery and vein
   D. Bulbourethral (Cowper's) glands
   E. Bulb of penis

8. Which of the following statements is not true of the muscles of the perineum?

   A. The central tendon serves as an attachment site for at least four muscles.
   B. The ischiocavernosus muscle covers the crus of the penis.
   C. The urogenital diaphragm and the levator ani are synonymous terms.
   D. The bulbocavernosus muscle is innervated by a branch of the pudendal nerve.
   E. The external urethral sphincter is part of the deep transverse perineus.

9. The portion of the uterine tube contained within the muscular wall of the uterus is termed the

   A. Ampulla
   B. Fimbria
   C. Isthmus
   D. Infundibulum
   E. Intramural

10. Regarding the anatomy of the female pelvis, which of the following statements is correct?

   A. The ureter passes inferior or external to the uterine artery.

B. The ureter passes superior or internal to the uterine artery.
C. The ureter penetrates the urogenital diaphragm to reach the urinary bladder.
D. The ureter courses in the infundibulopelvic ligament.

11. Contents of the spermatic cord include all of the following except the

A. Artery to the vas deferens
B. Testicular artery
C. Pampiniform plexus of veins
D. Vas deferens
E. Deep arteries of the penis

12. The ejaculatory duct is formed by the union of the

A. Duct of the seminal vesicle and urethra
B. Duct of the seminal vesicle and ductus deferens
C. Ductus deferens and urethra
D. Ductus epididymus and ductus deferens
E. Ductuli efferentia and ductus deferens

13. The pudendal nerve provides sensory innervation to the

A. Upper vagina
B. Vulva and perineum
C. Cervix, and lower uterus
D. Bladder and proximal urethra
E. Ovary, uterus, and vagina

14. All of the following nerves transmit sympathetic fibers except the

A. Lesser splanchnic
B. Greater splanchnic
C. Pelvic splanchnics
D. Least splanchnic
E. Thoracic splanchnics

15. The internal pudendal artery arises from the

A. Superior vesical
B. External iliac
C. Vaginal
D. Internal iliac
E. Uterine

16. The internal pudendal vessels and the pudendal nerve traverse the

A. Femoral canal
B. Alcock's canal
C. Canal of Nuck
D. Stensen's canal
E. Hunter's canal

17. The rectouterine space (pouch of Douglas) is contiguous to the

A. Posterior fornix of the vagina
B. Superficial perineal compartment
C. Urinary bladder
D. Ishiorectal fossa

18. In the female the _____ is homologous to the _____ in the male, which is located in the middle of the urethral crest of the prostatic urethra.

A. Vagina, utricle
B. Clitoris, penis
C. Ovary, testes
D. Labia minora, scrotum

19. The suspensory ligament of the ovary contains

A. Obliterated paramesonephric duct
B. Vaginal vessels
C. Ligamentum teres uterii
D. Ovarian vessels
E. Uterine vessels

20. Which of the following structures is (are) not found between the layers of the broad ligament?

A. Ovarian artery and vein
B. Uterine (fallopian) tube
C. Uterine arteries and veins
D. Ureter
E. Ligamentum teres (round ligament) uterii

21. Concerning the vestibule of the vagina all of the following are true except

A. It extends anteriorly to the glans clitoris
B. Both the urethra and the vagina open into it
C. It is flanked by erectile bodies
D. The pudendal cleft is a synonym for the vestibule

22. The _____ nerve(s) provide(s) motor innervation to the smooth muscle of the internal urethral sphinctor.

A. Pelvic splanchnic (nervi erigentes)
B. Pudendal
C. Obturator
D. Dorsal nerve of the penis/clitoris
E. Superior and inferior gluteal

23. A sperm cell does not pass through which of the following in its transit from the testis to the normal site of fertilization of the ovum?

A. Epididymis
B. Seminal vesicle
C. Urethra (male)
D. Uterus
E. Uterine tube

24. The terminal ends of the bulbourethral (Cowper's) glands in the male open into the

A. Penile (spongy or cavernous) urethra
B. Ejaculatory ducts
C. Ductus deferens
D. Prostatic urethra
E. Membranous urethra

25. Concerning the ischiorectal fat pad that fills the ischiorectal fossa, all of the following are true except

A. Anteriorly it extends between the pelvic and urogenital diaphragms
B. Posteriorly it extends to the sacrotuberous ligament deep to the gluteus maximus
C. Superiorly it reaches the level of the tendinous arch (arcuate line) of the levator ani
D. Inferiorly it protrudes into the superficial perineal compartment

26. The superior extent of the false pelvis is a line drawn between the

A. Coccyx and ischial spines
B. Two iliac crests
C. Sacral promentary posteriorly, pubic crest anteriorly, and pecten pubis laterally
D. Anterior superior and anterior inferior iliac spines
E. Pubic tubercles

27. Which of the following is not a portion of the levator ani component of the pelvic diaphragm?

A. Levator prostatae
B. Puborectalis
C. Pubococcygeus
D. Iliococcygeus
E. Coccygeus

28. Parasympathetic fibers in the inferior mesenteric plexus are branches of the

A. Vagus nerves
B. Pelvic splanchnic nerves
C. Least splanchnic nerve

D. White rami communicantes
E. Gray rami communicantes

29. Which of the following statements concerning the erectile bodies is not true?

A. These structures are adherent to the inferior layer of the UG diaphragm.
B. These structures are covered by skeletal muscles through their entire length.
C. The clitoris is formed solely by the union of two laterally placed erectile bodies (crura).
D. In the male the urethra travels through one of these erectile bodies, the corpus spongiosum penis.

30. A structure not present in the male urogenital triangle is the

A. Greater vestibular gland
B. Transverse perinei muscles
C. Bulbospongiosus muscle
D. Ischiocavernosus muscle
E. Branches of the internal pudendal artery

31. Rupture of the urethra in the bulb of the penis can cause urine to extravasate into the

A. Ischiorectal fossa
B. Anterior thigh
C. Peritoneal cavity
D. Superficial fascial cleft in the gluteal region
E. Superficial fascial cleft of the anterior abdominal wall

32. The pudendal (Alcock's) canal

A. Is a condensation of the internal investing layer of fascia on the medial surface of the obturator internus muscle
B. Transmits a terminal branch of external iliac artery
C. Is internal to the pelvic diaphragm (levator ani)
D. Extends from the lesser sciatic foramen to the posterior aspect of the urogenital diaphragm

33. During a bimanual pelvic examination, if the index finger is in the posterior fornix of the vagina and the middle finger is in the rectum, what would be held between the fingers?

A. Rectouterine pouch (of Douglas)
B. Body of the uterus in its normal position

C. Cervix
D. Neck of the urinary bladder
E. Cardinal (transverse cervical) ligaments

34. The isthmus of the uterus is

A. Between the external and internal os
B. Between the uterine cavity and internal os
C. Between the uterine cavity and uterine tube
D. At the attachment of the false ligament of the ovary

35. Branches of the sacral plexus that can be dissected and identified in the pelvic cavity include all of the following except the

A. Superior gluteal nerve
B. Inferior gluteal nerve
C. Pudendal nerve
D. Obturator nerve

36. All of the following are components of the pelvic wall except the

A. Coccygeus
B. Puborectalis
C. Deep transverse perineus
D. Piriformis
E. Obturator internus

37. Most of the lymphatic drainage from the ovaries and upper part of the uterus passes through all of the following except

A. The cysterna chyli
B. Para-aortic lymph nodes
C. The thoracic duct
D. Inguinal lymph nodes

38. Which of the following muscles have attachments within the true pelvis?

A. Psoas major
B. Iliacus
C. Piriformis
D. Pyramidalis
E. Psoas minor

39. The blood supply to the rectum is derived from branches of all of the following except the

A. Inferior mesenteric artery
B. Internal iliac artery
C. Internal pudendal artery
D. Obturator artery

40. Which of the following reproductive organs is(are) not found between the urinary bladder and the rectum?

A. Seminal vesicles
B. Uterus
C. Vagina
D. Epididymis

41. The round ligament of the uterus extends from the uterus to the

A. Vagina
B. Labia majora (majus)
C. Labia minora (minus)
D. Ovary
E. Cervix

42. The _____ nerve appears on the _____ aspect of the psoas major muscle and exits the pelvic cavity through the _____ to reach the thigh.

A. Genitofemoral, anterior, inguinal canal
B. Femoral, lateral, femoral canal
C. Obturator, lateral, femoral canal
D. Femoral, posterior, lesser sciatic foramen
E. Obturator, medial, obturator canal

43. Which of the following pairs of muscles are within the deep perineal compartment?

A. Superficial transverse perineus and external urethral sphincter
B. External urethral sphincter and levator ani
C. Deep transverse perineus and superficial transverse perineus
D. Deep transverse perineus and levator ani
E. External urethral sphincter and deep transverse perineus

44. Not under autonomic nervous system control is

A. Relaxation of the internal urethral sphincter
B. Contraction of the cremaster muscle
C. Release of seminal fluid from the seminal vesicle
D. Contraction of dartos muscle fibers
E. Filling of male and female erectile bodies

45. Relaxation of the _____ muscle will result in a decrease in the anterior–posterior curvature of the rectum.

A. External anal sphincter
B. External urethral sphincter
C. Puborectalis
D. Deep transverse perineus (in the female only)
E. Superficial transverse perineus

46. The deep dorsal vein of the penis/clitoris normally drains directly into the

A. Prostatic plexus of veins (uterine plexus in female)
B. Hepatic portal system
C. External iliac vein
D. Posterior division of the internal iliac vein
E. Femoral vein

47. Concerning the uterine artery all of the following are true except that it

A. Is a branch of the posterior division of the internal iliac
B. Courses between the layers of the broad ligament
C. Anastomoses with the ovarian artery
D. Gives branches to the vagina

48. Which of the following statements concerning the seminal vesicle is true?

A. It is medial to the ductus (vas) deferens as the ductus deferens terminates.
B. It lies in a horizontal position in the distended (full) bladder.
C. It is attached to the posterior aspect of the prostate gland.
D. It is situated between the pelvic diaphragm and the urogenital diaphragm.
E. Its duct joins the ductus deferens to form the ejaculatory duct.

49. The shortest portion of the male urethra is the

A. Prostatic urethra
B. Membranous urethra
C. Penile urethra
D. None of the above is correct because the prostatic portion is the same length as the membranous portion of the urethra

50. The central tendon (perineal body) of the perineum is a site of attachment for all of the following muscles except the

A. Bulbocavernosus
B. Superficial transverse perineus
C. External anal sphincter
D. Ischiocavernosus

SIX

Inferior Extremity

Inferior Extremity

At the completion of the study of the gluteal region and thigh the student should be able to

▶ Palpate the main surface features of the inferior extremity

▶ Label on a diagram the cutaneous nerves of the inferior extremity

▶ List the specializations of deep fasciae and delineate compartments of the thigh and leg

▶ Define the femoral triangle and give its contents

▶ List muscles of the three compartments of the thigh, and give their blood supply, nerve supply, and action

▶ Give the relationships of the gluteus maximus, medius, and minimus

▶ Label on a diagram the small lateral rotators of the thigh

▶ List the muscles of the posterior compartment

▶ Give the course and distribution of the superior gluteal, inferior gluteal and sciatic nerves, and the superior and inferior gluteal arteries

Surface Anatomy

The inferior extremity is subdivided into the hip, thigh, knee, leg, ankle, and foot. It is limited superiorly by the iliac crest, inguinal ligament, symphysis pubis, ischiopubic ramus, sacrotuberous ligament, and coccyx. The **iliac crest** is easily palpable in its entirety and extends superiorly as high as the level of the fourth lumbar vertebra. Anteriorly the **anterior superior iliac spine** is palpable and affords the lateral attachment for the **inguinal ligament,** whose position is marked by the inguinal fold (groin). Posteriorly the **gluteal fold** delineates the inferior border of the gluteus maximus muscle. The **ischial tuberosities** are easily felt when the thighs are flexed, as are also the **ischiopubic rami** on the medial aspect of the thighs. Just lateral to the midline, the **pubic tubercles** are readily palpable, especially in the thin individual. Approximately a hand's breadth below the crest of the ilium, the **greater trochanter** of the femur can be palpated about 5 cm posterior to the anterior superior iliac spine.

The massive **quadriceps femoris muscle** tapers inferiorly over the front of the thigh to terminate in the **suprapatellar tendon,** which attaches into the margins of the subcutaneous **patella** and continues as the **patellar ligament** to insert into the **tibial tuberosity.** Laterally the tendons of the **biceps femoris** can be palpated and, with the medially situated tendons of the **semimembranosus** and **semitendinosus,** form prominent cords at the posterior aspect of the knee. With the thigh flexed, abducted, and laterally rotated, the outline of the **sartorius muscle** is visible anteriorly as it crosses the thigh obliquely.

The subcutaneous lateral portions of the **condyles of the femur** give width to the knee. Inferiorly the **head of the fibula** is easily felt at the lateral side of the knee. The **anterior border** and **medial surface of the tibia** are subcutaneous throughout the length of the bone and are continuous proximally with the medial condyle of the tibia. The **malleoli,** formed laterally by the fibula and medially by the tibia, are readily recognized at the ankle. Posteriorly the prominent **calcaneal tendon** serves as the insertion of the soleus, plantaris, and gastrocnemius muscles into the tuberosity of the calcaneus (heel bone).

Fasciae

The **superficial fascia** has special features only in the thigh and on the sole of the foot. In the thigh it contains considerable adipose tissue and varies in thickness, being relatively thick in the inguinal region. Here is it disposed into a **superficial fatty** and a **deeper membranous layer,** with superficial lymph nodes, the great saphenous vein, and smaller blood vessels lying between them. The deeper layer is rather prominent on the medial side, where it blends with the deep fascia and covers the saphenous opening as the **cribriform fascia.** The superficial fascia fuses with the femoral sheath and lacunar ligament superiorly and with the deep fascia laterally. In the sole of the foot the superficial fascia is greatly thickened by fatty tissue disposed into pockets extending inwardly from the skin. These **fibrous fat pads** are especially thick at weight-bearing sites, such as on the heel, ball of the foot, and pads of the toes, where they protect deeply lying structures.

The **deep fascia** of the thigh, the **fascia lata,** varies considerably in thickness and strength. It attaches superiorly to the inguinal ligament, the body of the pubis, the pubic arch, and the ischial tuberosity; laterally to the iliac crest; medially to the sacrotuberous ligament, and posteriorly to the sacrum and the coccyx. The oval **saphenous opening** at the upper medial portion of the thigh presents a medial crescentic border. It is roofed by the cribriform fascia derived from the deep layer of the superficial fascia, which, in turn, is perforated by the great saphenous vein and other smaller vessels. At the medial side of the thigh the deep fascia is relatively thin, but it thickens laterally as the **iliotibial tract.** The latter is a wide, strong band that extends from the iliac crest to the lateral condyle of the tibia, the capsule of the knee joint, and the patellar ligament. The iliotibial tract affords insertion for the tensor fascia lata muscle and about three-fourths of the gluteus maximus muscle. With the body erect, the iliotibial tract serves as a powerful brace, which helps to steady the pelvis and keep the knee joint firmly extended. Internal extensions of the fascia lata attach deeply to the linea aspera of the femur as the **lateral, medial,** and **posterior intermuscular septa,** which separate the thigh into extensor, adductor, and flexor muscular compartments (Fig. 6-1).

Inferiorly the deep fascia attaches to the medial and lateral margins of the patella, the tibial tuberosity, condyles of both the tibia and the femur, and the head of the fibula. At the posterior aspect of the knee joint the fascia forms a roof for the popliteal fossa. The deep fascia of the leg is firmly attached anteromedially to the subcutaneous shaft of the tibia and sends deep **intermuscular septa,** which, with the interosseous membrane and tibia, divide the leg into extensor, peroneal, and flexor muscular compartments. The flexor (posterior) compartment is further subdivided by **transverse intermuscular septa** into deep, intermediate, and superficial portions. Inferiorly the deep fascia attaches at both malleoli and then continues onto the foot.

The deep fascia surrounding the ankle thickens

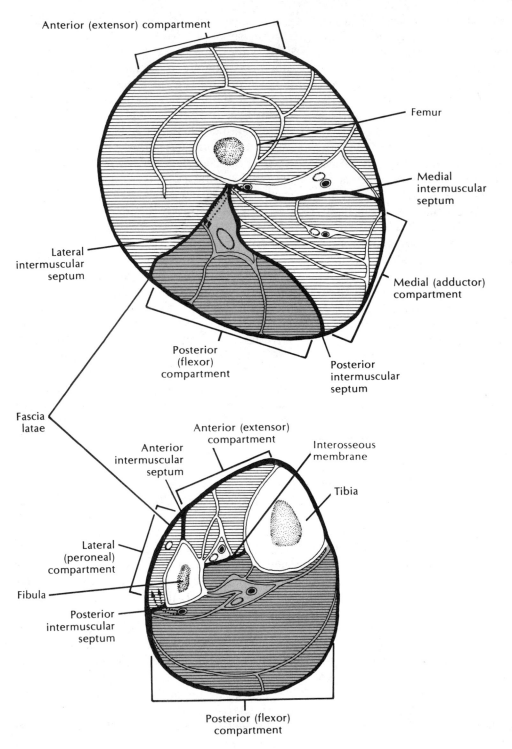

Figure 6-1. *Compartments of the thigh and leg.*

to form bands, the extensor, flexor, and peroneal retinacula. These bind or hold the tendons of the leg muscles in place as they extend into the foot. The retinacula prevent bow-stringing of the tendons during contraction. The **extensor retinaculum** is subdivided into a superior and inferior portion. The **superior extensor retinaculum** (transverse crural ligament) is a strong band 2 cm to 3 cm wide situated immediately above the ankle joint. It is attached laterally to the fibula and medially to the tibia. The **inferior extensor retinaculum** (cruciate ligament) is Y-shaped or V-shaped, with the Y lying on its side. The stem of the Y is attached laterally to the upper surface of the calcaneus and overlies the tendons of the extensor digitorum longus and peroneus tertius. The proximal band passes to the medial malleolus at the medial side of the ankle, while the distal band blends with the deep fascia over the dorsum of the foot. The **flexor retinaculum** (laciniate ligament) bridges the gap between the medial malleolus and the medial surface of the calcaneus and is firmly attached to both structures. Septa extend from the deep surface to the underlying bone and deltoid ligament to form osseofibrous tunnels transmitting structures from the posterior compartment of the leg. The **peroneal retinaculum** is subdivided into superior and inferior components and binds the peroneal tendons in place. Septa from the deep surface pass to the **peroneal trochlea,** forming two osseofibrous tunnels that separate the tendons of the peroneus longus muscle from the peroneus brevis. Both the superior and inferior peroneal retinacula extend from the lateral malleolus to the lateral surface of the calcaneus.

Cutaneous Innervation

Skin over the gluteal region is supplied by **lateral branches of the iliohypogastric** and **subcostal nerves** derived from the lumbar plexus and twelfth thoracic nerve (Fig. 6-2). Branches from the dorsal rami, the **superior** and **middle cluneal nerves,** supply the skin over the sacrum.

The **inferior cluneal nerves,** derived from the posterior femoral cutaneous nerve, pass superiorly over the lower border of the gluteus maximus to supply the skin of the inferior aspect of the buttocks (Fig. 6-3).

Cutaneous innervation of the anterior surface of the thigh is supplied by two branches of the femoral nerve, the **medial** and **intermediate femoral cutaneous nerves** (Fig. 6-4). The lateral aspect of the thigh is innervated by anterior and posterior branches of the **lateral femoral cutaneous nerve,** a direct branch of the lumbar plexus. The medial surface of the thigh is supplied by the medial femoral cutaneous and **cutaneous branches from the obturator and perineal nerves.** The **femoral branch** of the genitofemoral reaches a small area just below the inguinal ligament. The **genital branch of the genitofemoral,** together with the ilioinguinal nerve, distribute to the skin adjacent to the region of the superficial inguinal ring, the base of the penis and scrotum in the male, and the labia majora in the female. The posterior surface of the thigh is supplied by the **posterior femoral cutaneous nerve** from the sacral plexus.

Phantom Pain
Phantom pain is often felt by patients who have had a limb amputated. They experience pain or other sensations in the missing extremity as if the limb were still there. An explanation for this phenomenon is that sensory nerves that previously received impulses from the now-missing limb are being stimulated by irritation to the amputation stump. Stimuli from these nerves are interpreted by the brain as coming from the nonexistent (phantom) limb.

The femoral nerve continues into the leg as the **saphenous nerve.** It passes inferiorly to supply skin over the anteromedial and medial surfaces of the leg, the medial side of the foot, and a small area on the medial aspect of the sole. Its **infrapa-**
Text continues on page 249

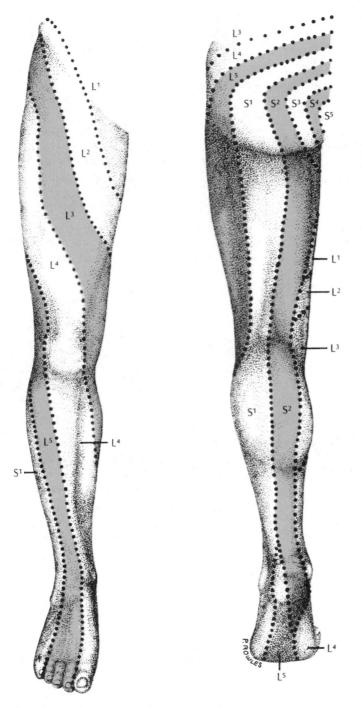

Figure 6-2. Dermatomes of lower extremity.

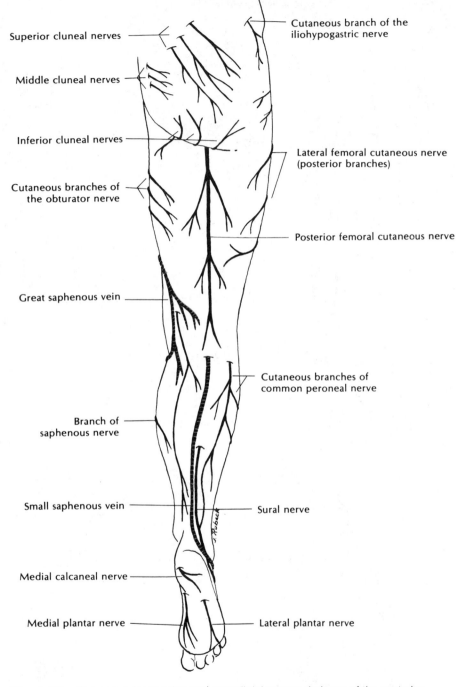

Superior cluneal nerves

Middle cluneal nerves

Inferior cluneal nerves

Cutaneous branches of
the obturator nerve

Great saphenous vein

Branch of
saphenous nerve

Small saphenous vein

Medial calcaneal nerve

Medial plantar nerve

Cutaneous branch of the
iliohypogastric nerve

Lateral femoral cutaneous nerve
(posterior branches)

Posterior femoral cutaneous nerve

Cutaneous branches of
common peroneal nerve

Sural nerve

Lateral plantar nerve

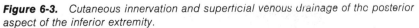

Figure 6-3. *Cutaneous innervation and superficial venous drainage of the posterior
aspect of the inferior extremity.*

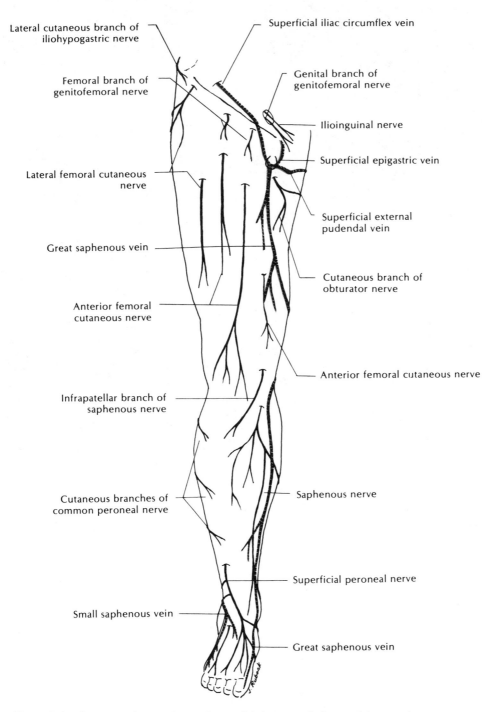

Lateral cutaneous branch of iliohypogastric nerve

Superficial iliac circumflex vein

Femoral branch of genitofemoral nerve

Genital branch of genitofemoral nerve

Ilioinguinal nerve

Lateral femoral cutaneous nerve

Superficial epigastric vein

Superficial external pudendal vein

Great saphenous vein

Cutaneous branch of obturator nerve

Anterior femoral cutaneous nerve

Anterior femoral cutaneous nerve

Infrapatellar branch of saphenous nerve

Cutaneous branches of common peroneal nerve

Saphenous nerve

Superficial peroneal nerve

Small saphenous vein

Great saphenous vein

Figure 6-4. *Cutaneous innervation and superficial venous drainage of the anterior aspect of the inferior extremity.*

tellar branch, arising in the thigh, pierces the lower end of the sartorious to ramify on the anterior surface of the leg immediately below the knee. The lateral surface of the upper leg receives its cutaneous innervation from the lateral sural branch of the **common peroneal.** The distal third of the anterior surface of the leg and the dorsum of the foot are innervated by a continuation of the **superficial peroneal nerve,** except for the adjacent sides of the great and second toe, which are supplied by a cutaneous branch of the **deep peroneal.** The **sural nerve,** derived from both tibial and common peroneal nerves, distributes to the posterior surface of the leg, the lateral side of the foot, the little toe, and a small portion of the posterolateral surface of the sole.

The **calcaneal branch of the tibial nerve** supplies most of the plantar surface of the heel, while the anterior two-thirds of the sole is innervated by the **medial** and **lateral plantar nerves,** both cutaneous branches of the tibial. Their areas of distribution to the sole are divided by an anteroposterior line passing through the midline of the fourth toe.

Venous Drainage

The **superficial venous drainage** of the inferior extremity is by way of two major vessels, the great and the small saphenous veins (see Figs. 6-3 and 6-4). The **great** (long) **saphenous vein** begins at the junction of the medial digital vein of the great toe and the dorsal venous arch of the foot. From its origin, it ascends anterior to the medial malleolus and along the medial border of the tibia to the knee, where it passes posterior to the medial epicondyle of the femur. It continues on the medial aspect of the thigh to the saphenous opening, where it pierces the cribriform fascia to empty into the femoral vein. In the inguinal region it receives the **superficial epigastric,** the **superficial iliac circumflex,** and the **superficial external pudendal veins.**

Intravenous Therapy
The long saphenous is readily available for intravenous entry. It is usually used only in emergency situations because of the greater risk of developing phlebitis when veins of the lower extremity are cannulated.

The **small** (short) **saphenous vein** begins laterally at the dorsal venous arch of the foot and ascends posterior to the lateral malleolus. It continues along the posterior aspect of the leg to the popliteal fossa, where it penetrates the deep fascia to empty into the **popliteal vein.**

The **deep venous drainage** is by way of the **venae comitantes** of the corresponding arteries. Numerous anastomoses occur between the great and small saphenous veins and between the superficial and deep venous drainage.

Varicose Veins of the Lower Extremity
One of the most common disorders of the vascular system is the dilation, elongation, and tortuosity of the superficial veins of the lower limb, called varicose veins. The principal cause of this affliction is an increased venous pressure with dilation of the veins, largely owing to valvular incompetence and gravity from the upright position. The increased venous pressure and varicosities cause edema of the skin and subcutaneous tissue with decreased blood flow and poor healing. This venous stasis and edema predisposes the skin to develp ulcers following minor skin abrasions or injury. If the deep veins are patent and can handle the venous return from the foot and leg, then surgical removal or stripping of the superficial varicosed veins decreases the venous pressure in the skin and subcutaneous tissue, reduces the edema, and allows the stasis ulcers to heal.

Gluteal Region

Muscles (Fig. 6-5; Table 6-1)

The gluteal region extends from the iliac crest superiorly to the gluteal fold inferiorly. The mass of the buttocks is formed by the **gluteus maximus,** the largest muscle of the body, which overlies the more deeply placed **gluteus medius** and **gluteus minimus muscles.** The gluteus maximus is the chief extensor and the most powerful lateral rotator of the thigh. The medius and minimus act as abductors and medial rotators. The small lateral rotator muscles of the thigh, the **piriformis, superior** and **inferior gemelli,** to-

gether with the tendons of the **obturator internus** and **externus,** and **quadratus femoris,** underlie the gluteus maximus, are related to the back of the hip joint, and insert into the greater trochanter of the femur.

Arteries and Nerves

The **superior gluteal artery** (Table 6-2), the largest branch of the internal iliac, courses posteriorly from its origin to pass between the lumbosacral trunk and the first sacral nerve. It leaves the

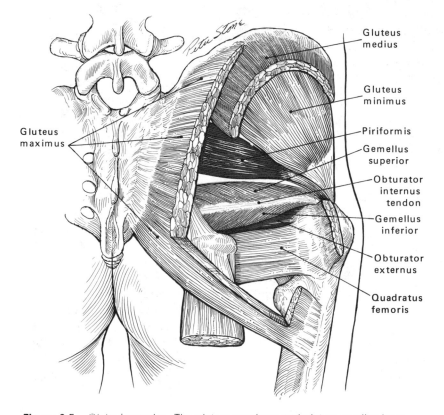

Figure 6-5. *Gluteal muscles. The gluteus maximus and gluteus medius have been cut to expose the muscles lying in the plane of the gluteus minimus.*

Table 6-1
Muscles of the Gluteal Region

Muscle	Origin	Insertion	Action	Nerve
Gluteus maximus	Upper portion of ilium, posterior aspect of sacrum and coccyx, and sacrotuberous ligament	Gluteal tuberosity and iliotibial tract	Chief extensor and powerful lateral rotator of thigh	Inferior gluteal
Gluteus medius	Ilium between middle gluteal line and iliac crest	Greater trochanter and oblique ridge of femur	Abducts and rotates thigh medially	Superior gluteal
Gluteus minimus	Ilium between middle and inferior gluteal lines	Greater trochanter and capsule of hip joint	Abducts and rotates thigh medially	Superior gluteal
Tensor fasciae latae	Iliac crest and anterior border of ilium	Iliotibial tract	Tenses fascia lata; assists in flexion, abduction, and medial rotation of thigh	Superior gluteal
Piriformis	Internal aspect of sacrum, greater sciatic notch, and sacrotuberous ligament	Upper part of greater trochanter	Rotates thigh laterally	Branches from S_1 and S_2
Gemelli superior and inferior	Superior, upper margin of lesser sciatic notch; inferior, lower margin of sciatic notch	Obliquely into either border of tendon of obturator internus and greater trochanter	Rotates thigh laterally	Branches from sacral plexus
Quadratus femoris	Lateral border of ischial tuberosity	Posterior aspect of greater trochanter and adjoining shaft of femur	Rotates thigh laterally	L_4, L_5, and S_1

pelvic cavity through the greater sciatic foramen above the level of the piriformis muscle (Fig. 6-6). It divides into a **superficial branch,** to supply the gluteus maximus from its deep surface, and a **deep branch** that passes between the gluteus medius and the gluteus minimus to supply both muscles, as well as the obturator internus, piriformis, levator ani, coccygeus muscles, tensor fasciae latae, and the hip joint.

The **inferior gluteal artery,** the larger of the two terminal branches of the internal iliac artery, passes posteriorly through the greater sciatic foramen below the level of the piriformis. It supplies the gluteus maximus and the lateral rotator muscles and gives a long slender branch that accompanies and supplies the sciatic nerve.

The **superior gluteal nerve** (L_4, L_5, and S_1) passes through the upper part of the greater sciatic foramen, courses anteriorly between the gluteus medius and the gluteus minimus to supply these two muscles, and the tensor fasciae latae. The **inferior gluteal nerve** (L_5, S_1, and S_2) leaves the pelvis by traversing the greater sciatic foramen, below the level of the piriformis, to supply the gluteus maximus (Table 6-3).

Text continues on page 254

Table 6-2
Arterial Supply to the Gluteal Region and Thigh

Artery	Origin	Course	Distribution	Anastomoses
Superior gluteal	Posterior division, internal iliac	Leaves greater sciatic foramen above level of piriformis	Gluteus maximus, medius, and minimus	Deep iliac circumflex, inferior gluteal, lateral femoral circumflex
Inferior gluteal	Anterior division, internal iliac	Leaves greater sciatic foramen below level of piriformis	Gluteus maximus, small lateral rotators of thigh	Superior gluteal, medial femoral circumflex, first perforating branch of profunda
Femoral	Continuation of external iliac artery distal to inguinal ligament	Passes through anterior compartment of thigh to adductor hiatus	Anterior compartment of thigh	No direct anastomoses
Deep external pudendal	Femoral	Courses on superficial aspect of pectineus muscle	Anterior compartment of thigh	Internal pudendal, medial femoral circumflex
Lateral femoral circumflex	Femoral	Passes laterally to divide into ascending, transverse and descending branches	Anterior compartment of thigh	Deep iliac circumflex, medial femoral circumflex, genicular branches of popliteal
Profunda femoral	Femoral	Descends in anterior compartment	Gives perforating branches to posterior compartment muscles	Medial femoral circumflex and genicular arteries
Medial femoral circumflex	Femoral	Passes between pectineus and adductor brevis to encircle proximal thigh	Anterior compartment of thigh	Lateral femoral circumflex, inferior gluteal, first perforating branch of profunda
Perforating branches	Profunda femoris	Four perforating branches pierce posterior intermuscular septum	Posterior compartment of thigh	First perforating with inferior gluteal; fourth perforating with genicular branches of popliteal
Obturator	Anterior division, internal iliac	Traverses obturator canal to reach medial compartment of thigh	Medial compartment of thigh	Superior geniculars

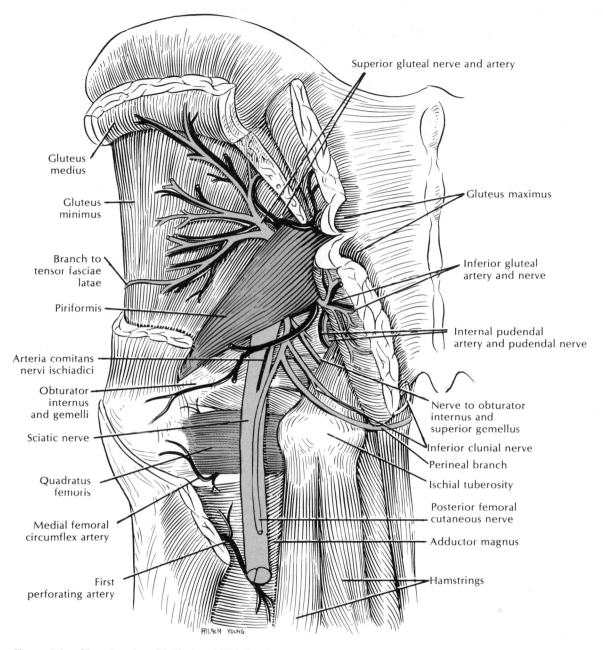

Gluteus medius

Gluteus minimus

Branch to tensor fasciae latae

Piriformis

Arteria comitans nervi ischiadici

Obturator internus and gemelli

Sciatic nerve

Quadratus femoris

Medial femoral circumflex artery

First perforating artery

Superior gluteal nerve and artery

Gluteus maximus

Inferior gluteal artery and nerve

Internal pudendal artery and pudendal nerve

Nerve to obturator internus and superior gemellus

Inferior clunial nerve

Perineal branch

Ischial tuberosity

Posterior femoral cutaneous nerve

Adductor magnus

Hamstrings

AILEEN YOUNG

Figure 6-6.   *Gluteal region. (Hollinshead WH: Textbook of Anatomy, 3rd ed. New York, Harper & Row, 1974)*

Table 6-3
Nerve Distribution to the Gluteal Region

Nerve	Origin	Course	Distribution
Cluneal nerves (superior, middle, inferior)	Dorsal rami of sacral nerves and twigs from posterior femoral cutaneous	Pass through posterior sacral foramina; inferior cluneal curves around gluteal fold	Skin over buttock
Posterior femoral cutaneous	Posterior and anterior divisions of sacral plexus	Passes through greater sciatic foramen, below level of piriformis; in thigh, lies on superficial aspect of biceps	Inferior cluneal branches to skin over buttock; skin over posterior aspect of thigh
Superior gluteal	Posterior divisions of sacral plexus	Passes through greater sciatic foramen, above level of piriformis, to course between gluteus medius and minimus	Gluteus medius, minimus, and tensor fasciae latae
Inferior gluteal	Posterior divisions of sacral plexus	Passes through greater sciatic foramen, below level of piriformis, to course on deep aspect of gluteus maximus	Gluteus maximus

Gluteal Injections

Since the gluteal region is a common site for injections, the sciatic nerve may be injured by a poorly placed intramuscular injection. Injury is usually avoided if the injection is made in the upper outer quadrant of the buttocks, which is far removed from the sciatic nerve and large blood vessels. Most nerve damage from injections affect the peroneal division of the sciatic nerve, which causes loss of power to dorsiflex the ankle or to extend the toes and is called foot drop. Other complications of parenteral injections include embolism, hematoma, abscess, intravascular injection of drugs, and sloughing of skin.

Thigh

Anterior Compartment of the Thigh

Femoral Sheath

The **femoral sheath** is a fasical funnel formed anteriorly by the endoabdominal fascia and posteriorly by the iliac fascia. It encloses the upper 3 cm to 4 cm of the femoral vessels and is situated deep to the inguinal ligament in a groove between the iliopsoas and the pectineus muscles. The interior of the sheath is divided into three compartments by two anteroposterior septa. The **medial compartment,** bounded medially by the crescentic base of the lacunar ligament, forms the **femoral canal** and contains only areolar connective tissue, a small lymph node, and lymphatic vessels. The

intermediate compartment encloses the **femoral vein,** and the **lateral compartment** encloses the **femoral artery** and the **femoral branch of the genitofemoral nerve.** Structures passing deep to the inguinal ligament, lateral to the femoral sheath, include the iliopsoas muscle and the femoral and lateral femoral cutaneous nerves.

Femoral Hernia

A femoral hernia results from the potential weakness of the lower abdominal wall at the femoral ring. This condition may allow entrance of a viscus, for example, a loop of small intestine, or a part of the omentum into the thigh. As the hernia sac passes through the femoral ring, which is medial to the femoral vein and just below the inguinal ligament, the hernia protrudes subcutaneously into the femoral triangle. If it continues to enlarge, it will often take a recurrent course upward over the inguinal ligament. The femoral hernia is always acquired and not congenital. It is formed most frequently in the female, probably due to the wider female pelvis and larger femoral canal. Loss of muscle tone of the abdominal musculature and stretching from multiple pregnancies are also predisposing factors in the higher incidence of femoral herniae in females.

Femoral Triangle

The **femoral triangle** covers the greater part of the upper third of the thigh (Fig. 6-7). Its superior **base** is formed by the inguinal ligament, and its **sides** by the medial border of the sartorius laterally and the medial border of the adductor longus medially. The above muscles meet at the **apex** inferiorly, where a narrow intermuscular cleft, the **adductor canal,** continues distally. The **roof** of the femoral triangle is formed by skin and superficial and deep fasciae. It contains superficial inguinal lymph nodes and vessels, the femoral branch of the genitofemoral nerve, superficial branches of the femoral artery, and the upper end of the great saphenous vein. The iliopsoas, adductor longus, and pectineus muscles form the **floor** of the triangle. Its **contents** include the **femoral artery** and **vein** passing from the base to the apex; the **deep external pudendal, profunda femoral,** and **lateral** and **medial femoral circumflex arteries;** the **femoral branch** of the genitofemoral, the **lateral femoral cutaneous** and **femoral nerves,** and several **inguinal lymph nodes.**

Femoral Vein Cannulation

The ease of access to the femoral vein makes it a preferred site for cannulation, especially during cardiopulmonary resuscitation (CPR) when, because of the chest compression, areas in the head and neck are difficult to stabilize. To find the vein the femoral artery is located as it extends into the thigh at the midpoint of the inguinal ligament. Placing your fingers on the pulsating artery, the needle may be introduced into the vein that lies medial to the artery as they traverse the femoral triangle.

Adductor Canal

At the apex of the femoral triangle, the femoral artery and vein continue inferiorly in the **adductor canal,** a muscular cleft between the vastus medialis and the adductors longus and magnus. It is deep to and covered by the sartorius muscle as it crosses the thigh obliquely. At the lowermost extent of the adductor canal, the posterior wall, formed by the adductor magnus, presents a deficiency, or opening, the **adductor hiatus,** which leads into the popliteal fossa. In addition to the femoral artery and vein, the canal contains the saphenous nerve and branches of the femoral nerve to the vastus medialis muscle.

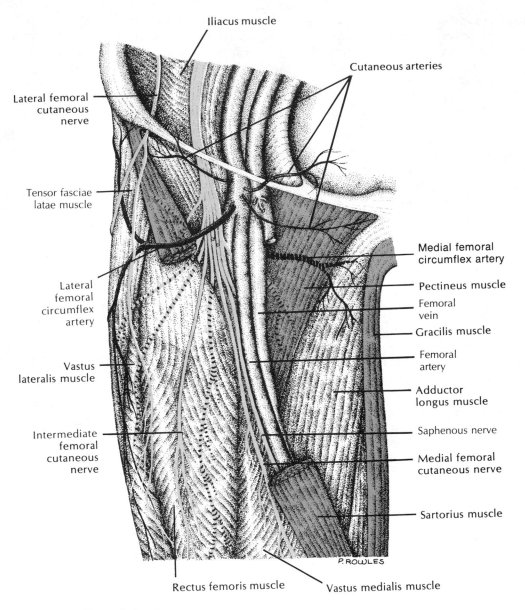

Iliacus muscle

Cutaneous arteries

Lateral femoral cutaneous nerve

Tensor fasciae latae muscle

Lateral femoral circumflex artery

Vastus lateralis muscle

Intermediate femoral cutaneous nerve

Medial femoral circumflex artery

Pectineus muscle

Femoral vein

Gracilis muscle

Femoral artery

Adductor longus muscle

Saphenous nerve

Medial femoral cutaneous nerve

Sartorius muscle

P. ROWLES

Rectus femoris muscle

Vastus medialis muscle

Figure 6-7. *Femoral triangle.*

Table 6-4
Muscles of the Anterior Compartment of the Thigh

Muscle	Origin	Insertion	Action	Nerve
Iliopsoas: Compound muscle formed from iliacus and psoas major				
Iliacus	Iliac fossa and lateral portion of sacrum	Lesser trochanter of femur by way of iliopsoas tendon	Flexes thigh	Femoral
Psoas major	Lumbar vertebrae	Lesser trochanter of femur by way of iliopsoas tendon	Flexes and rotates thigh medially	Second and third lumbar
Sartorius	Anterior superior iliac spine	Upper part of medial surface of tibia	Acts on both hip and knee joints; mainly flexes leg	Femoral
Quadriceps: Consists of rectus femoris and vasti muscles, the four muscles combining into an aponeurotic and tendinous insertion into tibial tuberosity, with patella interposed as sesamoid bone				
Rectus femoris	Straight head from anterior inferior iliac spine; reflected head from postero-superior aspect of rim of acetabulum	Tibial tuberosity	Extends leg and flexes thigh	Femoral
Vastus lateralis	Intertrochanteric line, greater trochanter, linea aspera, and lateral intermuscular septum	Tibial tuberosity	Extends leg	Femoral
Vastus medialis	Intertrochanteric line, spiral line, and medial intermuscular septum	Tibial tuberosity	Extends leg	Femoral
Vastus intermedius	Upper two-thirds of shaft of femur and distal one-half of lateral intermuscular septum	Tibial tuberosity	Extends leg	Femoral
Articularis genus	Variable slip of muscle on deep aspect of vastus intermedius that, in extension of knee, pulls synovial membrane out of the way of the articular surfaces.			

Muscles

The muscles of the anterior aspect of the thigh are the iliopsoas, the sartorius, and the quadriceps femoris (Table 6-4). The latter two muscles are contained within the **extensor (anterior) compartment,** which is limited by the lateral and medial intermuscular septa. The **sartorius** acts as a flexor of the leg and, together with the rectus femoris and iliopsoas, also flexes the thigh. The bulky **quadriceps femoris,** one of the largest and most powerful muscles in the body, consists of the **rectus femoris** and the **vastus lateralis, medialis,** and **intermedius.** As powerful extensors of the leg, the four muscles form an aponeurotic and tendinous insertion on the tibia, with the patella interposed as a sesamoid bone. Muscles in the extensor compartment are innervated by the femoral nerve and receive their blood supply from the femoral artery (Fig. 6-8).

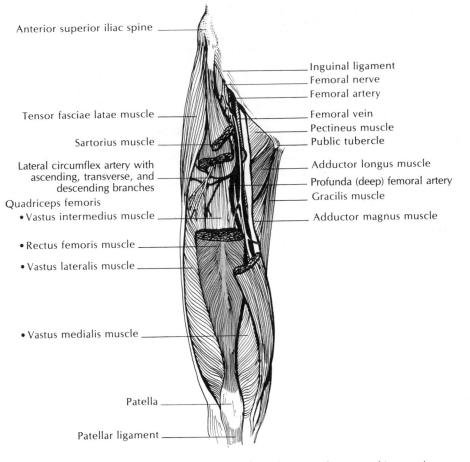

Anterior superior iliac spine

Inguinal ligament
Femoral nerve
Femoral artery

Tensor fasciae latae muscle

Femoral vein
Pectineus muscle
Public tubercle

Sartorius muscle

Lateral circumflex artery with ascending, transverse, and descending branches

Adductor longus muscle
Profunda (deep) femoral artery
Gracilis muscle

Quadriceps femoris

Adductor magnus muscle

• Vastus intermedius muscle

• Rectus femoris muscle

• Vastus lateralis muscle

• Vastus medialis muscle

Patella

Patellar ligament

Figure 6-8. *Anterior and medial views of thigh. Sartorious muscle removed to reveal contents of adductor canal; rectus femoris muscle resected to show principal arteries of anterior compartment.*

Arteries and Nerves

The **femoral artery,** the continuation of the external iliac, passes deep to the inguinal ligament to enter the femoral triangle (see Table 6-2). At the apex of the triangle it passes deep to the sartorius muscle to traverse the adductor canal. It becomes the **popliteal artery** at the adductor hiatus. Branches within the femoral triangle include the **superficial epigastric artery,** which pierces the femoral sheath to ascend superficial to the inguinal ligament and courses toward the umbilicus; the **superficial iliac circumflex,** which penetrates the femoral sheath to run in the subcutaneous tissue toward the anterior superior iliac spine; the **superficial external pudendal,** which supplies skin and muscles in the inguinal region and gives anterior scrotal or labial branches to the skin of the external genitalia, and the **deep external pudendal,** which lies on and supplies the pectineus and adductor longus muscles, and sends twigs to the external genitalia.

The large **profunda branch** of the femoral artery arises from its posterolateral aspect and passes inferiorly, posterior to the adductor longus. It gives rise to the medial and lateral circumflex arteries, then courses deeply to give origin to the **four perforating branches.** These perforating vessels supply the adductor and the flexor muscle groups. The **medial femoral circumflex artery** courses between the pectineus and the iliopsoas, then around the neck of the femur, where it divides into a **superficial** and a **deep branch** to supply muscles in the region. The **lateral femoral circumflex** passes laterally, deep to the sartorius and rectus femoris, where it divides into ascending, transverse, and descending branches. The **ascending branch** distributes to the gluteal region. The **transverse branch** encircles the femur to anastomose with the medial femoral circumflex, while the **descending branch** passes inferiorly to terminate in the anastomosis around the knee joint.

The **femoral nerve,** the largest branch of the lumbar plexus, emerges from the lateral side of the psoas major muscle just below the iliac crest to descend between the psoas and the iliacus muscles. It passes deep to the inguinal ligament, in the lateral neuromuscular compartment, to enter the thigh and divide into a number of branches in the femoral triangle. Its terminal and longest branch, the **saphenous nerve,** passes through the adductor canal to become superficial at the adductor hiatus and course between the sartorius and gracilis muscles to supply the skin of the leg and the medial side of the foot. Branches of the femoral nerve in the abdomen include the **nerve to the iliacus** and the **nerve to the psoas.** In the thigh, it gives **muscular branches** to the sartorius, rectus femoris, and vasti muscles, and twigs to the pectineus, as well as the **intermediate** and **medial femoral cutaneous branches.**

Adductor Compartment of the Thigh

Muscles (Table 6-5)

The adductor muscles, arranged in three layers, occupy the medial (adductor) compartment which is limited by the medial and posterior intermuscular septa (Fig. 6-9). The **pectineus** and **adductor longus** are the most superficial; the

adductor brevis lies intermediate, and the **adductor magnus** occupies the deepest stratum. The **gracilis** lies superficially on the medial aspect of the thigh. The two divisions of the obturator nerve are interposed between the three muscular layers and supply the muscles in the adductor compartment. All three adductors act in adduction and medial rotation of the thigh. The pectineus and gracilis, in addition to adduction, act as flexors. The pectineus aids in flexion of the thigh, and the gracilis in flexion of both the thigh and the leg. The gracilis also assists in medial rotation of the leg.

Arteries and Nerves

The **obturator artery** is a branch of the internal iliac that arises in the pelvis and accompanies the obturator nerve through the obturator canal. Here it divides into anterior and posterior branches that follow the margins of the obturator foramen deep to the obturator externus (Fig. 6-10). The **anterior branch** supplies the obturator externus, the pectineus, the adductors, and the gracilis; the **posterior branch** is distributed to muscles attaching to the ischial tuberosity and gives off a branch to supply the head of the femur.

The **obturator nerve** originates in the abdomen as a branch of the lumbar plexus. It emerges from the medial surface of the psoas major muscle to pass along the lateral wall of the pelvic cavity, through the obturator canal, and divide into anterior and posterior divisions as it enters the thigh. The divisions pass to either side of the adductor brevis muscle. The **anterior division** lies between the adductors longus and brevis, supplies these muscles, and sends branches to the gracilis and pectineus muscles. The **posterior division** passes between the adductors brevis and magnus, supplying these muscles as well as the obturator externus. The pectineus and the adductor magnus each have dual innervations. The pectineus receives additional innervation from the **femoral nerve.** The adductor magnus acquires a second nerve supply from the **tibial division** of the sciatic nerve.

Table 6-5
Muscles of the Adductor Compartment of the Thigh

Muscle	Origin	Insertion	Action	Nerve
Adductor longus	Body of pubis immediately below pubic crest	Linea aspera of femur	Adducts, flexes, and rotates thigh medially	Obturator
Adductor brevis	Body of pubis below origin of adductor longus	Between lesser trochanter and linea aspera and upper part of linea aspera	Adducts, flexes, and rotates thigh medially	Obturator
Adductor magnus	Side of pubic arch and ischial tuberosity	Extensive into linea aspera, medial supracondylar ridge, and adductor tubercle	Adducts, flexes, and rotates thigh medially; distal fibers aid in extension and lateral rotation of thigh	Obturator; distal portion by sciatic
Pectineus	Pectineal line and pectineal surface of pubis	Posterior aspect of femur between lesser trochanter and linea aspera	Adducts and assists in flexion of thigh	Obturator and femoral
Gracilis	Lower half of body of pubis	Upper part of medial surface of tibia	Adducts thigh; flexes knee joint and rotates leg medially	Obturator
Obturator externus	Margins of obturator foramen and obturator membrane	Posterior aspect of intertrochanteric fossa of femur	Flexes and rotates thigh laterally	Obturator

Posterior Compartment of the Thigh

Muscles

The posterior compartment of the thigh is located between the posterior and lateral intermuscular septa and contains the flexor, or hamstring, muscles (Table 6-6). Muscles of this compartment are innervated by the sciatic nerve; the short head of the biceps receiving its nerve supply from the common peroneal division, and the remaining muscles from the tibial division. The **biceps** arises by a long and short head; the **semimembranosus** has a long membranous origin, and the **semitendinosus** a long tendinous insertion. The long head of the biceps, the semimembranosus, and the semitendinosus span both hip and knee joint, and all act as flexors of the leg and extensors of the thigh. They also act in rotation of the leg and thigh, with the biceps rotating laterally and the semimembranosus and semitendinosus rotating medially.

Arteries and Nerves

No major artery courses through the posterior compartment (Fig. 6-11). The blood supply is derived from the previously described **perforating branches** of the large profunda femoris branch of the femoral artery.

The **sciatic nerve,** the principal branch of the sacral plexus, is the largest nerve of the body. It enters the thigh inferior to the piriformis muscle, passing through the greater sciatic foramen to de-

Text continues on page 264

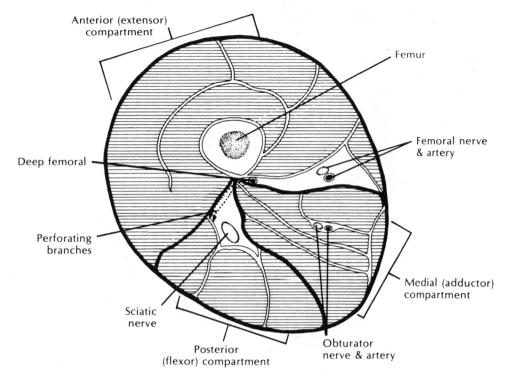

Figure 6-9. *Compartments of the thigh.*

Table 6-6
Muscles of the Posterior Compartment of the Thigh

Muscle	Origin	Insertion	Action	Nerve
Biceps femoris	Long head, common tendon with semitendinosus from ischial tuberosity; short head, linea aspera and upper half of supracondylar ridge of femur	Common tendon into head of fibula	Flexes knee; rotates leg laterally; long head extends hip	Long head by tibial portion, short head by peroneal portion of sciatic
Semitendinosus	In common with long head of biceps from ischial tuberosity	Upper part of medial surface of tibia	Flexes knee; rotates leg medially; extends hip joint	Tibial portion of sciatic
Semimembranosus	Ischial tuberosity	Medial condyle of tibia	Extends hip joint; flexes knee; rotates leg medially	Tibial portion of sciatic

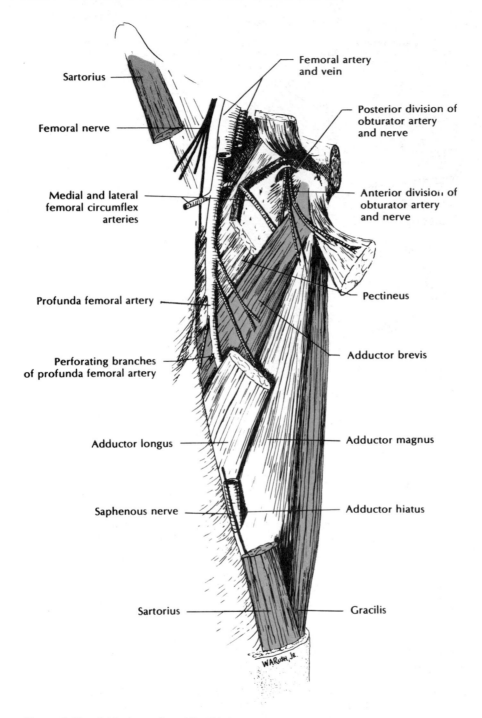

Sartorius

Femoral nerve

Medial and lateral
femoral circumflex
arteries

Profunda femoral artery

Perforating branches
of profunda femoral artery

Adductor longus

Saphenous nerve

Sartorius

Femoral artery
and vein

Posterior division of
obturator artery
and nerve

Anterior division of
obturator artery
and nerve

Pectineus

Adductor brevis

Adductor magnus

Adductor hiatus

Gracilis

WARush, Jr.

Figure 6-10. *Adductor region of the thigh.*

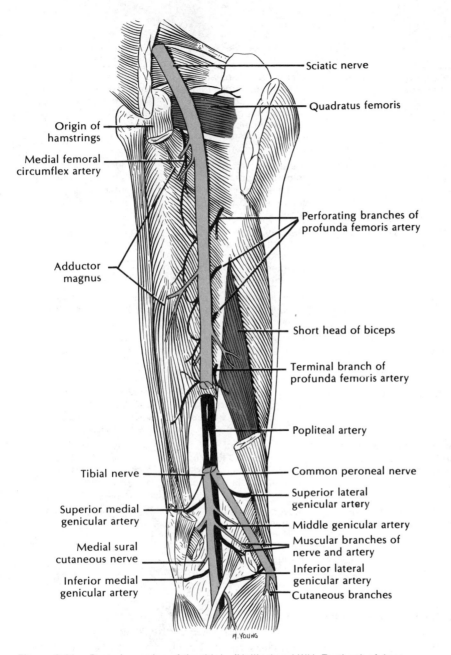

Sciatic nerve

Quadratus femoris

Origin of hamstrings

Medial femoral circumflex artery

Perforating branches of profunda femoris artery

Adductor magnus

Short head of biceps

Terminal branch of profunda femoris artery

Popliteal artery

Tibial nerve

Common peroneal nerve

Superior lateral genicular artery

Superior medial genicular artery

Middle genicular artery

Muscular branches of nerve and artery

Medial sural cutaneous nerve

Inferior lateral genicular artery

Inferior medial genicular artery

Cutaneous branches

A. YOUNG

Figure 6-11. *Posterior region of the thigh. (Hollinshead WH: Textbook of Anatomy, 3rd ed. New York, Harper & Row, 1974)*

Table 6-7
Nerve Distribution to Thigh

Nerve	Origin	Course	Distribution
Genitofemoral	Anterior divisions of lumbar plexus	Pierces and runs on anterior surface psoas major; genital branch passes through superficial inguinal ring, femoral branch courses with femoral artery in femoral sheath	Genital branch supplies cremaster and skin of scrotum (labium majus); femoral branch-skin over femoral triangle
Ilioinguinal nerve	Anterior divisions of lumbar plexus	Passes in cleft between internal abdominal oblique and transversus abdominis to traverse inguinal canal	Skin over femoral triangle, scrotum (labium majus), and pubic symphysis
Lateral femoral cutaneous	Posterior divisions of lumbar plexus	Crosses posterior abdominal wall to pass deep to inguinal ligament 2 to 3 cm medial to the anterior superior iliac spine	Skin of lateral aspect of thigh
Medial and intermediate femoral cutaneous	Femoral	Arise in femoral triangle to course in superficial fascia	Skin of medial and anterior aspect of thigh
Femoral	Posterior divisions of lumbar plexus	Crosses posterior abdominal wall to pass deep to inguinal ligament in neuromuscular compartment	Muscles of anterior (extensor) compartment
Obturator	Anterior divisions of lumbar plexus	From deep to psoas major, crosses lateral wall of pelvic cavity to traverse obturator canal	Muscles of medial (adductor) compartment, twigs to pectineus, and skin over medial aspect of thigh
Tibial	Anterior divisions of sacral plexus as component of sciatic	Traverses greater sciatic foramen below level of piriformis, lies on adductor magnus as it passes through thigh	Semimembranosus, semitendinosus, long head of biceps and twigs to adductor magnus
Common peroneal	Posterior divisions of sacral plexus as component of sciatic	Traverses greater sciatic foramen below level of piriformis, lies on adductor magnus as it passes through thigh	Short head of biceps

scend between the gluteus maximus muscle, and the gemelli muscles, the obturator internus tendon, and the quadratus femoris muscle. As it continues distally through the thigh, it lies on the adductor magnus and is crossed obliquely by the long head of the biceps femoris. It terminates by dividing into **tibial** and **common peroneal nerves.** This division may occur anywhere from its origin in the pelvis to the popliteal fossa, but usually takes place as the sciatic nerve enters the popliteal fossa. Branches from the common peroneal division supply the short head of the biceps. The tibial division supplies the long head of the biceps, the semimembranosus, thc semitendinosus, and the adductor magnus muscles (Table 6-7).

Herniated Disc

A protrusion of the nucleus pulposus with concomitant pressure on an emerging spinal nerve is the most common cause of pain in the lower back. The pain is due to a spasm of the intrinsic muscles of the back. The muscle spasm will result in a deviation of the vertebral column toward the affected side. The level of the disc involved may be ascertained by the deficit of sensation to the involved dermatome of the lower extremity or by loss of specific muscle reflexes.

Leg

Objectives

At the completion of the study of the leg and popliteal fossa the student should be able to

 Describe the retinacula of the knee and ankle and delineate the compartments of the leg

▶ *List the muscles in each of the compartments of the leg and give their actions*

▶ *Delineate the course and distributions of the common peroneal and tibial nerves, and give their branches*

▶ *Describe the distribution of the branches of the popliteal artery*

In the leg, as in the thigh, the muscles are located in fascial compartments. The **lateral (peroneal) compartment** is limited by the **anterior** and **posterior intermuscular septa;** the **anterior (extensor) compartment** is between the **anterior intermuscular septum** and the **tibia;** and the **posterior (flexor) compartment** is bounded by the **tibia** and the **posterior intermuscular septum.** Passing between the tibia and the fibula, the deeply lying **interosseous membrane** separates the extensor from the flexor compartment (Fig. 6-12). The latter is further divided by two **transverse intermuscular septa** into a deep subdivision containing the tib-ialis posterior muscle, a superficial subdivision containing the gastrocnemius and soleus muscles, and an intermediate compartment containing the remaining muscles, the posterior tibial artery, and the tibial nerve. Each of the three major compartments transmits a major nerve that supplies muscles within the compartment, but, as in the thigh, one compartment, the lateral, has no major artery coursing through it.

Popliteal Fossa

The **popliteal fossa** is a diamond-shaped area on the posterior aspect of the knee joint, extending from the lower third of the femur to the upper part of the tibia (Fig. 6-13). With the knee flexed this area presents a depression, but with the joint fully extended forms a slight posterior bulge. The fossa is **bounded** superiorly and laterally by the biceps femoris; superiorly and medially by the semitendinosus, semimembranosus, gracilis, sartorius, and adductor magnus, and inferiorly by the converging heads of the gastrocnemius and the laterally placed plantaris muscle. The **floor** is formed from superior to inferior by the popiteal surface of the femur, the oblique popliteal ligament, the popliteus muscle, and the upper part of the tibia. The fossa contains the common peroneal and tibial nerves, the popliteal artery and vein, the

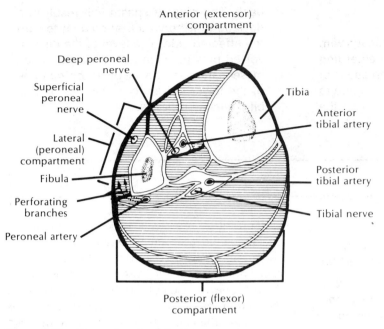

Figure 6-12. *Compartments of the leg.*

posterior femoral cutaneous nerve, the small saphenous vein, lymph nodes, and numerous synovial bursae.

The **popliteal artery** is the direct continuation of the femoral at the adductor hiatus. It descends through the popliteal fossa to terminate at the lower border of the popliteus muscle by dividing into the **anterior** and **posterior tibial arteries.** Branches of the popliteal artery within the fossa consist of several **genicular branches** that join in the anastomosis around the knee joint.

Anterior Compartment of the Leg

Muscles (Table 6-8)

Within the anterior compartment (Fig. 6-14) the superficially located **tibialis anterior muscle** lies along the lateral side of the tibia, adjacent to the interosseous membrane. It acts in dorsiflexion and inversion of the foot and, with the tibialis posterior and the peroneus longus, functions to maintain the longitudinal arch of the foot. The long, thin **extensor digitorum longus** is situated along the fibula and acts to extend the toes and secondarily to dorsiflex and evert the foot. Usually, the **peroneus tertius** can be distinguished from the extensor digitorum longus only by its insertion into deep fascia on the foot or to the bases of the fourth and fifth metatarsals. The **extensor hallucis longus,** deeply located between the tibialis anterior and the extensor digitorum longus, extends the great toe and aids in dorsiflexion of the foot. Muscles in the anterior compartment are innervated by the deep peroneal (anterior tibial) nerve and receive their blood supply from the anterior tibial artery. The tibialis anterior usually receives additional innervation directly from the common peroneal.

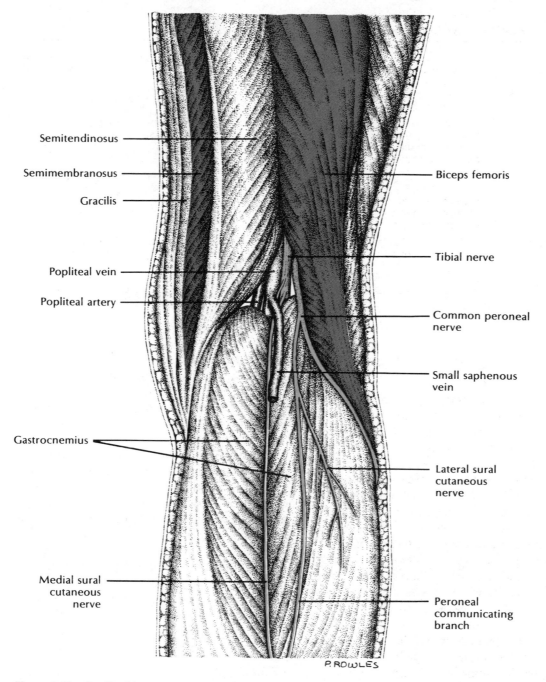

Semitendinosus

Semimembranosus

Gracilis

Popliteal vein

Popliteal artery

Gastrocnemius

Medial sural
cutaneous
nerve

Biceps femoris

Tibial nerve

Common peroneal
nerve

Small saphenous
vein

Lateral sural
cutaneous
nerve

Peroneal
communicating
branch

P. ROWLES

Figure 6-13. *Popliteal fossa.*

Table 6-8
Muscles of the Anterior Compartment of the Leg

Muscle	Origin	Insertion	Action	Nerve
Tibialis anterior	Lateral condyle of upper two-thirds of tibia and interosseous membrane	First cuneiform and first metatarsal	Dorsiflexes and inverts foot	Deep peroneal
Extensor digitorum longus	Lateral condyle of tibia, upper three-fourths of fibula, and interosseous membrane	By four tendons that form membranous expansions over metatarsophalangeal joints of second to fifth toes	Extends toes; continued action dorsiflexes and everts foot	Deep peroneal
Peroneus tertius	Distal one-fourth of fibula and interosseous membrane	Fifth metatarsal or deep fascia of foot	Dorsiflexes and everts foot	Deep peroneal
Extensor hallucis longus	Middle half of fibula and interosseous membrane	Base of distal phalanx of great toe	Extends great toe; aids in dorsiflexion and inversion of foot	Deep peroneal

Arteries and Nerves (Table 6-9)

At the bifurcation of the popliteal artery, its **anterior tibial branch** passes distally between the heads of origin of the tibialis posterior muscle. It then pierces the proximal portion of the interosseous membrane to course inferiorly on the anterior surface of the membrane, passing between the tibialis anterior and the extensor digitorum longus proximally, and the tibialis anterior and the extensor hallucis longus distally. The anterior tibial artery continues under the inferior extensor retinaculum as the **dorsalis pedis artery.** In addition to supplying the muscles of the anterior compartment, its branches include the **posterior tibial recurrent,** which passes superiorly between the popliteal ligament and the popliteus muscle to join the genicular anastomosis, and the **anterior tibial recurrent,** which crosses the interosseous membrane to pass superiorly and joins the genicular anastomosis. At the ankle the **medial anterior malleolar branch** passes deep to the tendon of the tibialis anterior to anastomose with branches of the posterior tibial artery, whereas the **lateral anterior malleolar** winds around the lateral malleolus deep to the extensor digitorum longus and peroneus tendons to join the malleolar branches of the posterior tibial and dorsalis pedis arteries.

The **deep peroneal** (anterior tibial) **nerve,** a branch of the common peroneal, courses around the head of the fibula to pass deep to the peroneus longus muscle, pierces the extensor digitorum longus, and descends on the interosseous membrane in the anterior compartment. It passes under the inferior extensor retinaculum to become the **medial** and **lateral digital branches.** It gives twigs to the peroneus longus, supplies muscles in the anterior compartment, the extensor digitorum brevis on the dorsum of the foot, and its most distal branch supplies the skin of the contiguous sides of the great and second toe.

Foot Drop

Disruption of the nerve supply to the extensor muscles in the leg will result in a

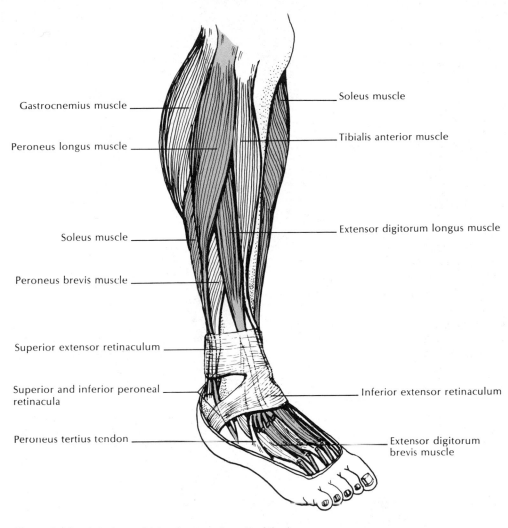

Gastrocnemius muscle

Peroneus longus muscle

Soleus muscle

Peroneus brevis muscle

Superior extensor retinaculum

Superior and inferior peroneal retinacula

Peroneus tertius tendon

Soleus muscle

Tibialis anterior muscle

Extensor digitorum longus muscle

Inferior extensor retinaculum

Extensor digitorum brevis muscle

Figure 6-14. *Anterior and lateral compartments of the leg.*

"foot drop." Upon examination, the patient will be unable to extend the foot against resistance. Usually this is due to a herniated disc or to trauma to the common peroneal nerve. The superficial position of the common peroneal as it winds around the neck of the fibula makes it vulnerable to injury. It is the most frequently damaged nerve in the lower extremity.

Lateral Compartment of the Leg

Muscles (Table 6-10)

Two muscles, the **peroneus longus** and the **peroneus brevis,** are located in the lateral compartment of the leg (see Fig. 6-14). Both are supplied by the superficial peroneal nerve, with the

Table 6-9
Arterial Supply to the Leg and Foot

Artery	Origin	Course	Distribution	Anastomoses
Popliteal	Continuation of femoral artery at adductor hiatus	Passes from adductor hiatus through popliteal fossa to leg	Superior, middle, and inferior geniculars to both lateral and medial aspect of knee	Obturator, lateral femoral circumflex, fourth perforating
Anterior tibial	Popliteal	Courses above interosseous membrane to anterior compartment of leg	Anterior compartment of leg	No direct anastomoses
Dorsalis pedis	Continuation of anterior tibial artery distal to inferior extensor retinaculum	Descends on extensor digitorum brevis to first interosseous space	Muscles on dorsum of foot; pierces first dorsal interosseous to contribute to formation of plantar arch	Lateral plantar
Posterior tibial	Popliteal	Passes through the posterior compartment of leg; distal to flexor retinaculum divides into medial and lateral plantars	Posterior and lateral compartment of leg	Peroneal
Peroneal	Posterior tibial	Descends in posterior compartment adjacent to posterior intermuscular septum	Posterior compartment; perforating branches supply lateral compartment of leg	Posterior tibial
Medial plantar	Posterior tibial	Courses along medial aspect of plantar surface of foot	Plantar surface of foot	Lateral plantar
Lateral plantar	Posterior tibial	Courses along lateral aspect of plantar surface of foot; adjacent to metacarpophalangeal joints joins dorsalis pedis to form plantar arch	Lateral aspect of plantar surface	Dorsalis pedis and dorsal digital branches
Plantar digitals	Plantar arch	Common plantar digitalis pass to webbings of the toe; bifurcate into proper plantar digitals	Muscles in anterior aspect of foot; tendons to the toes	Dorsal digitals

Table 6-10
Muscles of the Lateral Compartment of the Leg

Muscle	Origin	Insertion	Action	Nerve
Peroneus longus	Lateral condyle of tibia, head, and upper two-thirds of fibula	First metatarsal and first cuneiform	Everts and aids in plantar flexion of foot	Superficial peroneal
Peroneus brevis	Lower two-thirds of fibula	Base of fifth metatarsal	Everts and aids in plantar flexion of foot	Superficial peroneal

peroneus longus frequently receiving additional innervation from the common peroneal. Both muscles act as evertors and aid in plantar flexion of the foot. The tendons of the peroneus longus and the tibialis anterior form a sling by their manner of insertion and give support for the longitudinal arch of the foot.

Arteries and Nerves

No major artery courses in the lateral compartment of the leg (see Table 6-9). The muscles are supplied by perforating twigs of the **peroneal branch** of the posterior tibial artery located within the posterior compartment.

The **superficial peroneal** (musculocutaneous) **nerve,** a branch of the common peroneal, passes between the extensor digitorum longus and the peronei muscles to descend in the lateral compartment. Within the compartment it supplies the peroneus longus and peroneus brevis muscles. In the distal third of the leg it becomes superficial to supply skin on the anterior surface of the leg and the dorsum of the foot.

Posterior Compartment of the Leg

Muscles (Table 6-11)

The deeply placed **tibialis posterior muscle** is the principal invertor of the foot and is separated from the **flexor digitorum longus** and the **flexor hallucis longus muscles** by the deeper of two transverse intermuscular septa (Fig. 6-15). The tibialis posterior aids in plantar flexion of the foot, and both flexors assist in inversion of the foot. The **gastrocnemius** and the **soleus** are separated from the intermediate flexor muscles by the more superficial transverse intermuscular septum. Inserting by the common **calcaneal tendon** (of Achilles) into the calcaneus, the gastrocnemius and the soleus are the strongest of the plantar flexors of the foot.

Calcaneal Bursitis
Calcaneal bursitis is an inflammation of the bursa located between the calcaneal tendon and the upper part of the calcaneus. It is fairly common in long distance runners.

Two small muscles within the popliteal region complete the flexor group of muscles, namely, the **popliteus,** which performs the important initial "unlocking action" in flexion of the knee, and the relatively unimportant **plantaris,** which sends its long slender tendon to insert into the medial side of the calcaneal tendon.

Trick Knee
In the normal gait, the final action in extension of the leg (knee) is a lateral rotation of the tibia that "locks" the femur and tibia so that the extremity becomes a rigid, weight-bearing pillar. Intermittently in some patients, following knee injuries,

Table 6-11
Muscles of the Posterior Compartment of the Leg

Muscle	Origin	Insertion	Action	Nerve
Gastrocnemius	Lateral head, lateral condyle of femur; medial head, popliteal surface and medial condyle of femur	With soleus through calcaneal tendon into posterior surface of calcaneus	Plantar flexes foot and flexes knee	Tibial
Soleus	Upper one-third of fibula, soleal line on tibia	With gastrocnemius through calcaneal tendon into posterior surface of calcaneus	Plantar flexes foot	Tibial
Plantaris	Popliteal surface of femur above lateral head of gastrocnemius	Into medial side of calcaneal tendon	Plantar flexes foot	Tibial
Popliteus	Popliteal groove, lateral condyle of femur	Tibia above soleal line	With knee fully extended, rotates femur laterally	Tibial
Flexor digitorum longus	Middle one-half of tibia below soleal line	Tendon divides into four tendons, which insert into distal phalanges of four lateral toes	Flexes phalanges of four lateral toes; continued action plantar flexes and inverts foot	Tibial
Flexor hallucis longus	Lower two-thirds of fibula and intermuscular septa	Base of distal phalanx of great toe	Flexes great toe; continued action plantar flexes and inverts foot	Tibial
Tibialis posterior	Interosseous membrane and adjoining tibia and fibula	Into tuberosity of navicular with slips to cuneiforms, cuboid, and bases of second, third, and fourth metatarsals	Principal invertor of foot; plantar flexes foot	Tibial

this action is not finalized; the knee "gives way" when weight is placed on it, and the person falls.

Arteries and Nerves

The **posterior tibial artery** (see Table 6-9), the larger terminal branch of the popliteal, passes distally in the intermediate portion of the posterior compartment of the leg on the superficial aspect of the flexor digitorum longus. At the ankle it passes deep to the flexor retinaculum where it divides into the **lateral** and **medial plantar arteries.** At the knee it gives off the **fibular branch,** which passes laterally toward the head of the fibula, giving twigs to the peroneus longus and peroneus brevis muscles as it ascends to join the genicular anastomosis. The large **peroneal branch** courses laterally, between the flexor hallucis longus and the fibula, to give a nutrient branch to the fibula, a communicating branch to the posterior tibial, and perforating branches to

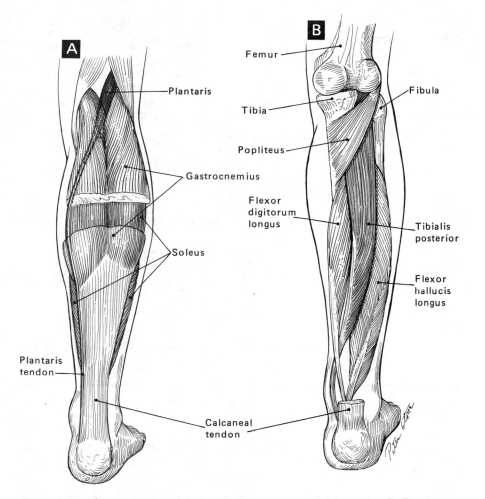

Figure 6-15. *Flexor muscles of the leg.* **A.** *The more superficial muscles.* **B.** *The deep flexors.*

the peroneal compartment. Its **branch to the tibia** is the largest nutrient artery in the body. At the ankle joint **medial** and **posterior malleolar** and **medial calcaneal branches** join, respectively, the malleolar and calcaneal anastomoses.

From the popliteal fossa the **tibial** (posterior tibial) **nerve** passes over the tendinous arch of the soleus to descend into the intermediate divi-

sion of the posterior compartment of the leg. Initially, it lies on the tibialis posterior, then on the flexor digitorum longus as it descends to divide under the flexor retinaculum to become the **medial** and **lateral plantar nerves.** It gives **muscular branches** to all the muscles of the posterior compartment as well as a **communicating branch to the sural nerve** (Table 6-12).

Table 6-12
Nerve Distribution to Leg

Nerve	Origin	Course	Distribution
Saphenous	Continuation of femoral in femoral triangle	Passes deep to sartorius to reach leg	Skin of medial aspect of leg and foot
Sural	Derived from both common peroneal and tibial	Tibial component passes between heads of gastrocnemius, joins common peroneal component in middle third of leg	Skin of posterior and lateral aspect of leg and foot
Tibial	Branch of sciatic in popliteal fossa	Courses through leg in intermediate portion of posterior compartment	Muscles in posterior (flexor) compartment, contributes to sural
Superficial peroneal	Branch of common peroneal in popliteal fossa	Winds around neck of fibula to course deep to peroneus longus	Muscles of lateral (peroneal) compartment, skin on distal third of leg
Deep peroneal	Branch of common peroneal in popliteal fossa	Courses through leg deep to extensor digitorum longus	Muscles of anterior (extensor) compartment

Foot

Objectives

At the completion of the study of the foot the student should be able to

▶ *Label on a diagram the cutaneous nerves of the foot*

▶ *List the muscles of the great toe, little toe, and intermediate group*

▶ *Describe the course and distribution of the lateral and medial plantar arteries and nerves*

Plantar Aponeurosis

The plantar aponeurosis is a sheet of deep fascia of great strength and importance. It is divided into medial, intermediate, and lateral portions, differentiated by their density and demarcated superficially by two shallow grooves that extend longitudinally along the foot. At the divisional lines septa pass from the deep aspect of the plantar aponeurosis to the deeper structures of the foot. The posterior end of the intermediate portion is narrow and attaches to the medial tubercle of the calcaneus. It widens as it extends forward and, near the heads of the metatarsals, divides into five slips, which pass to each of the five toes. This portion forms a strong tie-beam, especially for the great toe, between the calcaneus and each proximal phalanx.

Muscles (Table 6-13)

The muscles within the sole of the foot function basically as groups and are important in posture, locomotion, and support of the arches of the foot (Fig. 6-16). The **plantar muscles** are described in **four layers** and arranged in **three groups,** a medial group for the great toe, a lateral group for the small toe, and an intermediate group.

The **superficial layer** includes the **abductor hallucis, abductor digiti minimi,** and the

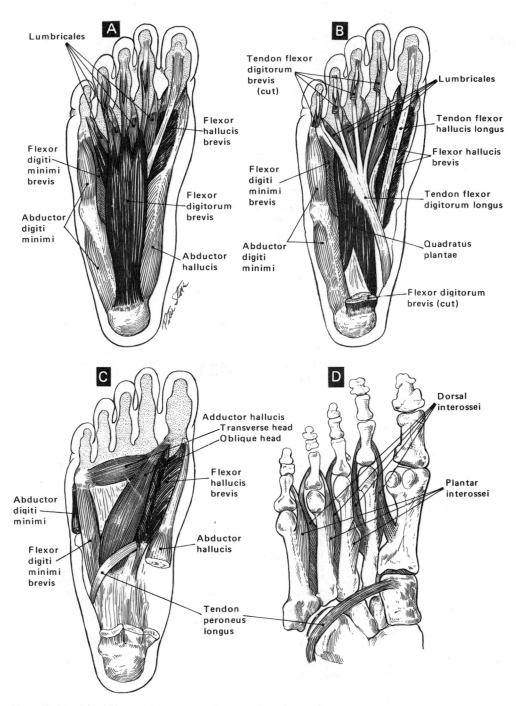

Figure 6-16. *Muscles of the foot. In the four drawings successively deeper layers of muscles are shown.*

Table 6-13
Muscles of the Foot

Muscle	Origin	Insertion	Action	Nerve
Extensor digitorum brevis (only muscle on dorsum of foot)	Dorsal surface of calcaneus	Divides into four tendons which insert into tendons of extensor digitorum longus of four medial toes	Dorsiflexes toes	Deep peroneal
Abductor hallucis	Medial tubercle of calcaneus	With medial belly of flexor hallucis brevis into proximal phalanx of great toe	Abducts and aids in flexion of great toe	Medial plantar
Flexor digitorum brevis	Medial tubercle of calcaneus and plantar fascia	Divides into four tendons that enter flexor sheath and split to admit passage of long flexor tendons before inserting into middle phalanx of lateral four toes	Flexes lateral four toes	Medial plantar
Abductor digiti minimi	Medial and lateral tubercles of calcaneus	Lateral side of proximal phalanx of little toe	Abducts little toe	Lateral plantar
Quadratus plantae	Medial head, medial side of calcaneus and plantar fascia; lateral head, lateral margin of calcaneus and plantar fascia	Tendons of flexor digitorum longus	Assists flexor digitorum longus	Lateral plantar
Lumbricals (4)	Tendons of flexor digitorum longus	Medial side of base of proximal phalanx of lateral four toes and extensor expansion	Aid interossei in flexion of metatarsophalangeal joints; extend distal two phalanges	First by medial plantar; lateral three by lateral plantar
Flexor hallucis brevis	Cuboid and third cuneiform	Divides into two tendons; medial, into base of proximal phalanx of great toe with abductor hallucis; lateral, with adductor hallucis	Flexes great toe	Medial plantar
Adductor hallucis Oblique head	Anterior end of plantar ligament and sheath of peroneus longus	With lateral belly of flexor hallucis brevis into proximal phalanx of great toe	Adducts and flexes great toe	Lateral plantar
Transverse head	Capsule of lateral four metatarsophalangeal joints	Joins oblique head to insert as above	Acts as tie for heads of metatarsals; adducts great toe	Lateral plantar
Flexor digiti minimi	Base of fifth metatarsal and plantar fascia	Lateral side of base of proximal phalanx of little toe	Flexes small toe	Lateral plantar

(Continued)

Table 6-13 (*Continued*)

Muscle	Origin	Insertion	Action	Nerve
Plantar interossei (3)	Medial side of third, fourth, and fifth metatarsals	Medial side of base of proximal phalanges of third, fourth, and fifth toes	Adduct lateral three toes toward second toe; flex proximal, and possibly extend distal phalanges	Lateral plantar
Dorsal interossei (4)	Lie in intermetatarsal space and arise from adjacent bones	Tendons pass forward as above and insert into proximal phalanges on either side of second toe and into lateral side of third and forth toes	Abduct second, third, and fourth toes from midline of second toe; flex proximal, and possibly extend distal, phalanges	Lateral plantar

flexor digitorum brevis muscles. In the **second layer** the **tendon of the flexor hallucis longus** grooves the under surface of the talus bone (sustentaculum tali) to pass medially under the tendon of the flexor digitorum longus and insert into the base of the terminal phalanx of the great toe. The **tendons of the flexor digitorum longus** enter the fibrous flexor sheaths at the middle of the foot, pass between the tendons of the flexor digitorum brevis, and insert into the bases of the terminal phalanges. In addition to the above tendons, the **quadratus plantae** and **lumbricales** are components of this layer.

The **third layer** is made up of the **flexor hallucis brevis,** the **adductor hallucis,** and the **flexor digiti minimi.** The flexor hallucis brevis lies on the first metatarsal along the lateral side of the abductor hallucis and is grooved by the tendon of the flexor hallucis longus to form two partially separated bellies. The adductor hallucis has two separate heads, the oblique and transverse, which may act as two distinct muscles.

Both muscles and tendons are present in the **fourth** and **deepest layer.** The **tendon of the peroneus longus** crosses the sole of the foot obliquely from the lateral to the medial side, and slips of the **tendon of the tibialis posterior** cross the sole in the opposite direction. These tendons form a sling for the foot to help maintain both the longitudinal and transverse arches. The muscles in this layer, the **three plantar** and **four dorsal interossei,** are thin and flattened, with the medial muscles more deeply situated. Both groups are more easily seen from the plantar than from the dorsal surface and lie between, and arise from, the metatarsal bones. The line of reference for the action of abduction and adduction for these muscles passes through the central axis of the second toe.

Arteries and Nerves

The **dorsalis pedis artery** is the continuation of the anterior tibial artery as it courses under the extensor retinaculum (see Table 6-9). It continues on the dorsum of the foot to the base of the first interosseous space, where it terminates by dividing into the **deep plantar** and the **first dorsal metatarsal arteries.** In its course, it gives rise to **lateral** and **medial tarsal branches** supplying the extensor digitorum brevis, cutaneous and osseous twigs in the ankle region, and the **arcuate artery** near its termination. The latter passes laterally across the bases of the metatarsal bones and gives rise to the second, third, and fourth **dorsal metatarsal arteries,** each of which subsequently divides into two **dorsal digital branches** to the sides of the toes. The deep plantar branch of the

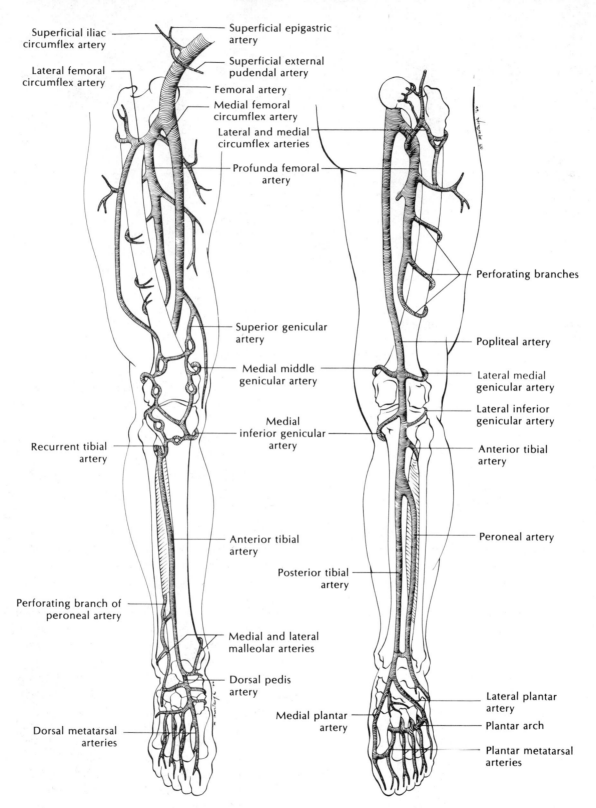

Figure 6-17. *Summary of arterial supply of the lower extremity.*

dorsalis pedis artery passes deeply through the first interosseous space to join with the lateral plantar artery to form the **deep plantar arch.**

Dorsalis Pedis Pulse
The dorsalis pedis artery lies in the subcutaneous tissue on the dorsum of the foot, between the tendons of the extensor digitorum longus and the extensor hallucis longus. It is easily palpable and may be used to determine the pulse rate if the radial pulse is not accessible.

The smaller of two terminal branches of the posterior tibial, the **medial plantar artery,** arises under cover of the flexor retinaculum. It courses deep to the abductor hallucis, then between the abductor hallucis and the flexor digitorum brevis to supply branches to the medial side of the great toe and gives muscular, cutaneous, and articular twigs along its course. The

larger of the two terminal branches of the posterior tibial is the **lateral plantar.** It passes forward between the first and second layers of muscles in the sole giving calcaneal, cutaneous, muscular, and articular branches as it continues toward the base of the fifth metatarsal. At the metatarsophalangeal joint it passes medially to join the deep plantar branch of the dorsalis pedis and together they form the deep plantar arch.

The **deep plantar arch** runs across the bases of the metatarsals to give a **plantar metatarsal artery** to each interosseous space, which subsequently divides into **plantar digital branches** to the adjacent sides of each toe. Separate branches supply the medial side of the great and the lateral side of the little toe. Perforating branches pass deeply in the interosseous space to join corresponding dorsal metatarsal vessels from the arcuate artery (Fig. 6-17).

The **medial plantar nerve,** the larger of the

Table 6-14
Nerve Distribution to Foot

Nerve	Origin	Course	Distribution
Saphenous	Femoral in thigh	Through superficial fascia of leg and anterior to medial malleolus	Skin on medial aspect of foot
Superficial peroneal	Common peroneal	Becomes superficial in distal third of leg	Skin on dorsum of foot
Deep peroneal	Common peroneal	Passes deep to extensor retinaculum to enter foot	Extensor digitorum brevis and skin on contiguous sides of great and second toe
Medial plantar	Tibial	Passes deep to flexor retinaculum to enter foot and course in medial compartment	Abductor hallucis, flexor digitorum brevis, flexor hallucis brevis, first lumbrical; skin of plantar surface medial to a line splitting the fourth toe (homologous to median of hand)
Lateral plantar	Tibial	Passes deep to flexor retinaculum to enter foot and course in lateral compartment	Supplies all muscles not supplied by medial plantar; skin lateral to a line splitting fourth toe (homologous to ulnar in hand)
Sural	Common peroneal and tibial	Descends in superficial fascia on back of leg	Lateral aspect of foot

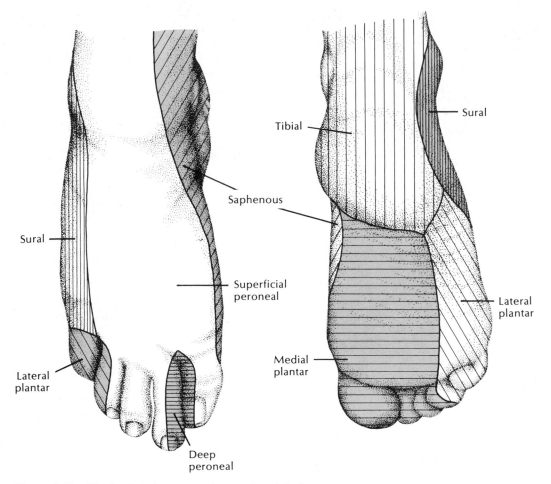

Figure 6-18. *Distribution of cutaneous innervation of the foot.*

two terminal branches of the tibial nerve, arises under the cover of the flexor retinaculum. It passes forward between the abductor hallucis and the flexor digitorum brevis, lateral to the medial plantar artery. It supplies the above muscles and sends **cutaneous branches** to the medial side of the sole. It terminates in four **digital nerves** that supply muscular branches to the flexor hallucis brevis, the first lumbrical, and cutaneous branches to the adjacent sides of the four medial toes and the medial side of the foot (Table 6-14).

The **lateral plantar nerve** passes laterally in the foot to course anteriorly, medial to the lateral plantar artery. It gives **muscular branches** to the quadratus plantae and abductor digiti minimi, and cutaneous branches to the lateral aspects of the sole. It terminates in superficial and deep branches. The **superficial branch** passes forward to supply the flexor digiti minimi, the interossei of the fourth intermetatarsal space, and skin on the lateral side of the sole, the fifth toe, and the lateral side of the fourth toe. The **deep branch** accompanies the lateral plantar artery and supplies the adductor hallucis, the remaining interossei, and the three lateral lumbricales muscles (Fig. 6-18).

Joints of the Inferior Extremity

Hip Joint

The hip joint is the best example of an enarthrodial (ball-and-socket) type of joint in the body (Fig. 6-19). The spheroidal **head** of the femur fits into a cuplike cavity, the **acetabulum,** in the innominate bone. The head of the femur is covered by articular cartilage, except for a small central area, the **fovea capitis femoris,** where the **ligamentum teres** attaches. Articular cartilage on the acetabulum forms an incomplete horse-shoe-shaped marginal ring, the **lunate surface,** with a central circular depression devoid of cartilage. In the fresh state, this is occupied by a mass of fat covered by the synovial membrane. A fibrocartilaginous rim, the **acetabular labrum,** deepens the articular cavity. It is triangular in cross section, with its base attached to the margin of the acetabulum and its free apex extending into the cavity.

The extensive **synovial membrane** of the hip joint passes from the margin of the articular cartilage of the head to cover the neck of the femur internal to the articular capsule. It reflects back onto the capsule and covers the acetabular labrum and the synovial fat pad, and ensheathes the ligamentum teres. It sometimes communicates with the bursa deep to the iliopsoas tendon.

The strong, dense, articular **capsule** of the hip joint is attached to the innominate bone just beyond the periphery of the acetabular labrum. Its femoral attachments are the intertrochanteric line anteriorly, the base of the neck superiorly, and just above the intertrochanteric crest posteriorly.

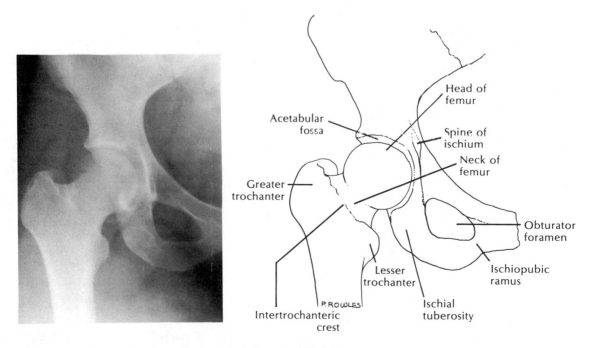

Figure 6-19. *Anteroposterior radiograph and diagram of hip joint.*

The capsule is composed of longitudinal collagenous fibers that are reinforced by accessory ligaments anteriorly and superiorly, and circular fibers (zona orbicularis) at the inferior and posterior aspect. The latter form a sling around the neck of the femur. The **iliofemoral ligament** (Y-shaped ligament of Bigelow), the strongest ligament in the body, reinforces the anterior aspect of the articular capsule. It is intimately associated with the capsule and is attached superiorly to the anterior inferior iliac spine; inferiorly it divides into two bands that attach to the upper and the lower parts of the intertrochanteric line. The triangular **ischiofemoral ligament** reinforces the posterior aspect of the capsule. It attaches superiorly to the ischium, below and behind to the acetabular margins, and inferiorly blends with the circular fibers of the capsule.

Dislocation of the Hip

Dislocation of the hip is quite rare because of: (1) the marked stability of the ball-and-socket joint; (2) the strong, tough, articular capsule; (3) the strength of the intrinsic ligaments; and (4) the extensive musculature over the joint. In congenital hip dislocation, which is most common in females, the cartilaginous lips (especially the upper lip) of the acetabular fossa allows the head of the femur to slip out of the fossa onto the gluteal surface of the ilium. Since the head of the femur in the fossa promotes normal development of the acetabulum, the hip dislocation must be reduced and a cast or traction is used to prevent recurrent dislocation.

The **ligamentum capitis femoris** (round ligament of the femur) is a flattened triangular band covered by synovial membrane attached by its apex to the fovea capitis femoris. Its base is attached by two bands to either side of the **acetabular notch.** In the interval between the osseous attachments of the round ligament to the acetabulum, the ligament blends with the transverse acetabular ligament as the latter extends across the acetabular notch to convert it into a foramen.

Knee Joint

The knee joint is usually classified as a ginglymus (hinge) joint, but is actually much more complex in function (Fig. 6-20 A and B). It combines the actions of three types of diarthrodial joints, namely, a **ginglymus** (hinge), a **trochoid** (pivot), and an **arthrodial** (gliding). Furthermore, it may be described as three separate articulations, the **femoropatellar** and the two **tibiofemoral joints.** The articular surfaces of the medial and lateral condyles of the femur, covered by articular cartilage and separated by a groove, diverge posteriorly and articulate with the two entirely separate condyles of the tibia. Each of the condyles is deepened by a meniscus and separated by the intercondylar crest. The **medial** and **lateral menisci** are two crescentic cartilaginous lamellae that cover the peripheral two-thirds of the articular surface of the tibia. They are triangular in cross section with their peripheral bases attaching by the **coronary ligaments** to the articular capsule and the tibia. Their free apices project into the articular cavity.

Aspiration of the knee

Aspiration of the knee joint may be necessary to relieve pressure, to evacuate blood, or to obtain a fluid sample for laboratory studies. In this procedure, the needle is inserted somewhat proximal and lateral to the patella through the tendinous part of the vastus lateralis muscle and is directed towards the middle of the joint. Through the same route the joint cavity may be irrigated and steroids are administered in the treatment of knee pathology.

The **synovial membrane** lining the knee joint is the largest and most extensive in the body. It begins at the superior border of the patella,

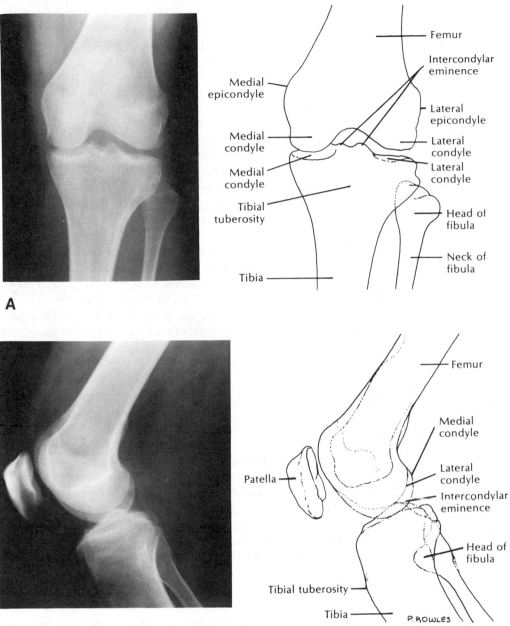

Figure 6-20. **A.** *Anteroposterior radiograph and diagram of knee joint.* **B.** *Lateral radiograph and diagram of knee joint.*

sends a blind sac deep to the quadriceps femoris muscle, and frequently communicates with the large **suprapatellar bursa** between the quadriceps femoris and the femur. Inferiorly the synovial membrane lies on either side of the patella, deep to the aponeurosis of the vasti muscles, and is separated from the patellar ligament by the **infrapatellar fat pad.** At the medial and lateral aspects of the patella, reduplications of the synovial membrane pass into the interior of the joint as the **alar folds.** At the tibia the synovial membrane attaches to the peripheries of the menisci and reflects over and ensheathes the cruciate ligaments.

Knee Injuries

The integrity of the knee joint depends upon the strength of its femoral–tibial ligaments and the tonus of the muscles playing over the joint, especially the quadriceps femoris. This muscle is capable of functionally preserving the joint, even if one or more of its ligaments is disrupted.

The most common football knee injury is the rupture of the medial collateral ligament, which is often accompanied by a tearing of the medial meniscus (medial semilunar cartilage). Since the ligament is firmly attached to the medial meniscus, these injuries usually occur together. (Injuries are less common to the lateral side of the knee because such an attachment is not present.) As the medial meniscus is torn, it may become wedged between the articular surfaces of the femur and tibia causing a "locking" of the joint.

A "clipping" injury is caused by a blow to the lateral side of the knee, damaging the anterior and posterior cruciate ligaments. Thus, in examining a patient for a knee injury, the physician is immediately concerned about the functional integrity of the three C's (collateral ligaments, cartilages, and cruciate ligaments).

The osseous arrangement of the knee joint as a weight-bearing structure is intrinsically unstable.

The stability of this joint is realized by compensating mechanisms of surrounding muscles and tendons, a strong articular capsule, strong internal and external ligaments, and modification of the joint surface by the menisci. The **patellar ligament** (ligamentum patellae), the central portion of the tendon of the quadriceps femoris, attaches superiorly to the apex and adjacent margins of the patella, and inferiorly to the tuberosity of the tibia. The posteriorly situated **oblique popliteal ligament** attaches superiorly to the femur above the condyles and to the upper margin of the intercondylar fossa; inferiorly it attaches to the posterior margin of the head of the tibia. This ligament forms part of the floor of the popliteal fossa, and the popliteal artery passes inferiorly on its surface. The **arcuate popliteal ligament** arches inferiorly from the lateral condyle of the femur to the posterior surface of the capsule, where two converging bands attach it to the styloid process of the fibula. The **tibial (medial) collateral ligament,** a broad, flat, membranous band, attaches superiorly to the medial condyle of the femur and inferiorly at the medial condyle of the tibia. In its lower part it is crossed by tendons of the sartorius, gracilis, and semitendinosus muscles; covers part of the semimembranosus tendon, and is intimately adherent to the medial meniscus. The **fibular (lateral) collateral ligament,** a strong, rounded fibrous cord, is attached superiorly to the lateral condyle of the femur. Inferiorly it splits the inserting tendon of the biceps femoris to attach to the head of the fibula. It has no attachment to the lateral meniscus.

Strong internal cruciate ligaments cross like the limbs on an X, are named from their position and attachment to the tibia, and are ensheathed by synovial membrane. The **anterior cruciate ligament** attaches in a depression anterior to the intercondylar eminence of the tibia and passes superiorly, posteriorly, and laterally to attach to the medial and posterior aspect of the lateral condyle of the femur. The **posterior cruciate** is stronger, shorter, and less oblique than the anterior. It attaches posterior to the intercondylar eminence and to the posterior extremity of the lateral meniscus. It passes superiorly, anteriorly, and medially to the medial condyle of the femur. Another internal structure, the posterior **menisco-femoral ligament (of Wrisberg),** a strong fasciculus arising from the posterior attachment of the lateral meniscus, attaches to the medial condyle of the femur immediately behind the attachment for the posterior cruciate ligament. The **transverse** and coronary ligaments of the knee connect the anterior margins of the menisci.

A large number of **synovial bursae** surround the knee joint. Anteriorly four bursae are present: one between the patella and the skin, another between the upper part of the tibia and the patellar ligament, one between the lower part of the tuberosity of the tibia and the skin, and a large bursa between the deep surface of the quadriceps femoris and the lower part of the femur. Laterally four bursae are noted: one between the lateral head of the gastrocnemius and the articular capsule, another between the fibular collateral ligament and the biceps tendon, a third between the fibular collateral ligament and the popliteus, and the last between the tendon of the popliteus and the lateral condyle of the femur. Five bursae are medially located: the first between the medial head of the gastrocnemius and the capsule; the second superficial to, and interposed between, the tibial collateral ligament and the sartorius, gracilis, and semitendinosus tendons; the third deep to, and interposed between, the tibial collateral ligament and the semimembranosus; the fourth between the semimembranosus and the head of the tibia; and the last between the semimembranosus and the semitendinosus. Any of these bursae may communicate with the synovial cavity of the knee joint.

tween the distal ends of the tibia and the fibula. Strong interosseous ligaments connect the roughened surfaces of adjacent portions of the bones to each other and are strengthened anteriorly and posteriorly by the **anterior** and **posterior tibiofibular ligaments.** The **talocrural joint** is between the tibia, the fibula, and the trochlea of the talus. The distal ends of the tibia and fibula form a deep socket, wider anteriorly than posteriorly, to receive the upper part of the talus. It is surrounded by the joint capsule, which is greatly thickened medially and laterally. The medial reinforcement, the **deltoid ligament**, is roughly triangular in shape and attaches superiorly to the medial malleolus and inferiorly to the talus, navicular, and calcaneus bones. Laterally three discrete ligaments, often referred to as the **lateral ligament**, include the **anterior talofibular ligament** passing from the lateral malleolus to the neck of the talus; the **posterior talofibular ligament** from the malleolar fossa to the posterior tubercle of the talus; and, between the two talofibular ligaments, the **calcaneofibular ligament** passing from the lateral malleolus to the lateral surface of the calcaneus.

Pott's Fracture

Pott's fracture is a serious fracture in which the lower part of one or both of the tibia and fibula are broken. If only one of the bones is fractured, the unbroken bone acts as a splint, resulting in minimal displacement of the limb. A complication of this injury is necrosis of the distal end of the fibula because of its poor blood supply. If this occurs, the stability of the ankle joint is impaired.

Ankle Joint

The ankle presents two joints, the tibiofibular and the talocrural. The strength of the ankle joint is largely dependent upon the integrity of the **tibiofibular syndesmosis,** a strong fibrous union be-

Joints of the Foot

Tarsal and metatarsal bones are bound together by ligaments to form the longitudinal and transverse arches of the foot (Fig. 6-21). These arches are maintained partly by the shape of the bones,

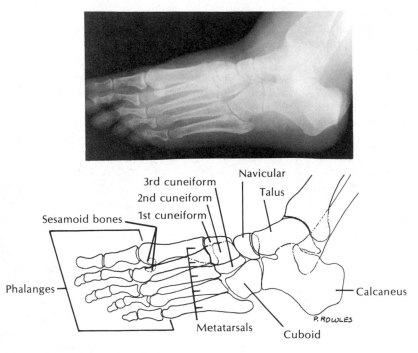

Figure 6-21. *Bones of the foot.*

partly by the tension of the ligaments and the plantar aponeurosis, but most importantly by the bracing of muscle tendons attached to the foot. The **longitudinal arch** presents a greater height and wider span on the medial side of the foot, with the talus forming the "keystone" at the summit of the arch. The short, solid **posterior pillar** is formed by the calcaneus; the much longer **anterior pillar** is supported by a medial and a lateral column. The **medial column** is formed by the navicular, three cuneiforms, and three medial metatarsal bones; the **lateral column** by the cuboid and lateral two metatarsal bones. The weight of the body is transmitted to the talus at the summit. The important ligaments that prevent flattening of the arch lie in the plantar concavity. They include the **plantar calcaneonavicular (spring) ligament,** which completes the socket formed by the navicular and the calcaneus; the **long plantar ligament** attaching to most of the plantar surface of the calcaneus and passing for-

ward to attach to the tuberosity of the cuboid; the **plantar calcaneocuboid (short plantar) ligament,** placed deep to the long plantar and consists of a strong wide band 2 cm to 3 cm passing from the anterior part of the plantar surface of the calcaneus to a ridge behind the groove in the cuboid; and the **plantar aponeurosis,** which passes from the anterior to the posterior pillars and acts as a tie beam.

Arches of the Foot
The entire body weight is supported by the feet. Normally, longitudinal and transverse arches of the foot result in contact at three points, the heel, and the first and fifth metatarsophalangeal joints. The integrity of these arches is maintained by the shape of tarsal bones, longitudinal and transverse ligaments, and the leg muscles. If the arches are impaired by injury, excessive fatigue, illness, over-

weight, or congenital defects, the muscles are weakened, the ligaments are stretched, the arches flattened, and pain is produced. Foot ailments that produce symptoms usually result from changes in the integrity of these arches or an excessive strain on the muscles or ligaments. The shape of the foot at birth may be flatfoot (pes planus), normal arch, or claw foot (pes cavus), but these congenital foot shapes usually do not produce symptoms by themselves. The common "flatfoot" is a foot with a flattened medial longitudinal arch, laterally displaced toes, and an everted foot. Claw foot describes a foot with an unusually high medial longitudinal arch that is caused by the powerful pull of the short intrinsic muscles of the sole of the foot. Postural strain, which usually involves such things as fatigue, overweight or poorly fitting shoes, usually produces pain in the foot or lower leg.

The **talocalcaneonavicular joint** lies anterior to the tarsal canal and forms part of the **transverse tarsal joint.** The head of the talus fits into a socket formed by the navicular above and the calcaneus below. The considerable interval between the navicular and the calcaneus is occupied by the spring ligament. The **calcaneocuboid joint** completes the transverse tarsal joint. The **bifurcate ligament,** which reinforces the capsule, attaches in a depression on the upper part of the calcaneus, where it then divides to pass to the navicular and cuboid bones.

The **tarsometatarsal joints** are formed by the cuneiforms and cuboid and the metatarsals. Individually there is little movement at the above joints, but working together they give elasticity and allow for twisting of the foot.

Hammer Toe

Hammer toe is a common deformity of the second or third toe. It may be congenital, or acquired by wearing poorly fitting shoes. In the typical case, the metatarsophalangeal joint and the distal interphalangeal joint are hyperextended, but the proximal interphalangeal joint is acutely flexed. A painful callus and small bursa often develop over the flexed joint.

The **metatarsophalangeal** and **interphalangeal joints** are similar. Between the heads of the metatarsal and the base of the proximal phalanx, the joint capsule is strengthened by **collateral ligaments.** The plantar portion of the capsule is thickened as the **plantar ligament** and is firmly fixed to the bases of the phalanges to allow the flexor tendons to glide freely over them. Plantar ligaments are connected by strong transverse fibers of the **deep transverse metatarsal ligament.** Aided by the transverse head of the adductor hallucis, this ligament helps to hold the heads of the metatarsals together with the tendons of the interossei passing deep, and the lumbricals superficial, to this ligament.

MAJOR ANATOMIC AND CLINICAL POINTS

Thigh

□ The saphenous nerve, a branch of the femoral nerve arising in the femoral triangle, continues to the foot. It supplies cutaneous innervation to the medial aspect of the leg and foot.

□ The thigh is divided into functional compartments with a major nerve to each compartment: femoral for extensors, obturator for adductors, sciatic for flexors.

□ One muscle in each compartment has dual innervation: pectineus by the femoral and the obturator, adductor magnus by the obturator and the tibial, short head of biceps by the common peroneal, and the long head by the tibial.

☐ Major arteries traverse only two compartments, the femoral in the extensor, and the obturator in the adductor. Blood supply to flexor compartment is by way of perforating branches of the deep femoral artery.

☐ In femoral herniae a loop of gut or mesentery passes deep to the inguinal ligament, continues through the femoral ring, and presents a mass in the proximal thigh.

☐ In varicosities it may be necessary to remove ("strip") the greater and lesser saphenous veins. The functional state of the deep veins must be assessed before this procedure is performed.

☐ Damage to the femoral nerve results in an inability to extend the knee, thus impairing a normal gait.

☐ Locating the femoral pulse aids in placing an intravenous needle or catheter in the femoral vein. With the fingers aligned with the long axis of the femoral artery the vein will lie medial to the artery.

Gluteal Region

☐ The piriformis muscle fills the greater sciatic foramen. Of the structures traversing the foramen only the superior gluteal vessels and nerve pass above the level of the piriformis.

☐ The series of short muscles deep to the gluteal muscles are all lateral rotators of the thigh.

☐ The semitendinosus, semimembranosus, and long head of the biceps are all two-joint muscles. They act to extend the hip and to flex the knee.

☐ The tibial and common peroneal components of the sciatic nerve are functional entities from their origin. The connective tissue ensheathment (epineurium) may be absent; if so, they will be anatomic entities in the gluteal region and hence no sciatic nerve would be demonstrable.

☐ Intramuscular injections into the gluteal mass should be made in the upper outer quadrant of the buttock to avoid hitting the sciatic nerve.

☐ Loss of the gluteus maximus does not noticeably impair gait on a level surface. However, in climbing stairs the patient must take one step at a time; the affected side cannot extend the hip to lift the weight of the body.

☐ Damage to the sciatic nerve would eliminate all cutaneous sensation to the leg and foot, except for a slender strip on the medial aspect supplied by the saphenous nerve.

Leg

☐ The leg is divided into three functional compartments with a major nerve to each compartment: deep peroneal to extensors, superficial peroneal to peroneal muscles, tibial to flexors.

☐ Major arteries traverse only two compartments: anterior tibial to extensor, posterior tibial to flexor. The peroneal compartment is supplied by perforating branches of the peroneal artery, which is in the posterior compartment.

☐ The tibia is subcutaneous along its entire length. The fibula does not participate in formation of the knee joint.

☐ The popliteus muscle performs a crucial action in initiating flexion of the knee. It acts to rotate the femur laterally ("unlocks the knee"), which must be accomplished before the knee can be flexed.

☐ The common peroneal nerve is highly vulnerable to trauma as it winds around the neck of the fibula. At this site it is just deep to the skin.

☐ The external investing layer of deep fascia is especially taut over the extensor muscles of the leg. Hypertrophy or a muscular spasm of the extensor muscles may compress the anterior tibial artery and cause ischemia; if severe, a fasciotomy may be necessary to circumvent necrosis.

☐ The knee must be fully extended and "locked" before the leg can bear the weight of the body. In a patient with a "trick knee" full extension on occasion does not occur and the knee "gives away."

☐ Numerous bursae surround the knee joint. Inflammation of these closed serous "sacs" is extremely painful. "Housemaid's knee" is involvement of the prepatellar bursa; "clergyman's knee" is involvement of the infrapatellar bursa.

Foot

- As the anterior tibial artery passes distal to the extensor retinaculum it becomes the dorsalis pedis.
- The medial plantar nerve is similar to the median nerve in the hand, it supplies the muscles of the great toe. The lateral plantar, similar to the ulnar nerve (in the hand), supplies most of intrinsic muscles of foot.
- Abduction and adduction of the toes is relative to a line passing through the long axis of the second toe.
- As in the hand each digit has two flexor tendons; the long flexor (similar to the deep flexor in forearm) is in the calf, the short flexor (superficial flexor in forearm) is in the sole of the foot.
- A "footdrop" results from injury to the deep peroneal nerve.
- The dorsalis pedis is a convenient artery to record the pulse in a recumbent patient.
- Medial and lateral movement at the articulation of the talus with the tibia and fibula is minimal. Excessive impact from either direction (inversion or eversion) will result in a "sprained," dislocated, or fractured ankle.
- "Flatfeet" involve the arches of the foot and may be due to malformation of the tarsal bones; stretching or elongation of supporting ligaments; or a fatigue or loss of tonicity of muscles, especially the peroneus longus, tibialis anterior, and tibialis posterior.

QUESTIONS FOR REVIEW

1. Structures traversing the adductor hiatus include all of the following except the
   A. Femoral artery
   B. Femoral vein
   C. Saphenous nerve
   D. Lymphatics

2. Landmarks of surface anatomy are helpful when describing the location of various dermatologic lesions, traumatic insults, and so forth. Approximately a hand's breadth below the crest of the ilium and two inches posterior to the anterior superior iliac spine can be palpated the
   A. Lesser trochanter of the femur
   B. Ischial spine
   C. Ischial tuberosity
   D. Greater trochanter of the femur
   E. Anterior inferior iliac spine

3. Which of the following statements concerning the obturator artery is false?
   A. It is a branch of the internal iliac
   B. It enters the thigh through the anterosuperior aspect of the obturator foramen
   C. It is accompanied through most of its course by the obturator nerve.
   D. In the thigh it lies deep to the adductor magnus.

4. The fascia lata extends inward to the linea aspera to divide the thigh into anterior, medial, and posterior muscular compartments. Muscles of the anterior compartment include all of the following except the
   A. Biceps femoris
   B. Sartorius
   C. Rectus femoris
   D. Vastus lateralis

5. Which of the following muscles does not act on two joints
   A. Sartorius
   B. Rectus femoris
   C. Biceps femoris
   D. Adductor longus

6. The popliteal fossa is a diamond-shaped area on the posterior aspect of the knee joint extending from the lower third of the femur to the upper part of the tibia. Its contents include the
   A. Common peroneal nerve
   B. Femoral nerve

C. Peroneal artery
D. Saphenous nerve
E. Obturator nerve

7. At the insertion of the small lateral rotators of the thigh, the tendon of the obturator externus is situated

A. Inferior to the quadratus femoris
B. Between the inferior gemellus and the quadratus femoris
C. Between the inferior gemellus and superior gemellus
D. Between the piriformis and superior gemellus

8. _____ is the tendinous cord that can be palpated on the superolateral border of the popliteal fossa when the knee joint is in partial flexion.

A. Biceps femoris muscle
B. Semitendinosus muscle
C. Semimembranosus muscle
D. Rectus femoris muscle
E. Vastus intermedius muscle

9. A muscle which plantar flexes the ankle joint is the

A. Tibialis anterior
B. Peroneus brevis
C. Extensor hallucis longus
D. Gastrocnemius

10. The prominent bump on the lateral aspect of the leg just distal to the knee joint is the

A. Tibial tuberosity
B. Head of the fibula
C. Lateral condyle of the femur
D. A lateral extension of the tibia

11. The superficial vein that terminates in the popliteal vein is the

A. Long (great) saphenous
B. Short (small) saphenous
C. Sural
D. Obturator

12. The nerve supply to the muscles of the anterior compartment of the leg is the

A. Tibial
B. Lateral plantar
C. Deep peroneal
D. Superficial peroneal

13. The plantar interossei of the foot are active in

A. Adduction only
B. Abduction only
C. Both abduction and adduction
D. Eversion

14. The ligaments of the foot are most likely to be stretched and strained

A. During prolonged quiet standing because in this instance the ligaments support the foot alone without the aid of the muscles
B. When walking because the muscles tire, forcing the ligaments to bear most of the arch support
C. When running because this is an unnatural action that produces abnormal strains on all ligaments and muscles
D. When sitting in a chair with the feet propped up.

15. The artery that enters the foot by passing posterior to the medial malleolus is the

A. Anterior tibial
B. Peroneal
C. Posterior tibial
D. Dorsalis pedis

16. Muscles that invert the foot include the

A. Tibialis posterior, tibialis anterior, and peroneus longus
B. Tibialis posterior, peroneus longus, and peroneus brevis
C. Peroneus anterior, peroneus longus, and peroneus brevis
D. Tibialis posterior, and tibialis anterior

17. All of the following muscles are supplied by the superior gluteal nerve except the

A. Gluteus maximus
B. Gluteus medius
C. Gluteus minimus
D. Tensor fasciae latae

18. The muscles of the posterior compartment of the thigh are primarily supplied by the

A. Obturator nerve
B. Sciatic nerve
C. Inferior gluteal nerve
D. Femoral nerve

19. The primary abductor of the hip joint (thigh) is the

    A. Gluteus medius
    B. Piriformis
    C. Pectineus
    D. Sartorius
    E. Obturator externis

20. The tendon of the biceps femoris muscle inserts into the

    A. Medial condyle of the tibia
    B. Lateral condyle of the tibia
    C. Tibial tuberosity
    D. Head of the fibula
    E. Patella

21. After the sciatic nerve has emerged from the pelvis inferior to the _____ muscle, it passes between the greater trochanter of the femur and the ischial tuberosity.

    A. Superior gemellus
    B. Quadratus femoris
    C. Gluteus minimus
    D. Obturator internus
    E. Piriformis

22. Of the following lists, which one includes muscles that take origin (at least in part) from the ischiopubic ramus (inferior ramus of pubis and ramus of ischium)?

    A. Semitendinosus, semimembranosus, and biceps femoris
    B. Sartorius, rectus femoris, and obturator externis
    C. Gracilis, adductor brevis, and obturator internis
    D. Vastus lateralis, tensor fasciae latae, and vastus medialis.

23. The gluteus medius and minimus muscles can perform all of the following actions except

    A. Medial rotation of the femur
    B. Abduction of the femur
    C. Prevention of excessive pelvic tilt to the opposite side when standing upon one extremity
    D. Powerful extension of the femur

24. Which of the following muscles originates from the anterior superior iliac spine?

    A. Semimembranosus
    B. Semitendinosus

C. Biceps femoris
D. Sartorius

25. Muscles inserting into the proximal portion of the medial surface of the tibia and forming the "pes anserinus" are all of the following except the

    A. Gracilis
    B. Semimembranosus
    C. Semitendinosus
    D. Sartorius

26. In order to avoid injecting into the sciatic nerve, intramuscular injections in the gluteal region should be made

    A. In the lower lateral quadrant
    B. In the upper lateral quadrant
    C. Midway between the greater trochanter and the ischial tuberosity
    D. Over the sacrotuberous ligament
    E. Less than 3 cm from the anterior superior iliac spine

27. The saphenous nerve is the terminal branch of which one of the following nerves?

    A. Femoral nerve
    B. Lateral femoral cutaneous nerve
    C. Obturator nerve
    D. Posterior femoral cutaneous nerve

28. The primary nerve supply to the medial (adductor) compartment of the thigh is the

    A. Femoral
    B. Obturator
    C. Tibial
    D. Common peroneal

29. Sensory innervation to the skin overlying the belly of the rectus femoris muscle is provided by means of the branches of the _____ nerve.

    A. Intermediate femoral cutaneous
    B. Common peroneal
    C. Lateral femoral cutaneous
    D. Sural
    E. Inferior cluneal

30. A deep cut into the subsartorial canal could cause all of the following except

    A. Profuse bleeding caused by the severence of the femoral artery
    B. Profuse bleeding caused by the cutting of the femoral vein

C. Paralysis of the muscles that produce dorsiflexion (extension) at the ankle joint (foot)
D. Loss of sensation on the medial aspect of the leg and foot

31. Classically, the largest sesamoid bone of the body is found in the tendon of a muscle innervated by the _____ nerve.

A. Medial plantar
B. Lateral plantar
C. Tibial
D. Common peroneal
E. Femoral

32. The major distribution of the perforating branches of the profunda femoris artery is to

A. Muscles of the posterior compartment of the thigh
B. Muscles of the anterior compartment of the thigh
C. Muscles of the medial compartment of the thigh
D. The popliteal fossa

33. Which of the following muscles insert into the lesser trochanter of the femur?

A. Obturator internis
B. Obturator externis
C. Iliopsoas
D. Pectineus

34. Inability to dorsiflex the foot, resulting in a gait characterized by the "footdrop" (toes striking the ground before the heel), would most likely be due to damage of the

A. Superficial peroneal nerve
B. Medial plantar nerve
C. Tibial nerve
D. Deep peroneal nerve

35. Which of the following pairs is incorrectly matched?

A. Peroneus tertius, deep peroneal nerve
B. Tibialis anterior, deep peroneal nerve
C. Extensor digitorum longus, superficial peroneal nerve
D. Peroneus brevis, superficial peroneal nerve
E. Peroneus longus, superficial peroneal nerve

36. Which of the following structures passes superficial to the neck of the fibula?

A. Saphenous nerve
B. Sural nerve
C. Popliteal artery
D. Common peroneal nerve
E. Short saphenous vein

37. The important "unlocking action" in flexion of the knee joint is performed by which muscle?

A. Popliteus
B. Plantaris
C. Piriformis
D. Pectineus
E. Peroneus tertius

38. The superficial peroneal nerve supplies all of the following, except the

A. Peroneus longus
B. Peroneus brevis
C. Peroneus tertius
D. Skin on the dorsum of the foot

39. Concerning the posterior tibial artery, all of the following are true except that it

A. Is a branch of the popliteal artery
B. Supplies structures in the posterior compartment of the leg
C. Gives a large branch, the peroneal, that pierces the posterior intermuscular septum to course through the lateral compartment of the leg
D. Divides into medial and lateral plantar branches at the ankle joint

40. The superior, lateral, and medial genicular arteries contribute to the circulation around the

A. Dorsum of the foot
B. Talocalcaneal joint
C. Knee joint
D. Shoulder joint
E. Hip joint

41. The most medial of the muscles of the anterior compartment of the leg is the

A. Extensor hallucis longus
B. Extensor digitorum longus
C. Tibialis anterior
D. Peroneus tertius

42. Each of the following muscles contributes directly to the stability of the knee joint except the
 A. Gastrocnemius
 B. Biceps femoris
 C. Quadriceps femoris
 D. Semitendinosus
 E. Soleus

43. The flexor digitorum brevis is innervated by the
 A. Superficial peroneal nerve
 B. Deep peroneal nerve
 C. Lateral plantar nerve
 D. Medial plantar nerve

44. Which of the following is not correct?
 A. The major contributor to the plantar arterial arch is the medial plantar artery.
 B. The interossei muscles of the foot are supplied by the lateral plantar nerve.
 C. The first lumbrical muscle in the foot is supplied by the medial plantar nerve.
 D. The lumbrical muscles of the foot arise from the tendons of the flexor digitorum longus.

45. The medial and lateral plantar arteries are branches of
 A. The anterior tibial artery
 B. The posterior tibial artery
 C. The peroneal artery
 D. The dorsalis pedis artery

46. The anterior tibial artery terminates at the ankle by becoming the
 A. Medial plantar artery
 B. Deep plantar artery
 C. Lateral plantar artery
 D. Dorsalis pedis artery

47. Concerning the flexor digitorum brevis all of the following are true, except
 A. The flexor accessorius (quadratus plantae) muscle inserts into its tendon
 B. It lies superficial to the tendons of the flexor digitorum longus
 C. It is innervated by the medial plantar nerve
 D. Its tendons insert into the middle (intermediate) phalanges of the toes

48. Concerning the flexor hallucis longus all of the following are true, except
 A. Its belly is located in the posterior compartment of the leg
 B. Its tendon courses between sesamoid bones of the metatarsophalangeal joint of the great toe
 C. It inserts into the base of the distal phalanx
 D. It is innervated by the medial plantar nerve

49. Which of the following are mismatched?
 A. First layer of plantar muscles, abductor hallucis
 B. Second layer of plantar muscles, 1st (medial) lumbrical
 C. Second layer of plantar muscles, quadratus plantae
 D. Third layer of plantar muscles, flexor hallucis brevis
 E. Fourth layer of plantar muscles, interossei

50. All of the following structures course behind the medial malleolus as they pass from the leg to the foot except the
 A. Tendon of tibialis posterior
 B. Posterior tibial artery
 C. Peroneal artery
 D. Flexor hallucis longus tendon
 E. Tibial nerve

SEVEN
Head and Neck

Head and Neck

Objectives

At the completion of study of the neck the student should be able to

▶ Delineate palpable surface features of the neck

▶ Demarcate subdivisions (i.e., triangles) of the neck

▶ Label on a schematic drawing the fasciae of the neck

▶ Identify on a cadaver the suprahyoid and infrahyoid muscles

▶ List the branches of the external carotid and give their course and distribution

▶ Describe the course and distribution of branches of the hypoglossal and vagus nerves

▶ Define the ansa cervicalis complex

▶ List branches and extensions of the cervical portion of the sympathetic trunk

▶ Describe the anatomic features relating to the scalene gap

▶ Describe the anatomy of the thyroid gland

Surface Anatomy of Head and Neck

Viewed from the front, the most prominent midline surface feature of the neck is the bulging, laryngeal prominence (Adam's apple), which is formed by the V-shaped thyroid cartilage of the larynx. At its superior border the laryngeal notch is palpable, while above, the **hyoid bone** can be felt. Immediately below the larynx is an important landmark, the **cricoid cartilage,** which is joined inferiorly by the **trachea.** The latter is reinforced by a series of palpable, horseshoe-shaped cartilaginous bars with their open ends directed posteriorly.

At the base of the neck the superior border of the manubrium, located between the **sternal heads** of the sternocleidomastoid muscles, forms an obvious concave depression, the **suprasternal space,** or **jugular notch.** Here the clavicles can be seen projecting laterally towards the tips of the shoulders, which are formed by the acromion processes of the scapulae. The insertion of the sternocleidomastoid muscle is the large, rounded **mastoid process** easily palpated behind the ear.

The chin is formed by the **mental protuberance** of the mandible, while at the posterior border of the bone the prominent **angle** is continued superiorly as the **ramus.** When the jaw is clenched, the **masseter muscle** is easily demonstrable superficial to the angle and the ramus. The bony prominence of the cheek, formed by the zygomatic bone, is continuous anteriorly and posteriorly with the zygomatic processes of the maxillary and temporal bones, respectively, and superiorly with the zygomatic process of the frontal bone. The freely movable **cartilaginous portion** of the nose can be followed superiorly to the stationary **nasal bones,** which form the bridge of the nose. At the lips, the skin of the face and the mucous membrane of the oral cavity are continuous.

The **rima palpebrarum,** a slitlike orifice at the free margins of the eyelid, fuses laterally and medially, as the **inner** and **outer canthi.** The skin of the lid is continuous at the free margin with the conjunctivum covering the inner surface of the eyelid. The eyelashes curve outward from the free border, while the ducts of the **tarsal (meibomian) glands** open onto the free surface of the lid. **Supraciliary** (brow) **ridges,** covered partially by the eyebrows, indicate the anterior bulging of the frontal air sinuses. The external portion of the ear, consisting of the **auricula,** or **pinna,** and the **external acoustic meatus,** will be described with the ear.

The nostrils, or **external nares,** are oval openings separated from each other by the lower movable cartilaginous part of the nasal septum. They have stiff hairs (vibrissae) projecting into the vestibule that screen particulate matter carried into the nasal cavity by air currents.

Fasciae

The **superficial fascia** of the head and neck is continuous with the superficial fascia of the pectoral, deltoid, and back regions (Fig. 7-1). The muscles of facial expression are embedded within this layer.

The **deep fasciae** consist of the external investing layer, the middle cervical layer, the visceral and the prevertebral fasciae. The **external investing layer (external cervical fascia)** completely invests the neck like a stocking and extends from the clavicle over the mandible to the zygomatic bone, where it blends with the fascia enclosing the masseter muscle. Posteriorly this layer fuses with the ligamentum nuchae and attaches superiorly to the external occipital protuberance and the superior nuchal line. In the anterior triangle this layer is bound to the hyoid bone and is subdivided into the suprahyoid and infrahyoid portions. The **suprahyoid portion** at-

taches to the inferior margin of the mandible, covers the submandibular gland, and sends a strong membranous process deep to the gland that attaches to the hyoid bone and angle of the mandible. This deep extension separates the submandibular and parotid glands and then splits to enclose the latter.

The **infrahyoid portion** of the external investing layer splits inferiorly to attach to the anterior and posterior aspect of the manubrium, where it forms the **suprasternal space.** This space is limited by the sternal heads of the sternocleidomastoideus and contains the lower portion of the anterior jugular veins, their communications across the midline, and some lymph nodes. Passing laterally this fascial layer opens to invest the sternocleidomastoideus, then fuses as it crosses the posterior triangle of the neck, separates again to ensheath the trapezius, and then fuses posteriorly in the midline with the ligamentum nuchae.

The **middle cervical fascia** is composed of two layers, with the stronger, more superficial layer enclosing the sternohyoideus and omohyoideus, and fusing superficially with the outer investing layer. The delicate deeper layer encloses the thyrohyoideus and sternothyroideus, and contributes to the formation of the carotid sheath before it fuses laterally with the external investing fascia.

The **prevertebral fascia** covers the anterior aspect of the cervical vertebrae, passes laterally to enclose the longus colli and scaleni muscles, then continues posteriorly to surround the levator scapulae muscle. Between the anterior and middle scaleni, this layer is prolonged into the axilla as the **cervicoaxillary sheath,** which surrounds the brachial plexus and the axillary artery. The prevertebral fascia is continuous inferiorly with the endothoracic fascia of the thoracic cavity. At the cervicothoracic aperture this fascial layer expands over the apex of the lung as the **cervical diaphragm** or **Sibson's fascia.** A potential cleft between the prevertebral fascia and the visceral fascia, the **retropharyngeal space,** is limited su-

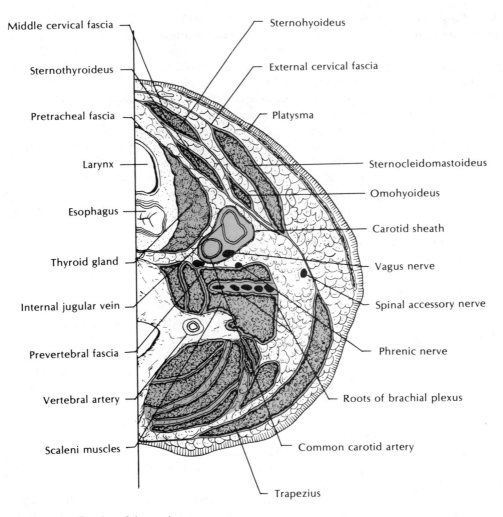

Figure 7-1. *Fasciae of the neck.*

periorly by the base of the skull and laterally by the attachment of the prevertebral fascia to the middle cervical fascia. Inferiorly the retropharyngeal space communicates with the posterior mediastinum.

The **visceral compartment** of the neck is located between the prevertebral fascia and the middle cervical fascia. It contains the major arteries and nerves within the neck, the cervical portions of the digestive and respiratory systems, and the thyroid and parathyroid glands. The **visceral**

(pretracheal) fascia is a tubular prolongation into the neck of the visceral fascia of the mediastinum, where it is continuous with the fibrous pericardium. It encloses the esophagus, trachea, pharynx, and larynx, and contributes laterally to the formation of the carotid sheath. That portion covering the superior constrictor muscles is called the **buccopharyngeal fascia** and is attached superiorly to the pharyngeal tubercle at the base of the skull, and anteriorly to the pterygoid hamulus and the pterygomandibular raphe.

Neck

Anterior Triangle of the Neck

The **anterior triangle** of the neck is bounded posteriorly by the anterior border of the sternocleidomastoideus and anteriorly by the midline of the neck; the base is formed by the lower border of the mandible, and the apex is at the sternum. For descriptive purposes it is subdivided into three paired and one common triangle (Fig. 7-2). The **muscular triangle,** delineated by the superior belly of the omohyoideus, the sternocleidomastoideus, and the midline, contains the infrahyoid muscles and thyroid gland. The superior belly of the omohyoid, the posterior belly of the digastric muscle, and the sternocleidomastoideus bound the **carotid triangle,** which contains the vagus and hypoglossal cranial nerves, the common, external, and internal carotid arteries, the internal jugular vein, and the hyoid bone. The **submandibular** (digastric) **triangle** is limited by the two bellies of the digastric muscle and the mandible. It contains the submandibular gland, nerves to the anterior belly of the digastric and mylohyoideus, the hypoglossal nerve, and the lingual and facial arteries. The unpaired **submental triangle** is between the anterior bellies of the digastric muscles and the hyoid bone. It contains the anterior jugular veins and lymph nodes.

Muscles (Fig. 7-3)

An extensive thin sheet of muscle, the **platysma,** covers the entire anterior aspect of the neck from the lower border of the mandible to the clavicle (Table 7-1). This superficial muscle of facial expression attaches to the skin and mandible. It is innervated by the facial nerve and acts to tense the skin over the neck. The **sternocleidomastoideus,** extending obliquely from the sternoclavicular joint to the mastoid process, divides the neck into anterior and posterior triangles. It has two origins, a **sternal head,** arising as a rounded tendon from the sternoclavicular joint, and a flattened **clavicular head,** from the superior aspect of the medial third of the clavicle. It lies superficial to the great vessels of the neck and the cervical plexus, with cutaneous branches of the latter (the lesser occipital, greater auricular, transverse cervical, and supraclavicular nerves), all emerging at about the midpoint of its posterior border. This muscle is innervated by the spinal accessory nerve and twigs from the second and third cervical nerves.

A relatively extensive flat sheet of muscle, the two **mylohyoid** muscles, forms the floor of the submental and part of the digastric triangles, as well as the muscular floor of the mouth. From either side of the neck, these muscles meet in the midline to insert into the median raphe. Situated deep to the mylohyoideus, the small bandlike **geniohyoideus** muscle passes from the genial tubercle on the internal aspect of the mandible to the hyoid bone. The mylohyoideus is innervated by the mandibular division of the trigeminal nerve, and the geniohyoideus by twigs from the first loop of the cervical plexus passing with the hypoglossal nerve. Both muscles act in depressing the mandible and elevating the hyoid bone.

The **digastric muscle,** consisting of two bellies attached to an **intermediate tendon,** demarcates with the inferior border of the mandible, the digastric subdivision of the anterior triangle. The intermediate tendon serves as a focal point in the relationships of the anterior triangle and is bound by a slip of fascia to the hyoid bone. The digastric is innervated by two nerves, the facial to the posterior belly and the mandibular division of the trigeminal to the anterior belly. It acts to elevate the hyoid bone and assists in depressing the mandible.

Text continues on page 302

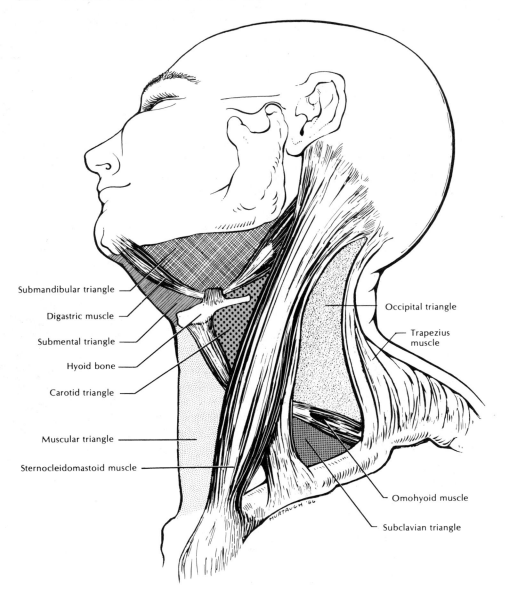

Submandibular triangle

Digastric muscle

Submental triangle

Hyoid bone

Carotid triangle

Muscular triangle

Sternocleidomastoid muscle

Occipital triangle

Trapezius muscle

Omohyoid muscle

Subclavian triangle

Figure 7-2. *Triangles of the neck.*

Table 7-1
Muscles of the Anterior Triangle of the Neck

Muscle	Origin	Insertion	Action	Nerve
Platysma	Superficial fascia of upper pectoral and deltoid regions	Skin and facial muscles overlying mandible and border of mandible	Depresses mandible and lower lip; tenses and ridges skin of neck	Facial
Sternocleidomastoideus	Manubrium and medial one-third of clavicle	Mastoid process and lateral half of superior nuchal line	Singly rotates and draws head to shoulder; together flex cervical column	Spinal accessory, C_2 and C_3
Omohyoideus	Medial lip of suprascapular notch	Lower border of body of hyoid	Steadies hyoid; depresses and retracts hyoid and larynx	C_2 and C_3 from ansa cervicalis
Digastric	Mastoid notch of temporal bone	Mandible near symphysis	Raises hyoid and base of tongue; steadies hyoid; depresses mandible	Posterior belly by facial; anterior belly by mandibular division of trigeminal
Mylohyoideus	Mylohyoid line of mandible	Median raphe and hyoid bone	Elevates hyoid and base of tongue; depresses mandible; raises floor of mouth	Mylohyoid branch of inferior alveolar
Geniohyoideus	Genial tubercle of mandible	Body of hyoid	Elevates hyoid and base of tongue	C_1 and C_2 coursing with hypoglossal nerve
Sternohyoideus	Posterior surface of manubrium and medial end of clavicle	Lower border of body of hyoid	Depresses hyoid and larynx	C_1, C_2, and C_3 from ansa cervicalis
Sternothyroideus	Posterior surface of manubrium	Oblique line of thyroid cartilage	Depresses thyroid cartilage	C_1, C_2, and C_3 from ansa cervicalis
Thyrohyoideus	Oblique line of thyroid cartilage	Lower portion of body and greater horn of hyoid	Depresses hyoid; elevates thyroid cartilage	C_1 and C_2 coursing with hypoglossal nerve

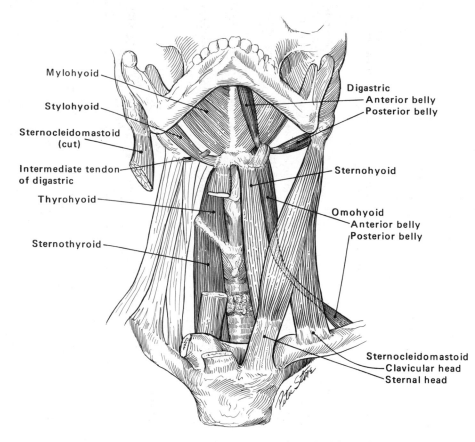

Mylohyoid

Stylohyoid

Sternocleidomastoid
(cut)

Intermediate tendon
of digastric

Thyrohyoid

Sternothyroid

Digastric
Anterior belly
Posterior belly

Sternohyoid

Omohyoid
Anterior belly
Posterior belly

Sternocleidomastoid
Clavicular head
Sternal head

Figure 7-3. *The suprahyoid and infrahyoid muscles. Suprahyoid muscles are those above the hyoid bone; infrahyoid muscles are those below it.*

The four straplike muscles attaching either to the hyoid bone or the thyroid cartilage make up the **infrahyoid muscles.** The superficially placed **sternohyoideus** passes from the sternum to the hyoid bone. It lies superficial to the **thyrohyoideus.** The latter extends from the oblique line of the thyroid cartilage to the hyoid bone and may be considered the superior extension of the **sternothyroideus.** The **omohyoideus** consists of two bellies with an intermediate tendon. The muscle forms a wide V in passing from the superior notch of the scapula to the hyoid bone. The intermediate tendon attaches by deep fascia to the manubrium and the first rib or costal cartilage. All

the infrahyoid muscles are innervated by the ansa cervicalis except for the thyrohyoideus, which receives a small branch from cervical components traveling with the hypoglossal nerve.

Arteries and Nerves

Arteries within the anterior triangle include the common carotid, the internal carotid, and the external carotid and its branches (Fig. 7-4; Table 7-2). The **common carotid artery** courses superiorly from behind the sternoclavicular articulation to the level of the superior border of the thyroid cartilage, where it bifurcates into the internal and external carotid arteries.

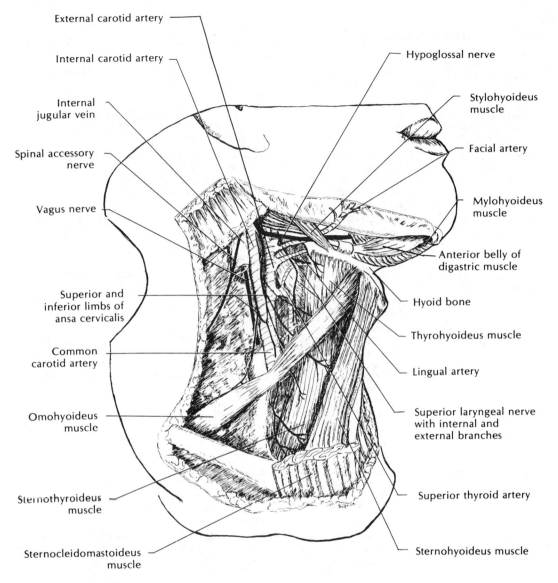

External carotid artery

Internal carotid artery

Internal jugular vein

Spinal accessory nerve

Vagus nerve

Superior and inferior limbs of ansa cervicalis

Common carotid artery

Omohyoideus muscle

Sternothyroideus muscle

Sternocleidomastoideus muscle

Hypoglossal nerve

Stylohyoideus muscle

Facial artery

Mylohyoideus muscle

Anterior belly of digastric muscle

Hyoid bone

Thyrohyoideus muscle

Lingual artery

Superior laryngeal nerve with internal and external branches

Superior thyroid artery

Sternohyoideus muscle

Figure 7-4. *Contents of the anterior triangle of the neck.*

Table 7-2
Arterial Supply to Anterior Triangle of the Neck

Artery	Origin	Course	Distribution	Anastomoses
Common carotid	Brachiocephalic on right; arch of aorta on left	Anterior triangle within the carotid sheath	Divides into internal and external carotids	No direct anastomoses
Internal carotid	Common carotid at superior border of thyroid cartilage	Ascends within the carotid sheath to reach carotid canal	Major blood supply to the brain and orbit; no branches in neck	No direct anastomoses in the neck
External carotid	Common carotid at superior border of thyroid cartilage	Ascends in anterior triangle to neck of mandible; terminates as superficial temporal and maxillary	Six arteries arise from external carotid in anterior triangle; supply face, neck, and a portion of scalp	No direct anastomoses
Superior thyroid	Anterior aspect of external carotid	Arches inferiorly to deep aspect of thyroid gland	Infrahyoid muscles, thyroid gland, laryngeal branch to internal aspect of larynx	Inferior thyroid, ascending pharyngeal, across midline
Lingual	Anterior aspect of external carotid	At greater cornua of hyoid bone pierces hyoglossus to reach deep aspect of tongue	Primary supply to tongue	Across midline
Facial	Anterior aspect of external carotid	Follows a tortuous course deep to submandibular gland → inferior border of mandible → angles towards the medial canthus of eye to terminate as angular	Suprahyoid muscles, submandibular gland and anterior aspect of face via inferior and superior labials, lateral nasal, and angular	Septal and infraorbital from maxillary, transverse facial, ophthalmic branches at orbit and across midline
Ascending pharyngeal	Medial aspect of external carotid	Ascends on lateral aspect of pharynx	Pharynx	Superior and inferior thyroids
Occipital	Posterior aspect of external carotid	Follows inferior border of posterior belly of digastric to posterior aspect of scalp	Superficial fascia and skin over occipital bone, occipitalis, posterior belly of digastric, and stylohyoideus	Posterior auricular and supraorbital
Posterior auricular	Posterior aspect of external carotid	Follows superior border of posterior belly of digastric to posterior aspect of ear	Pinna, posterior belly of digastric, stylohyoid and auricular muscles	Occipital, supraorbital, and superficial temporal

Carotid Sinus and Carotid Body

At the bifurcation of the common carotid, the internal carotid artery is dilated as the carotid sinus. At this location the internal carotid has abundant nerve endings supplied by the glossopharyngeal nerve. The carotid sinus is sensitive to the changes in blood pressure and serves as a reflex pressoreceptor that assists in the regulation of the blood pressure in vessels of the brain. Posterior to the bifurcation in the wall of the internal carotid, there is a small ovoid structure, the carotid body. It is a highly vascular, epithelioid body that is also well supplied by the glossopharyngeal nerve. It is sensitive to chemical changes in the blood, such as hypoxia. The carotid body acts as a chemoreceptor and, as a reflex, causes an increase in heart rate, blood pressure, and respiratory rate, to eliminate the hypoxia.

The **internal carotid** has no branches in the neck, but ascends within the carotid sheath with the internal jugular vein and the vagus nerve. It passes into the skull through the carotid canal as the principal blood supply to the brain and orbital cavity.

Six branches arise from the **external carotid artery** in the anterior triangle. Its two terminal branches, the superficial temporal and internal maxillary, bifurcate at the neck of the mandible. The six branches in the neck radiate from the area of the intermediate tendon of the digastric muscle to the various structures they supply.

Ligation of the External Carotid Artery
Ligation of the external carotid artery controls bleeding from a surgically inaccessible branch of this vessel, such as in the lingual, ascending pharyngeal, middle meningeal, or sphenopalatine. Collateral circulation is quickly established by anastomoses between the superior thyroid with the inferior thyroid arteries, and by branches crossing the midline of the face.

The **superior thyroid branch** arises from the anterior aspect of the external carotid opposite the thyrohyoid membrane and arches inferiorly to supply the thyroid gland. In its course it is accompanied by the external laryngeal nerve that innervates the cricothyroideus muscle. The **lingual artery** also arises from the anterior aspect of the external carotid, opposite the greater horn of the hyoid bone, and passes deep to the posterior belly of the digastric and mylohyoideus to supply the tongue. In its course it is crossed superficially by the hypoglossal nerve. The **facial** (external maxillary) **artery,** also from the anterior aspect of the external carotid, follows a sigmoid course to pass deep to the posterior belly of the digastric. It grooves the posterior and superior borders of

the submandibular gland, then curves over the inferior border of the mandible onto the face. Its distribution will be considered with the discussion of the face. From the medial aspect of the external carotid, the **ascending pharyngeal branch** ascends, with the internal carotid, to supply prevertebral muscles, the pharynx, and the palatine tonsil. Arising posteriorly, the **occipital branch** follows the inferior border of the posterior belly of the digastric muscle to pass superiorly and posteriorly to the mastoid process, where it ramifies on the back of the head. The **posterior auricular branch** courses along the superior border of the posterior belly of the digastric to pass to the notch between the external auditory meatus and the mastoid process. It supplies the area behind the ear. Deep to the parotid gland, the external carotid terminates by bifurcating into the **superficial temporal artery,** which ascends anterior to the ear and supplies the side of the face and head, and the **maxillary** (internal maxillary) **branch,** which will be described with the infratemporal fossa.

Collateral Circulation
Collateral circulation of the face and neck is extensive as many arteries of these regions anastomose freely with each other. In the neck, these channels are especially plentiful between arteries supplying the neck muscles and thyroid gland. Anastomoses may also occur between the terminal branch of the facial artery and the ophthalmic artery. This establishes communication between the external and the internal carotid.

Nerves and their branches encountered in the anterior triangle of the neck include the glossopharyngeal, vagus, hypoglossal, cervical plexus, and cervical sympathetic chain (Table 7-3).

The **vagus nerve,** the longest of the cranial nerves, leaves the skull through the jugular foramen in company with the glossopharyngeal and the spinal accessory nerves, and the internal jugular vein. In the neck it is enclosed within the carotid sheath and enters the thoracic cavity by pass-

Table 7-3
Nerves in Anterior Triangle

Nerve	Origin	Course	Distribution
Vagus nerve (X)	Brain stem	Traverses jugular foramen to pass through neck in carotid sheath with carotid artery and internal jugular vein	Branches in neck include superior and inferior cardiac branches to cardiac plexus; superior and inferior laryngeal nerves to larynx; pharyngeal branches
Superior laryngeal	Vagus (X)	Divides into internal and external branches; internal laryngeal pierces thyrohyoid membrane to reach inside of larynx; external laryngeal follows superior thyroid artery	Internal laryngeal sensory to larynx above level of vocal folds; external laryngeal supplies cricothyroid
Inferior (Recurrent) laryngeal	Vagus (X)	On right side, hooks around subclavian artery; on left side, hooks around arch of aorta; nerves then run in tracheoesophageal groove to reach larynx	Motor to intrinsic muscles of larynx except cricothyroid (by external laryngeal), sensory below level of vocal folds
Ansa cervicalis	First and second loops of cervical plexus	Twig from first loop joins hypoglossal nerve, most fibers leave hypoglossal near intermediate tendon as superior limb of ansa; inferior limb from second loop of cervical plexus; limbs join to form loop (ansa) on superficial aspect of internal jugular	From ansa branches pass to sternohyoid, sternothyroid and omohyoid muscles
Nerve to thyrohyoid	Element of ansa cervicalis complex leaves hypoglossal just distal to superior limb of ansa cervicalis	Passes directly to thyrohyoid muscle	Thyrohyoid
Nerve to geniohyoid	Element of ansa cervicalis complex leaves hypoglossal nerve in floor of oral cavity	Passes directly to geniohyoid muscle	Geniohyoid
Hypoglossal (XII)	Brain stem	Leaves cranial cavity through hypoglossal canal, courses deep to posterior belly and intermediate tendon of digastric; at intermediate tendon, passes forward on hyoglossus and mylohyoid to reach floor of oral cavity	All intrinsic and extrinsic tongue muscles except palatoglossus

ing anterior to the subclavian artery. It has two **sensory ganglia,** and **superior** (jugular) at the jugular foramen and the **inferior** (nodose) 2 cm to 3 cm below. Initial branches of the vagus include the **recurrent meningeal,** which re-enters the cranial cavity by way of the jugular foramen to supply dura in the posterior cranial fossa; the **auricular branch**, cutaneous to the posterior aspect of the pinna of the ear and the floor of the external auditory meatus; the **pharyngeal branches,** which join branches of the glossopharyngeal and superior cervical ganglion to form the pharyngeal plexus, and the **nerve to the carotid body.**

The **superior laryngeal branch** of the vagus arises at the inferior ganglion, passes deep and medial to the internal and external carotid arteries, and divides into an internal and an external branch. The **internal laryngeal nerve,** together with the laryngeal branch of the superior thyroid artery, pierces the thyrohyoid membrane to supply sensory fibers to the mucous membrane of the larynx above the level of the vocal folds, and parasympathetic fibers to glands of the epiglottis, base of the tongue, and the upper larynx.

The **external laryngeal nerve,** a long slender branch, accompanies the superior thyroid artery. It passes deep to the sternothyroideus muscle and the thyroid gland to supply the cricothyroideus muscle.

Injury to the External Laryngeal Nerve
Injury to the external laryngeal nerve may occur during thyroidectomy. It is a branch of the superior laryngeal nerve that lies next to the superior thyroid artery, often ligated in thyroid surgery. If the nerve is included in the ligature for the superior thyroid artery, the nerve supply to the cricothyroid muscle is interrupted and the vocal cord cannot be lengthened. With this loss of tension on one cord, the voice becomes weak, hoarse, and easily fatigued.

Two or three **superior cardiac branches** arise at the level of the inferior ganglion and may join a sympathetic cardiac branch from the superior cervical ganglion as a common nerve to the heart. The **inferior cardiac branch** arises either from the vagus or from the recurrent laryngeal nerve at the root of the neck. All vagal cardiac branches from the right side and the superior cardiac branches from the left side pass behind the major vessels of the heart to terminate in the deep cardiac plexus. The inferior cardiac branch from the left side descends anterior to the arch of the aorta to end in the superficial cardiac plexus. Stimuli from cardiac branches of the vagus nerve decrease the rate and force of the heart beat.

Arising low in the neck, the **inferior (recurrent) laryngeal nerve** is sensory to the larynx below the level of the vocal folds and motor to all the intrinsic muscles of the larynx, except the cricothyroideus. On the right side the inferior laryngeal nerve hooks around the subclavian artery, on the left around the arch of the aorta. Both nerves then pass posteriorly to ascend in the tracheoesophageal groove. At the larynx they penetrate the cricothyroid membrane to reach the interior of the larynx.

Inferior Laryngeal Nerve Injury
In performing a thyroidectomy, this nerve is vulnerable to injury because of its intimate relation to the thyroid gland and the inferior thyroid artery. The surgeon must carefully identify and displace the nerve away from the thyroid before removing the gland. Permanent postoperative hoarseness results from damage or division of this nerve. If only one nerve is involved, speech is not greatly affected because the other vocal cord is still functional. Bilateral cutting of the nerves causes the vocal cords to become fixed in a neutral abduction–adduction position, resulting in loss of speech and impaired breathing.

The **glossopharyngeal nerve** leaves the cranium through the jugular foramen together with the vagus and spinal accessory nerves, and the internal jugular vein. Initially, it courses between the internal jugular vein and internal carotid artery, then passes forward between the internal carotid and the styloid process. It presents two **sensory ganglia,** the small **superior** and the larger **inferior (petrosal)** ganglion. From the inferior ganglion the **tympanic nerve** passes through the temporal bone to join the tympanic plexus on the promontory of the middle ear cavity. It carries parasympathetic fibers destined for the parotid gland, and sensory fibers to the middle ear cavity. Filaments from the **promontory plexus** are reconstituted to form the **minor petrosal nerve,** which passes through a small canal in the temporal bone to emerge into the middle cranial fossa and then continues through the fissure between the temporal and the sphenoid bones to reach the otic ganglion. A second branch of the glossopharyngeal, the **nerve to the carotid sinus,** descends from the main trunk at the jugular foramen and joins with fibers from the vagus to pass to the anterior surface of the internal carotid and supply pressor receptor fibers to the carotid sinus. Three or four **pharyngeal branches** join with branches from the vagus and the superior cervical sympathetic ganglion to form the pharyngeal plexus. The **nerve to the stylopharyngeus** is the only named muscular branch of the glossopharyngeal nerve. **Tonsillar branches** supply sensation to the palatine tonsil, soft palate, and fauces, and communicate with the lesser palatine nerve. Two **lingual branches** supply the posterior third of the tongue with both general and special (taste) sensation.

Cervical Sympathetic Trunk

The **cervical portion of the sympathetic trunk** is embedded within the connective tissue posterior to the carotid sheath and anterior to the prevertebral muscles. Unlike the sympathetic chain in the thorax, where typically one ganglion occurs for each spinal nerve, only three ganglia are present. In the neck the first four cervical spinal nerves receive gray rami communicantes from the superior cervical ganglion, the fifth and sixth cervical nerves from the middle cervical ganglion, and the seventh and eighth cervical nerves from the inferior cervical ganglion. The latter frequently unites with the first thoracic ganglion to form the stellate ganglion. **No white rami communicantes** are present in the cervical region. Each of the cervical spinal nerves receives at least one gray ramus communicans from their respective fused ganglia.

The **superior cervical ganglion,** the largest ganglion of the sympathetic chain, is located between the internal carotid artery and the longus capitis muscle at the level of the second and third cervical vertebrae. It is spindle-shaped, about 2 cm to 3 cm long, and sends twigs superiorly along the internal carotid artery as the internal carotid plexus. It also sends **communicating twigs** to the glossopharyngeal, vagus, spinal accessory, and hypoglossal cranial nerves; **gray rami** to the first four cervical spinal nerves; **pharyngeal branches** to the pharyngeal plexus, and a **cardiac branch,** which on the left side passes to the superficial cardiac plexus, and on the right to the deep cardiac plexus.

Horner's Syndrome

Horner's syndrome results from disruption of the cervical sympathetic nerves. The injury may result from pressure on the cervical chain or ganglia by a malignant tumor in the neck or upper lung, surgery, or penetrating injuries to the neck. Symptoms include constriction of the pupil, narrowing of the palpebral fissure, recession of the eyeball (enophthalmos), drooping of the eyelid (pseudoptosis), loss of sweating (anhidrosis), and flushing of the face (vasodilation).

The smaller **middle cervical ganglion** is located opposite the summit of the loop of the inferior thyroid artery. It gives off **gray rami** to the fifth and sixth cervical nerves, a **thyroid branch,** which forms a plexus around the inferior thyroid artery, **cardiac branches** to the deep cardiac plexus, and the **ansa subclavia** as filaments that descend anteriorly to hook around the subclavian artery and join the inferior cervical ganglion.

The **inferior cervical ganglion** is small, irregular in shape, located behind the common carotid artery, and frequently joins with the ganglion of the first thoracic nerve to form the **stellate ganglion.** Its branches include **gray rami** to the seventh and eighth cervical spinal nerves, branches to the **ansa subclavia,** contributions to the **subclavian** and **vertebral plexuses,** and a **cardiac branch** to the deep cardiac plexus.

Cervical Plexus

The **cervical plexus** is formed by the **ventral rami** of the first four cervical nerves (Fig. 7-5). Each of these ventral rami divides into an **ascending** and a **descending division.** The ascending limb of the first cervical nerve passes into the skull to supply sensation to the meninges. The descending limb of the fourth cervical nerve contributes to the brachial plexus. The remaining ascending and descending divisions unite to form **three loops** opposite the first four cervical vertebrae. The cervical plexus receives postganglionic sympathetic fibers from the superior cervical sympathetic ganglion. Branches of the plexus include the cutaneous nerves, to be described later with the posterior triangle (the **lesser occipital, greater auricular, transverse cervical,** and the three **supraclavicular nerves**), the ansa cervicalis, segmental muscular branches to the prevertebral muscles, and the **phrenic nerve.**

The **ansa cervicalis,** which innervates the infrahyoid muscles, is derived from the first and second loops of the cervical plexus. Twigs from the first loop (C_1 and C_2) pass to the hypoglossal nerve, travel with it for 1 cm or 2 cm, where most of the fibers leave the hypoglossal as the **supe-**

rior ramus (descendens hypoglossi) of the ansa cervicalis, and descends to form the ansa with a branch from the second loop (C_2 and C_3), the **inferior ramus** (descendens cervicalis). Fibers from the first and second cervical nerves, remaining with the hypoglossal nerve, branch off as independent twigs to supply the thyrohyoideus and geniohyoideus muscles. The superior ramus of the ansa cervicalis sends twigs to the superior belly of the omohyoideus. The inferior ramus supplies the inferior belly of the omohyoideus, and from the loop, branches pass to the sternohyoideus and sternothyroideus muscles.

Segmental muscular branches from the cervical plexus supply the prevertebral muscles. Twigs from the first loop of the cervical plexus pass to the rectus capitis lateralis, longus capitis, and rectus capitis anterior; from the second loop to the longus capitis and longus colli, and from the third loop to the middle scalenus and levator scapulae. Contributions from the cervical plexus to the spinal accessory nerve give additional innervation to the sternocleidomastoideus (C_2 and C_3) and the trapezius (C_3 and C_4). The **phrenic nerve** is derived from the fourth cervical nerve with contributions from the third and fifth cervical nerves. It courses inferiorly to cross the scalenus anterior obliquely, passes deep to the transverse cervical and suprascapular arteries to pass through the thorax, where it innervates the diaphragm.

The Hiccup

The hiccup is an involuntary, spasmodic contraction of the diaphragm, which momentarily draws air into the lungs. At that instant, the glottis is closed causing the characteristic sound. The cause is unknown but may be due to the stimulation of nerve endings in the digestive tract or the diaphragm. Inhalation of air with about 5% CO_2 usually stops hiccuping. If the problem continues unabated, section of the phrenic nerve may be indicated.

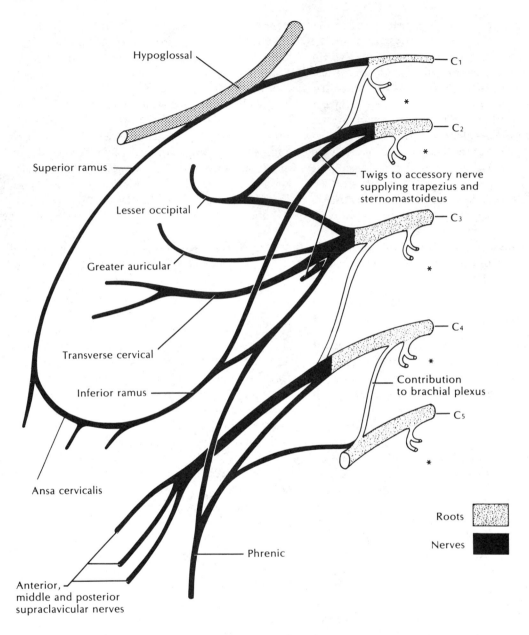

Hypoglossal

Superior ramus

Lesser occipital

Greater auricular

Transverse cervical

Inferior ramus

Ansa cervicalis

Anterior,
middle and posterior
supraclavicular nerves

Phrenic

C₁

C₂

Twigs to accessory nerve
supplying trapezius and
sternomastoideus

C₃

C₄

Contribution
to brachial plexus

C₅

Roots

Nerves

*Segmental branches to rectus capitis lateralis,
rectus capitis anterior, longus capitis, longus colli,
scaleni muscles and levator scapulae.

Figure 7-5. *Cervical plexus.*

Prevertebral Region

The **prevertebral muscles** form a longitudinal muscular mass anterior to the vertebral column, which includes the **anterior, middle,** and **posterior scaleni;** the **longus capitis** and **colli,** and the **rectus capitis anterior** and **lateralis muscles.** The scaleni, longus capitis, superior portion of the longus colli, and rectus capitis lateralis muscles all arise from transverse processes of the cervical vertebrae. The inferior portion of the longus colli arises from the bodies of cervical vertebrae, and the rectus capitis anterior from the lateral mass of the atlas. The scaleni insert into the first two ribs, the rectus capitis anterior and lateralis into the occipital bone, and the colli muscles into the bodies or transverse processes of the cervical vertebrae (Table 7-4). All the muscles are ensheathed by **prevertebral fascia,** are segmentally innervated, and act as a group in flexion and rotation of the head and neck.

Structures associated with the prevertebral region include the **vertebral artery, the carotid sheath** and its contents, the **cervical portion of the sympathetic trunk,** the **spinal nerves** emerging through the intervertebral foramina, and the **cervical plexus.**

Table 7-4
Prevertebral Muscles

Muscle	Origin	Insertion	Action	Nerve
Scalenus anterior	Transverse processes of third to sixth cervical vertebrae	Scalene tubercle on first rib	Bilaterally stabilize neck; unilaterally inclines neck to side	Twigs from cervical plexus and C_5 through C_7
Scalenus medius	Transverse processes of lower five cervical vertebrae	Upper surface of first and second ribs	As above	Cervical plexus and C_4 through C_8
Scalenus posterior	Transverse processes of fifth and sixth cervical vertebrae	Outer surface of second rib	As above	Twigs from C_7 or C_8
Longus capitis	Transverse processes of third through sixth cervical vertebrae	Basilar portion of occipital bone	Flexes and rotates head	Twigs from C_1 through C_4
Longus colli	Transverse processes and bodies of third cervical to third thoracic vertebrae	Anterior tubercle of atlas, bodies of second to fourth cervical vertebrae, and transverse processes of fifth and sixth cervical vertebrae	Flexes and rotates head	Twigs from C_2 through C_6
Rectus capitis anterior	Lateral mass of atlas	Basilar portion of occipital bone	Flexes and rotates head	Twigs from C_1 and C_2
Rectus capitis lateralis	Transverse process of atlas	Jugular process of occipital bone	Bends head laterally	Twigs from C_1 and C_2

Posterior Triangle of the Neck

The **posterior triangle** of the neck is bounded anteriorly by the posterior border of the sterno-cleidomastoideus and posteriorly by the anterior border of the trapezius (Fig. 7-6). The **apex** is at the junction of the above muscles on the superior nuchal line, and the **base** is formed by the middle third of the clavicle. The posterior belly of the omohyoideus muscle divides the posterior triangle into a small **subclavian** and a larger **occipital triangle.** Six muscles form the **floor** of the triangle. From superior to inferior, they are the semi-spinalis capitis, splenius capitis, levator scapulae, and the posterior, middle, and anterior scaleni.

Central Venous Pressure

When lying on an examination table all patients have distended neck veins. If the patient changes to a sitting position, the central venous pressure (CVP) can be estimated. This is done by observing how many centimeters the external jugular vein is distended above the level of the clavicle.

Important structures within the posterior triangle include the spinal accessory, phrenic, lesser occipital, greater auricular, transverse cervical, and supraclavicular nerves (Table 7-5); the roots of the brachial plexus; the subclavian, suprascapular, and transverse cervical arteries; and the subclavian and **external jugular veins** (Table 7-6).

Text continues on page 315

Table 7-5
Nerves in Posterior Triangle

Nerve	Origin	Course	Distribution
Spinal accessory (XI)	Brain stem and C_{1-5}	Exits cranial cavity via jugular foramen, crosses posterior triangle to reach trapezius	Sternocleidomastoid and trapezius
Phrenic	Cervical plexus	Crosses anterior surface of scalenus anterior to enter inlet of thoracic cavity	Thoracic diaphragm
Lesser occipital	Second loop of cervical plexus	From midpoint of posterior border of sternocleidomastoid courses along posterior border to area behind ear	Skin behind ear
Greater auricular	Second loop of cervical plexus	From midpoint of posterior border of sternocleidomastoid crosses muscle to angle to mandible	Skin between sternocleidomastoid and ramus of mandible
Transverse cervical	Second loop of cervical plexus	Crosses sternocleidomastoid transversely	Skin over anterior triangle
Supraclaviculars	Third loop of cervical plexus	From midpoint of posterior border of sternocleidomastoid fan out inferiorly	Skin over shoulder and upper chest to second intercostal space
Segmental branches	Ventral rami of spinal nerves	Pass directly to muscles supplied	Prevertebral muscles and muscles in floor of posterior triangle

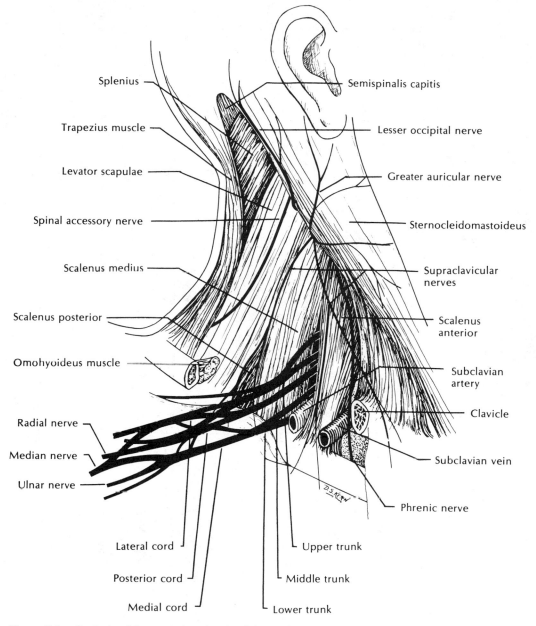

Splenius

Trapezius muscle

Levator scapulae

Spinal accessory nerve

Scalenus medius

Scalenus posterior

Omohyoideus muscle

Radial nerve

Median nerve

Ulnar nerve

Semispinalis capitis

Lesser occipital nerve

Greater auricular nerve

Sternocleidomastoideus

Supraclavicular nerves

Scalenus anterior

Subclavian artery

Clavicle

Subclavian vein

Phrenic nerve

Lateral cord

Posterior cord

Medial cord

Upper trunk

Middle trunk

Lower trunk

Figure 7-6. *Contents of the posterior triangle of the neck.*

Table 7-6
Arterial Supply to the Posterior Triangle and Root of Neck

Artery	Origin	Course	Distribution	Anastomoses
Subclavian	Brachiocephalic on right; aortic arch on left	Traverse root of neck; at first rib becomes axillary	Root of the neck, posterior triangle and scapular region, vertebral canal, brain, and thorax	No direct anastomoses
Vertebral	Subclavian	Traverses transverse foramina C_6–C_1; passes through foramen magnum to form basilar with opposite vertebral	Deep muscles of neck, vertebral canal, spinal cord and brain	With opposite vertebral; vertebral canal branches with other twigs to vertebral canal, and deep cervical
Thyrocervical trunk	Subclavian	Ascends to divide into inferior thyroid, suprascapular, transverse cervical and ascending cervical branches	Thyroid gland, deep muscles of neck; muscles of scapular region	No direct anastomoses
Inferior thyroid	Thyrocervical trunk	Passes in a sigmoid course to reach deep aspect of thyroid gland	Thyroid gland; laryngeal branch to interior of larynx	Superior thyroid
Suprascapular	Thyrocervical trunk	Passes superficial to anterior scalene; crosses posterior triangle to reach superior border of scapula	Supraspinatus and infraspinatus branches supply respective muscles on dorsum of scapula	Dorsal scapular and scapular circumflex
Transverse cervical	Thyrocervical trunk	Passes superficial to anterior scalene to cross posterior triangle; divides into ascending and dorsal scapular branches*	Ascending cervical supplies trapezius; dorsal scapular supplies levator scapulae and rhomboids	Suprascapular branches, and scapular circumflex
Ascending cervical	Thyrocervical trunk (or inferior thyroid)	Ascends on superficial aspect of anterior scalene; parallels course of phrenic nerve	Deep muscles of neck	Deep cervical
Costocervical trunk	Subclavian artery	Trunk divides into a deep cervical and superior intercostal	Deep muscles of neck, and upper two intercostal spaces	No direct anastomoses
Deep cervical	Costocervical trunk	Ascends in neck on deep muscles	Deep muscles of neck	Ascending cervical and vertebral
Superior intercostal	Costocervical trunk	Descends over neck of first rib to provide posterior intercostals to upper two intercostal spaces	Muscles and overlying skin of first two intercostal spaces	Anterior intercostals of first two intercostal spaces

* The dorsal scapular artery may arise as a direct branch from the third part of the subclavian artery, rather than as a branch of the transverse cervical.

The latter, formed by the union of the retromandibular and posterior auricular veins, descends vertically from the angle of the mandible to pass obliquely across the sternocleidomastoideus, where it pierces the deep fascia about 2 cm to 3 cm above the clavicle and empties into the subclavian vein.

Intravenous Cannulation

When other peripheral veins are collapsed or otherwise unavailable, as is often the case with old, obese, or chronically ill patients, the external jugular vein is a readily available alternative for cannulation. The risk of complications is relatively small compared to subclavian or internal jugular entry. Intravenous cannulation is frequently used in a Code Blue (emergency) situation.

The **spinal accessory nerve** emerges from the posterior border of the sternocleidomastoideus to cross the triangle obliquely, dividing it into equal parts. It courses deep to the trapezius muscle and supplies the trapezius and sternocleidomastoid. The **phrenic nerve,** arising from the third, **fourth,** and fifth cervical nerves, crosses the scalenus anterior obliquely to descend into the thorax where it supplies the thoracic diaphragm.

Drooped Shoulder

A superficial wound in the posterior triangle can sever the relatively superficial spinal accessory nerve. Injury to this nerve denervates the trapezius muscle. This would result in asymmetry of the lateral slope of the neck from atrophy of the muscle and a downward displacement or drooping of the shoulder.

Inferior to the spinal accessory nerve, four cutaneous branches of the cervical plexus emerge from the posterior border of the sternocleidomastoideus. The **lesser occipital nerve** (C_2 and C_3) follows the posterior border of the sternocleidomastoideus cephalad to pierce the deep fascia near the mastoid process, where it supplies the scalp above and behind the ear, and the medial surface of the pinna. The **greater auricular nerve** (C_2 and C_3) passes obliquely across the sternocleidomastoideus to ascend behind the ear, where it gives a **mastoid branch** to the mastoid process, an **auricular branch** to both surfaces of the pinna, and **facial branches** to the skin in front of the ear and over the parotid gland. The **transverse cervical nerve** (C_2 and C_3) crosses the sternocleidomastoideus horizontally to divide into a **superior** and an **inferior branch.** It supplies the skin over the anterior triangle of the neck. A large branch from C_3 and C_4 divides into the **medial, intermediate,** and **lateral supraclavicular nerves.** They pierce the platysma near the clavicle to supply the skin of the chest as low as the second intercostal space and the skin over the upper portion of the shoulder.

Local Anesthetic Block

Infiltration of a local anesthetic agent at the midpoint of the posterior border of the sternocleidomastoid can anesthesize most of the skin of the neck. This injection blocks the four cutaneous branches of the cervical plexus that emerge behind the sternocleidomastoideus.

Within the posterior triangle, the **trunks of the brachial plexus** emerge through the **scalene gap** between the anterior and the middle scaleni muscles to continue through the cervicoaxillary canal into the axilla.

Four branches of the brachial plexus arise in the posterior triangle. These are referred to as **supraclavicular branches** of the plexus because they originate proximal to the clavicle. One branch, the **dorsal scapular,** arises from the ventral ramus of C_5. It pierces the scalenus medius to cross the posterior triangle and supply the levator scapulae and rhomboids. The **long thoracic** arises from ventral rami C_5, C_6, and C_7. It descends on the lateral thoracic wall to run on the superficial aspect of the serratus anterior, which it supplies. Two branches, the nerve to the subclavius and the suprascapular, arise from the upper trunk. The **nerve to the subclavius** supplies the subclavius muscle. The **suprascapular** crosses the posterior triangle to reach the superior border of the scapula. It passes through the scapular notch, deep to the transverse scapular ligament, to terminate in branches to the supraspinatus and infraspinatus muscles.

Scalenus Anticus Syndrome (Cervical Rib Syndrome)

This syndrome results from compression of the brachial plexus and subclavian artery. It may be due to a hypertrophy of the scalene muscles, which narrows the "scalene gap" through which these structures pass to enter the posterior triangle. It may also be caused by the presence of a supernumerary (cervical) rib. In this condition there may be pain, numbness, and weakness of the upper extremity. If this is not corrected it can result in muscular atrophy and reflex disorders.

The **subclavian artery** crosses the first rib to pass through the scalene gap in company with, but anterior to, the brachial plexus, while the **subclavian vein** passes anterior to the scalenus anterior as it crosses the first rib. Two branches of the subclavian artery, the **suprascapular** and the **transverse cervical,** cross the scalenus anterior from medial to lateral to pin down the phrenic nerve as they course toward the scapular region.

Subclavian Vein Catheterization

At the scalene gap the subclavian vein passes anterior to the scalenus anterior. The scalenus anterior inserts into the scalene tubercle on the first rib. When using the supraclavicular approach to insert a needle or catheter into the subclavian vein, palpation of the scalene tubercle safeguards the underlying subclavian artery and brachial plexus. The latter structures course deep to the scalenus anterior as they pass through the scalene gap.

Root of the Neck

The **root of the neck** is the junctional area between the neck proper and the thorax and is limited laterally by the first rib, anteriorly by the manubrium, and posteriorly by the first thoracic vertebra (Fig. 7-7). It transmits all the structures passing between the neck and the thorax.

Thyroid Gland

The **thyroid** is an encapsulated, endocrine, butterfly-shaped gland, located in front and to the sides of the trachea at the level of the fifth through seventh cervical vertebrae. It consists of a **right** and a **left lobe** connected by a narrow **isthmus** of glandular tissue located at the level of the third or fourth tracheal ring. Each lobe is somewhat conical in form and consists of a base situated at the level of the fifth or sixth tracheal ring and an apex resting against the side of the thyroid cartilage extending to its superior border. Its full, rounded lateral (superficial) surface is covered by infrahyoid muscles, and the medial (deep) surface is molded by the structures on which it lies, namely, the cricoid and thyroid cartilages, and the cricothyroideus and inferior constrictor muscles superiorly, and the trachea and esophagus inferiorly. There may be a slender extension of the thyroid gland, the **pyramidal lobe,** passing either from the isthmus or the left lobe superiorly to the hyoid bone. A narrow slip of muscle, the **levator glandulae thyroideae,** is sometimes attached to this lobe.

Thyroglossal Duct

The thyroid gland develops from a downgrowth of a solid column of cells in the floor of the primitive pharynx at the foramen caecum of the tongue. During the caudal migration of the thyroid into the neck region, the cell column becomes canalized as the thyroglossal duct. The distal cells of the duct give rise to the thyroid. Although the fully developed thyroid

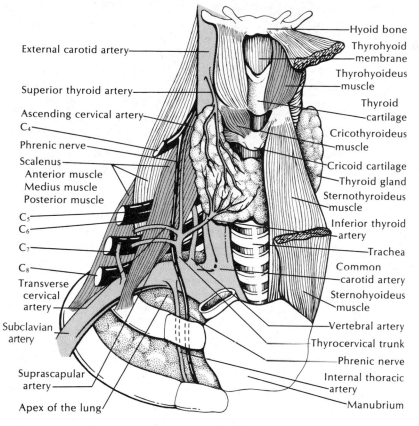

External carotid artery

Superior thyroid artery

Ascending cervical artery

C4

Phrenic nerve

Scalenus
Anterior muscle
Medius muscle
Posterior muscle

C5
C6

C7

C8

Transverse
cervical
artery

Subclavian
artery

Suprascapular
artery

Apex of the lung

Hyoid bone
Thyrohyoid membrane
Thyrohyoideus muscle
Thyroid cartilage
Cricothyroideus muscle
Cricoid cartilage
Thyroid gland
Sternothyroideus muscle
Inferior thyroid artery
Trachea
Common carotid artery
Sternohyoideus muscle
Vertebral artery
Thyrocervical trunk
Phrenic nerve
Internal thoracic artery
Manubrium

Figure 7-7. *Root of the neck.*

has no duct, remnants of the duct may be found along its course. Cysts develop within such vestiges, especially in the sublingual region and in the midline above and below the thyroid cartilage. If they enlarge or become infected, these cysts must be removed surgically. Operative cure of a thyroglossal duct cyst requires removal of the cyst and the duct by removing the middle portion of the hyoid bone and ligating the duct close to its origin, at the base of the tongue.

The thyroid gland is a highly vascularized organ supplied by two pairs of relatively large arteries. The **superior thyroid arteries,** branches from the external carotids, supply the apices, and the **inferior thyroid arteries,** from the thyrocervical trunks, supply the base and the deep surface. A single, inconstant, small branch, the **thyroidea ima artery,** may arise from the brachiocephalic, the left common carotid, or the arch of the aorta. These vessels anastomose freely within each lobe but little communication occurs across the midline. **Venous drainage** is accomplished by three pairs of veins that form a superficial plexus; however, most of the blood is drained from the deep surface. The **superior thyroid veins** drain the upper portion of the gland and either cross the common carotids to join the internal jugular veins, or follow the superior thyroid arteries to end in the common facial veins. The

middle thyroid veins arise near the lower portion of the gland and cross the common carotids to empty into the internal jugulars. The **inferior thyroid veins** are the largest veins draining the thyroid. They begin as a plexus over the isthmus and pass inferiorly to empty into the brachiocephalic veins.

Parathyroid Glands

Two or more pairs of small endocrine glands, the **superior** and **inferior parathyroids,** are usually embedded in the posterior aspect of the thyroid gland. The superior glands are more constant in position and usually lie at the level of the middle of the thyroid, while the inferior glands are situated near the base of the thyroid. Their vascular supply is usually derived from twigs of the inferior thyroid artery.

Parathyroid Tetany

Parathyroid tetany results from the accidental removal of all or most of the parathyroids during a total thyroidectomy. Surgeons, therefore, are careful to identify and leave intact the parathyroid glands (usually four in number). The hormone of the parathyroid regulates calcium metabolism and plasma calcium concentrations. With removal of the parathyroids, the plasma calcium is decreased, causing increased neuromuscular activity, such as muscular spasms, weakness, and nervous hyperexcitability, a condition called tetany. Tetany from removal of the parathyroid glands can cause death, unless adequate calcium or vitamin D is provided, or exogenous parathyroid hormone is administered.

Arteries in the Root of the Neck

The arteries in the root of the neck originate from the arch of the aorta and include the brachiocephalic (innominate) on the right side and the common carotid and subclavian arteries on the left (see Table 7-6). The **brachiocephalic** arises from the arch of the aorta to pass superiorly and divides behind the sternoclavicular joint into right common carotid and right subclavian arteries. Anteriorly the brachiocephalic artery is covered by the sternoclavicular joint and the sternohyoideus and sternothyroideus muscles; medially it rests on the trachea; laterally it is related to the right brachiocephalic vein, and posteriorly it is separated from the apical pleura by fat and connective tissue.

On the right side the **subclavian artery** originates behind the sternoclavicular joint as a terminal branch of the brachiocephalic; on the left, it is a direct branch of the arch of the aorta and enters the root of the neck by passing behind the sternoclavicular joint. Both arteries arch laterally, groove the pleura and lungs, and pass between the anterior and middle scaleni muscles to become the axillary arteries at the lateral border of the first rib. The subclavian artery is arbitrarily divided into three parts by the anterior scalenus muscle. Branches from the first part (proximal to the muscle) include the vertebral, the thyrocervical trunk, and the internal thoracic arteries; the second part (deep to the muscle) gives off the costocervical trunk, while the dorsal scapular artery may arise as a single branch of the third part (distal to the muscle).

The **vertebral artery** arises from the posterosuperior aspect of the subclavian. It ascends vertically along the lateral border of the longus colli muscle to the level of the cricoid cartilage. It passes between the scalenus anterior and the longus colli muscles to enter the foramen transversarium of the sixth cervical vertebra. It traverses similar foramina in each of the cervical vertebrae to finally enter the cranial cavity through the foramen magnum. It joins the opposite vertebral artery to form the basilar artery, which contributes to the formation of the circle of Willis, which supplies the brain.

Subclavian Steal Syndrome

The subclavian steal syndrome refers to a reversal of blood flow from the internal carotid through the basilar artery (that supplies the brain stem), down the vertebral to the subclavian artery. It can only occur when the subclavian artery is occluded proximal to the origin of the vertebral artery. When a patient with subclavian steal syndrome increases the demand for blood flow to the upper extremity by vigorous exercise, for example, the extra blood is "stolen" from the basilar–vertebral system, which can produce syncope or stroke.

Arising from the anterior aspect of the subclavian at the medial margin of the anterior scalenus muscle, the short, wide **thyrocervical trunk** gives origin to four branches. The largest branch, the **inferior thyroid artery,** follows an S-shaped course, ascending along the medial border of the scalenus anterior to pass medially between the carotid sheath and vertebral vessels at the level of the cricoid cartilage. It then descends along the posterior border of the thyroid gland, which it supplies from its deep aspect. The **ascending cervical artery** is a small, constant branch of the trunk, which ascends obliquely across the scalenus anterior to give twigs to prevertebral muscles and the vertebral canal. Coursing laterally across the scalenus anterior, anterior to the phrenic nerve, the **transverse cervical branch** traverses the posterior triangle of the neck. It courses deep to the omohyoid muscle to reach the anterior border of the trapezius. It divides into a **superficial branch** (ascending cervical), ramifying on the deep surface of the trapezius and supplying it, and a **deep** (descending) **branch** that passes deep to the levator scapulae and rhomboids to follow the medial border of the scapula and to supply these muscles. The **suprascapular artery** follows the transverse cervical artery to pass to the scapular notch, where it meets the suprascapular nerve. At the notch the artery passes superficially and the nerve deep to the transverse scapular ligament. Both structures divide into **supraspinatus** and **infraspinatus branches** to supply muscles in their respective fossae on the dorsum of the scapula.

The **internal thoracic artery,** arising from the inferior aspect of the subclavian, passes inferomedially to enter the thorax behind the first costal cartilage. It lies between the pleura and ribs and the intercostal muscles. It is usually crossed anteriorly by the phrenic nerve. This vessel continues inferiorly along the lateral border of the sternum to give **anterior intercostal branches** to the first six intercostal spaces, and then terminates by dividing into the **superior epigastric** and **musculophrenic arteries.**

The short **costocervical trunk** arises from the posterior aspect of the subclavian to pass superoposteriorly over the pleura to the neck of the first rib. Opposite the first intercostal space it gives a **deep cervical branch** that ascends between the neck of the first rib and the transverse process of the seventh cervical vertebra to supply deep muscles of the neck. The **superior intercostal branch** of the trunk courses inferiorly, anterior to the neck of the first rib, to give posterior intercostal branches to the first and second intercostal spaces.

As a **variation** of the branches of the subclavian artery, the **dorsal scapular artery** arises, in about 50% of the cases, from the third part of the subclavian to course parallel to the medial border of the scapula. It supples the levator scapulae and the rhomboids and replaces the deep (descending) branch of the transverse cervical artery from the thyrocervical trunk. When the dorsal scapular artery is present, the transverse cervical artery supplies only the trapezius muscle and is called the **superficial cervical** artery.

Deep Back

Objectives

At the completion of the study of the deep back the student should be able to

▶ *Palpate surface landmarks of the back*

▶ *Locate the intrinsic muscles of the back and name the muscles in each group.*

▶ *Identify the major processes of a vertebra and differentiate vertebrae from cervical, thoracic, and lumbar regions.*

▶ *Define cervical enlargement, dura mater, subarachnoid space, filum terminale, and dentate ligament*

▶ *Describe the blood supply of the spinal cord*

Deep Muscles (Table 7-7)

The deep muscles of the back are arranged in two groups: a **longitudinal group** consisting of the **erector spinae,** or **sacrospinalis (iliocostalis, longissimus,** and **spinalis);** and a **transverse group,** including the **semispinalis,** the **multifidi,** the **rotatores,** the **interspinales,** and the **intertransversarii** (Fig. 7-8). (Current terminology lists the term erector spinae for all the deep muscles of the back.)

The longitudinal group spans the interval from the spines of the vertebrae to the angles of the ribs and acts to extend the vertebral column. In width this group forms a mass of muscle about as broad as the palm of a hand. These muscles extend vertically from the fourth segment of the sacrum to the mastoid process of the temporal bone and are placed side by side like three fingers.

The muscles of the transverse group extend from the spines of the vertebrae to the tips of the

transverse processes and are present between the fourth sacral vertebra and the occipital bone. The muscles of the transverse group are placed one on top of the other, like the layers of a sandwich, and act primarily to twist the vertebral column. The deep muscles are innervated segmentally by dorsal rami of the spinal nerves, except for some of the intertransversarii, which are supplied by ventral rami.

Fascia of the deep muscles forms thin muscular envelopes, except in the lumbar region, where it thickens and is disposed in three anteroposterior layers, or lamina, as the **thoracolumbar fascia.** The thick strong **posterior layer** covers the sacrospinalis and continues superiorly onto the thorax. Inferiorly it attaches to the iliac crest and sacrum. Medially it fuses with the periosteum of the vertebral spines, and superiorly with the ligamentum nuchae. Laterally it joins the middle and anterior layers at the lateral border of the erector spinae. The **middle** and **anterior layers** sheathe the quadratus lumborum muscle. At the lateral border of this muscle they fuse with the posterior layer to give origin to the three muscles of the anterolateral abdominal wall (Fig. 7-9.)

Suboccipital Triangle (Table 7-8)

The muscles that bound the small **suboccipital triangle** are medially, the **rectus capitis posterior major,** which overlies the **minor;** laterally, the **obliquus capitis superior;** and inferiorly, the **obliquus capitis inferior** (Fig. 7-10). The **semispinalis capitis** and the **longissimus capitis,** lying deep to the flattened, relatively extensive **splenius muscle,** form the roof of the triangle. The floor is formed by the posterior

Text continues on page 325

Table 7-7
Deep Muscles of the Back

Muscle	Origin	Insertion
Erector spinae (longitudinal group)	Series of muscles forming mass that extends from sacrum to skull. Acting unilaterally, they bend vertebral column to that side; bilaterally they extend vertebral column. They are segmentally innervated by dorsal rami of spinal nerves, as are all muscles of back listed below.	
Iliocostalis lumborum	Iliac crest and sacrospinal aponeurosis	Lumbodorsal fascia and tips of transverse processes of lumbar vertebrae and angles of lower six or seven ribs
Iliocostalis thoracis	Superior borders of lower seven ribs medial to angles	Angles of upper seven ribs and transverse process of seventh cervical vertebra
Iliocostalis cervicis	Superior borders at angles of third to seventh ribs	Transverse processes of fourth, fifth, and sixth cervical vertebrae
Longissimus thoracis	Sacrospinal aponeurosis, sacroiliac ligaments, transverse processes of lower six thoracic and first two lumbar vertebrae	Transverse processes of lumbar and thoracic vertebrae and inferior borders of ribs lateral to their angles
Longissimus cervicis	Transverse processes of upper five or six thoracic vertebrae	Transverse processes of second through sixth cervical vertebrae
Longissimus capitis	Transverse processes of first four cervical vertebrae and articular processes of last four cervical vertebrae	Mastoid process of temporal bone
Spinalis thoracis	Spines of upper two lumbar and lower two thoracic vertebrae	Spines of second through ninth thoracic vertebrae
Spinalis cervicis	Spines of upper two thoracic and lower two cervical vertebrae	Spines of second through fourth cervical vertebrae

Muscle	Origin	Insertion	Action
Semispinalis capitis	Transverse processes of upper six thoracic and seventh cervical vertebrae	Between superior and inferior nuchal lines	Extends and inclines head laterally
Semispinalis thoracis	Transverse processes of lower six thoracic vertebrae	Spines of upper six thoracic and lower two cervical vertebrae	Extends and inclines head laterally
Semispinalis cervicis	Transverse processes of upper six thoracic vertebrae	Spines of second through sixth cervical vertebrae	Extends and inclines head laterally
Multifidus	Sacrum and transverse processes of lumbar, thoracic, and lower cervical vertebrae	Spinous processes of lumbar, thoracic, and lower cervical vertebrae	Abducts, rotates, and extends vertebral column
Rotatores	Transverse processes of second cervical vertebra to sacrum	Lamina above vertebra of origin	Rotate and extend vertebral column
Interspinales	Superior surface of spine of each vertebra	Inferior surface of spine of vertebra above vertebra of origin	Extend and rotate vertebral column
Intertransversarii	Extend between transverse processes of cervical, lumbar, and lower thoracic vertebrae. Unilaterally, they bend vertebral column laterally; bilaterally they stabilize column.		
Splenius cervicis	Spinous processes of third through sixth thoracic vertebrae	Transverse process of first three cervical vertebrae	Inclines and rotates head and neck
Splenius capitis	Ligamentum nuchae and spinous processes of upper five thoracic vertebrae	Mastoid process and superior nuchal line	Inclines and rotates head and neck

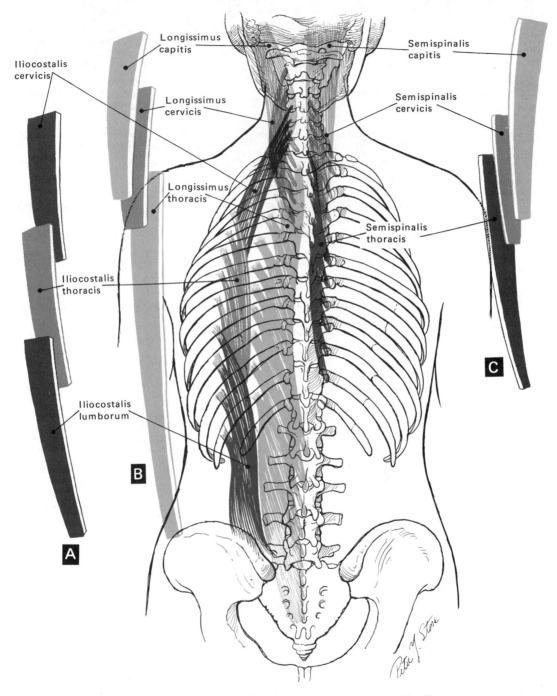

Iliocostalis
cervicis

Longissimus
capitis

Longissimus
cervicis

Longissimus
thoracis

Iliocostalis
thoracis

Iliocostalis
lumborum

B

A

Semispinalis
capitis

Semispinalis
cervicis

Semispinalis
thoracis

C

Figure 7-8. *Intrinsic muscles of the back.* **A.** *The overlapping arrangement of the iliocostalis muscles.* **B.** *The arrangement of the longissimus muscles.* **C.** *The arrangement of the semispinalis muscles.*

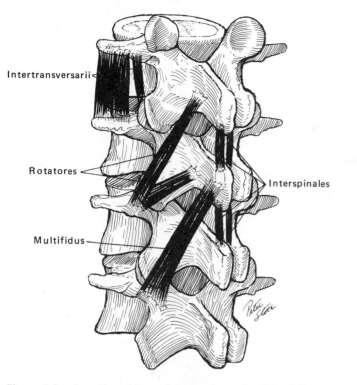

Figure 7-9. *A section of the vertebral column showing relation of the transverse muscles of the back to the vertebrae. These muscles are intrinsic muscles of the back.*

Table 7-8
Muscles of the Suboccipital Triangle

Muscle	Origin	Insertion	Action	Nerve
Rectus capitis posterior major	Spine of axis	Lateral half of inferior nuchal line	Extends and rotates head	Dorsal ramus of C_1 (suboccipital nerve)
Rectus capitis posterior minor	Posterior tubercle of atlas	Occipital bone below medial portion of inferior nuchal line	Extends head	As above
Obliquus capitis superior	Transverse process of atlas	Occipital bone between superior and inferior nuchal lines	Extends head	As above
Obliquus capitis inferior	Spine of axis	Transverse process of atlas	Rotates and extends head	As above

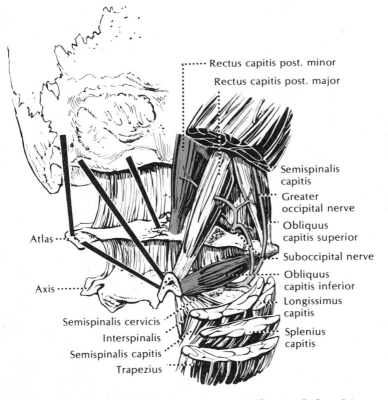

Rectus capitis post. minor
Rectus capitis post. major
Semispinalis capitis
Greater occipital nerve
Obliquus capitis superior
Suboccipital nerve
Obliquus capitis inferior
Longissimus capitis
Splenius capitis

Atlas
Axis
Semispinalis cervicis
Interspinalis
Semispinalis capitis
Trapezius

Figure 7-10. *Suboccipital triangle and contents. (Gardner E, Gray DJ, O'Rahilly R: Anatomy. Philadelphia, Saunders, 1975)*

Table 7-9
Nerve Distribution to Suboccipital Triangle and Back

Nerve	Origin	Course	Distribution
Suboccipital	Dorsal ramus of C_1	Passes between skull and first cervical vertebra to reach suboccipital triangle	Muscles of suboccipital triangle
Greater occipital	Dorsal ramus of C_2	Emerges inferior to obliquus capitis inferior muscle and ascends to reach back of scalp	Skin over occipital bone
Least occipital	Dorsal ramus of C_3	Passes directly to skin	Dermatome of C_3
Dorsal rami	Spinal nerves	Pass segmentally to muscles and skin	Intrinsic muscles of back (erector spinae) and skin over back of lower neck and back

atlantooccipital membrane and the posterior arch of the atlas. The triangle contains the **vertebral artery** as it passes into the skull through the foramen magnum, and the **suboccipital nerve** (C_1), which supplies the muscles of the triangle. The **greater occipital nerve** (C_2) arches around the inferior border of the obliquus capitis inferior to supply sensation to the back of the scalp (Table 7-9). The muscles of the triangle may act as extensors and rotators of the head, but function chiefly as postural muscles.

Vertebral Column

The **vertebral column** forms the central portion of the axial skeleton of the body. It presents two **primary curvatures,** the thoracic and the sacral, which are convex posteriorly, and two **secondary curvatures,** the cervical and the lumbar, which are concave posteriorly. It is composed of seven cervical, twelve thoracic, five lumbar, five fused sacral, and three to five fused coccygeal vertebrae.

Scoliosis
This abnormal curvature of the spine is a condition in which a lateral bending of the vertebral column, usually in the thoracic region, is present. This is the most common of the abnormal vertebral curvatures. It may be congenital owing to an absence of the lateral half of a vertebra (hemivertebra) or acquired from a persistent severe sciatica. In the latter the trunk is usually bent away from the painful side, thus reducing the pressure on the sciatic nerve fibers as they emerge through the intervertebral foramina. Poliomyelitis may cause scoliosis by the paralysis of muscles on one side of the body, which produces a lateral deviation of the trunk towards the unaffected side.

Kyphosis
This abnormal curvature of the spine is an exaggeration of the convex curvature of the thoracic region (hunchback) of the vertebral column. In tuberculosis of the spine, vertebral bodies may partially collapse, causing an acute angular bending of the vertebral column, which is called gibbus. In the elderly, degeneration of the intervertebral discs leads to senile kyphosis. "Round shouldered" is an expression of mild kyphosis.

Lordosis
This abnormal curvature of the spine is an exaggerated convex curvature of the vertebral column in the lumbar region (sway back). It is present in congenital double dislocation of the hip, in which the support of the pelvis occurs posterior to the acetabulum. It may be caused from increased weight of abdominal contents, as in pregnancy or extreme obesity.

The component parts of a **typical vertebra** are the body and a number of processes that surround the centrally located vertebral foramen (Fig. 7-11). The massive **body** gives strength to the vertebral column and is separated from adjacent vertebral bodies by the intervertebral discs. The **transverse processes** project laterally from the junction of the pedicles and the laminae. In the cervical region they contain a foramen for the passage of the vertebral artery, and in the thoracic region they afford articulation for the ribs. The short, slightly rounded **pedicles** project posteriorly from the body and are joined by the flattened **laminae** to form the arch. The pedicles, laminae, and posterior surface of the body form the vertebral foramen. Adjacent vertebral foramina form the vertebral canal. At the junction of the laminae the **spinous processes** project posteriorly. At the junction of the pedicle with the lamina the **superior** and **inferior processes** bear articular facets that form synovial joints with

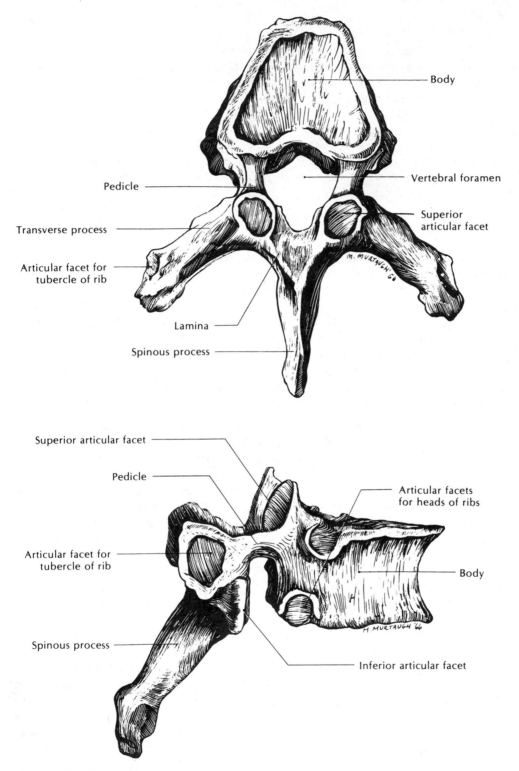

Body

Vertebral foramen

Pedicle

Superior
articular facet

Transverse process

Articular facet for
tubercle of rib

Lamina

Spinous process

M. MURTAUGH
'66

Superior articular facet

Pedicle

Articular facets
for heads of ribs

Articular facet for
tubercle of rib

Body

Spinous process

H. MURTAUGH '66

Inferior articular facet

Figure 7-11. *Typical thoracic vertebra.*

adjacent vertebrae. The deep **vertebral notch** on the inferior border and the shallow notch on the superior border of each pedicle, together with the intervertebral discs, form the **intervertebral foramina** between adjacent vertebrae. They transmit spinal nerves, vessels, and meninges.

Intervertebral discs unite adjacent, unfused vertebral bodies. They consist of an outer layer of fibrocartilage, and **anulus fibrosus,** while centrally they present a relatively soft gelatinous mass, the **nucleus pulposus.**

Slipped Disc

Slipped disc is a layman's term for a herniated intervertebral disc. Discs are subject to pathologic changes where the strain is greatest, for example, the lower lumbar region. If the compression force is excessive, a rupture of the disc occurs, and the semisolid nucleus pulposus herniates partially or completely through the anulus fibrosus to impinge on spinal nerves that emerge from the vertebral column adjacent to the discs in this region. Such nerve compression causes a painful neuralgia (sciatica) down the back, and lateral side of the leg, and into the sole of the foot. Traction, bed rest, and analgesia usually relieve the pain. If such conservative treatment is ineffective, then surgical decompression of the spinal nerves by laminectomy or removal of some of the nucleus pulposus may be necessary to relieve pain.

Regional characteristics differentiate vertebrae of the cervical, thoracic, and lumbar regions. Cervical vertebrae have foramina within their transverse processes, the **foramina transversaii,** which transmits the vertebral artery and vein. The transverse processes present anterior and posterior tubercles, while the spinous processes of the third through sixth vertebrae are bifid. The first and second cervical vertebrae are atypical. The first, or **atlas,** lacks both a body and a spinous process. Instead of the usual articular processes,

the lateral mass of the atlas articulates superiorly with the occipital condyles of the skull and inferiorly with the second cervical vertebra. These lateral masses are joined by the anterior and posterior arches, which present, in the midline, anterior and posterior tubercles. The **odontoid process** projects superiorly from the upper part of the body of the second cervical vertebra, or **axis,** and represents the transposed body of the first vertebra. **Thoracic vertebrae** present **articular facets for the ribs** on the body for articulation with the head of the rib, and on the transverse process for the rib tubercle. The long slender spinous processes of thoracic vertebrae are directed inferiorly. The **lumbar vertebrae** are differentiated by their lack of foramina transversarii and rib facets. They present massive bodies, broad and somewhat quadrilateral horizontal spinous processes, and long slender transverse processes.

Ligaments of the Vertebral Column

The broad, thick **anterior longitudinal ligament** passes over the anterior surface of the vertebral bodies from the skull to the coccyx and is firmly attached to the bodies and the intervertebral discs. The **posterior longitudinal ligament** extends over the posterior surface of the vertebral bodies and, therefore, along the inner anterior surface of the vertebral canal. The **supraspinous ligament** attaches along the tips of the spinous processes, where it enlarges as the ligamentum nuchae in the cervical region. **Interspinous ligaments** pass between adjacent superior and inferior borders of spinous processes and fuse with the supraspinous ligament. The **ligamentum flavum** extends from the anterior aspect of one lamina to the posterior aspect of the lamina below. **Intertransverse ligaments** form small bands between adjacent transverse processes.

Lumbar Puncture

Lumbar puncture is a procedure performed to: (1) obtain a sample of cerebrospinal fluid for diagnostic studies, (2) relieve intracranial pressure, (3) administer drugs, or (4) induce anesthesia. Since the spinal cord, in the adult, does not extend inferior to the second lumbar vertebra, the interspace between L_4 and L_5 is usually chosen for the puncture. At this level there is no danger of damage to the spinal cord. In its course the needle pierces the skin, superficial fascia, supraspinous ligament, interspinous ligament, epidural space, dura, and arachnoid to enter the subarachnoid space.

Spinal Cord and Meninges

As an extension of the brain stem, the **spinal cord** is located in the vertebral canal between the foramen magnum and the level of the first or second lumbar vertebra. Below this level nerve rootlets and the filum terminale, a prolongation of pia mater, form the **cauda equina,** which occupies the vertebral canal. The cylindric spinal cord is slightly flattened anteroposteriorly and has **cervical** and **lumbar enlargements** at the levels of origin of the nerves to the upper and lower extremities. The cord is grooved anteriorly by the **anterior median fissure,** and posteriorly by the **posterior median sulcus.** The designation of a **spinal cord segment** or **spinal level** refers to that portion of the cord associated with the origin of ventral and dorsal roots of a specific spinal nerve.

Relation of Spinal Cord Segments to Vertebrae

The spinal cord is considerably shorter than the vertebral column, extending inferiorly only to the first or second lumbar vertebra. Thus, the relationship between origin of the nerves from the spinal cord and their emergence through intervertebral foramina of the vertebral canal does not coincide. In the lower cervical and upper thoracic region the spinal cord segment is approximately two vertebrae superior to the vertebral level. Thus, the spinous process of C_6 overlies spinal cord segment C_8; in the lower thoracic and upper lumbar regions, T_{11} and T_{12} spinous processes overlie the five lumbar spinal cord segments, while L_1 spinous process overlies the five sacral segments.

Such knowledge is critical to the surgeon in locating a tumor in a patient showing signs of spinal cord compression, for example, loss of sensation over the thumb, an area innervated by C_6. Therefore, to expose this spinal cord segment a laminectomy would be performed on vertebra C_4.

The **blood supply** of the spinal cord is from the **vertebral arteries,** supplemented by segmental **spinal branches** of deep cervical, intercostal, lumbar, and sacral arteries. As the vertebral artery enters the cranial cavity, just prior to its union to the artery of the opposite side to form the basilar artery, it gives off **anterior** and **posterior spinal branches.** The branches from each side unite to form a single trunk anteriorly while posteriorly paired trunks extend inferiorly. These arterial trunks course along the entire length of the cord and receive segmental contributions from the above-named intervertebral branches.

The spinal cord, bathed in cerebrospinal fluid, is invested by three membranes (meninges): the dura, arachnoid, and pia mater (Fig. 7-12). The outer, tough, dense **dura mater** is a fibrous tube that extends from the foramen magnum, where it is continuous with dura of the brain, to the coccyx. Below the level of the second sacral vertebra it narrows considerably and blends with the connective tissue covering the posterior aspect of the coccyx. As each spinal root emerges from the vertebral canal, it carries with it, for a short distance, a prolongation of the dura as the **dural sleeve.** The **epidural** space, between the dura and walls of the vertebral canal, contains fat and a plexus of thin-walled veins. The **subdural space** is a capil-

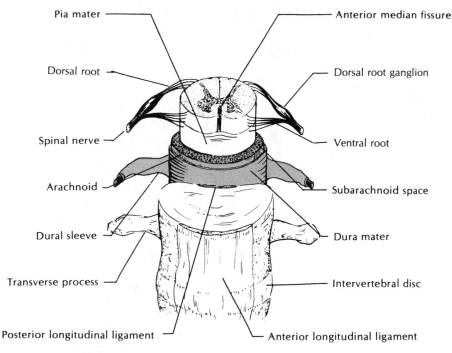

Pia mater

Dorsal root

Spinal nerve

Arachnoid

Dural sleeve

Transverse process

Posterior longitudinal ligament

Anterior median fissure

Dorsal root ganglion

Ventral root

Subarachnoid space

Dura mater

Intervertebral disc

Anterior longitudinal ligament

Figure 7-12. *Cross section of the spinal cord and meninges.*

lary interval between the dura and the arachnoid, containing a small quantity of fluid.

The avascular, delicate, transparent **arachnoid mater,** coextensive with the dura mater, is also continuous through the foramen magnum with the arachnoid covering the brain. Filamentous extensions pass through the subarachnoid space to fuse with the pia mater. Laterally this membrane is prolonged a short distance with the sheath for the spinal nerve. The **subarachnoid space,** a relatively extensive trabeculated interval between the arachnoid and the pia, contains the **cerebrospinal fluid** that serves as a protective liquid cushion for the cord.

Cerebrospinal Fluid Pressure
Cerebrospinal fluid pressure is measured by a manometer attached to a needle inserted into the subarachnoid space in a lumbar puncture procedure. With the patient in a lateral recumbent position, the manometer should register about 120 mm of water.

The **pia mater** is a delicate, vascular membrane intimately adherent to the spinal cord. From the termination of the cord at the first or second lumbar vertebra the pia mater continues inferiorly as the **filum terminale (internum). Denticulate ligaments** are serrated lateral extensions from the pia that anchor and limit torsion or twisting of the cord within the dural sac. The medial edge of this thin ligament has a continuous attachment to the cord midway between the anterior and posterior nerve roots, while the serrated lateral border attaches to the dura at intervals between emerging spinal nerves.

Face

Objectives

At the completion of the study of the face, scalp, and parotid gland the student should be able to

▶ *Delineate dermatomes of the three divisions of the trigeminal nerve and name their terminal branches that supply the skin*

▶ *Label on an illustration the muscles of facial expression*

▶ *List the branches of the facial nerve and facial artery*

▶ *List the layers of the scalp and locate the major sutures of the skull*

▶ *Describe the danger area of the face and scalp and discuss its importance*

▶ *Describe the morphological characteristics, location, and relationships of the parotid gland*

▶ *List the structures coursing through the parotid gland, giving their origin, course, and distribution*

Cutaneous Innervation

The cutaneous innervation to the face is supplied by all three divisions of the **trigeminal,** or fifth cranial, nerve (Fig. 7-13). The **ophthalmic division** (V_1) divides into three branches, the lacrimal, frontal, and the nasociliary nerves, which supply cutaneous innervation above the level of the eyes and on the dorsum of the nose. The **lacrimal nerve** supplies the conjunctiva and skin of the upper eyelid. The **frontal nerve** divides into a **supratrochlear branch** that supplies skin of the forehead and upper eyelid, and a **supraorbital branch** passing through the supraorbital foramen to supply the upper eyelid and the scalp as far posteriorly as the lambdoidal suture. The naso-

ciliary nerve gives **long ciliary branches** to the eye that pierce the lamina cribrosa, and a **posterior ethmoidal** branch that supplies posterior and middle ethmoidal air cells. The nasociliary nerve then bifurcates to give an **anterior ethmoidal** branch to the anterior ethmoidal air cells, the anterosuperior aspect of the nasal cavity, and the skin over the dorsum of the nose, and an **infratrochlear** branch that supplies skin of the upper eyelid.

Trigeminal Neuralgia

Trigeminal neuralgia, or tic douloureux, is associated with excruciating pain over the face, especially areas innervated by the mandibular and maxillary divisions of the fifth cranial nerve. The etiology is unknown, but it is most common in the middle aged and elderly. Alcohol injections into the trigeminal ganglion or around the divisions of the nerves as they emerge from the skull may relieve trigeminal neuralgia.

The **maxillary division** (V_2) divides into the infraorbital and zygomatic nerves, which supply on oblique area between the mouth and the eyes. The **infraorbital nerve** appears on the face at the infraorbital foramen and divides into an **inferior palpebral branch** to the skin and conjunctiva of the lower lid, an **external nasal branch** to the side of the nose, and a **superior labial branch** to the upper lid. The **zygomatic nerve** terminates in **zygomaticofacial** and **zygomaticotemporal branches** to the skin over the zygomatic bone and the temporal region.

The **mandibular division** (V_3) supplies the remainder of the face. Its sensory branches are derived from both the anterior and posterior divisions of the nerve. The **buccal branch,** from the anterior division, breaks into a series of twigs to

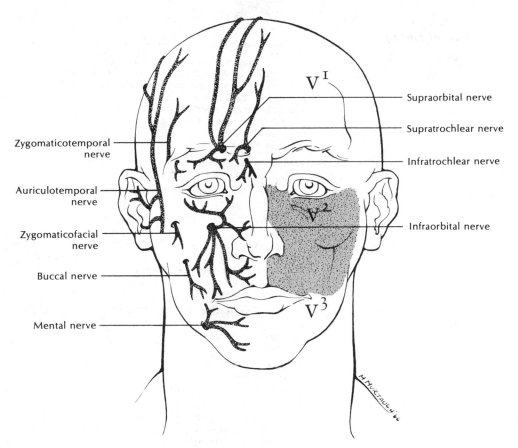

Figure 7-13. *Cutaneous innervation of the face.*

supply the mucosa of the cheek. The **auriculo-temporal nerve,** from the posterior division, divides into an auricular branch to the external auditory meatus, tympanic membrane, skin of the tragus, and the upper and outer part of the pinna, and a **superficial temporal branch,** which supplies most of the temporal region. A terminal branch of the mandibular division is the **inferior alveolar nerve.** It passes through the mandibular foramen to course in the mandibular canal. It emerges from the canal at the mental foramen to divide into the **mental branch** supplying the skin of the chin, and the **inferior labial branch** to the lower lip (Table 7-10).

Nerve Block

Nerve block is a term signifying loss of sensation in a region, such as in local dental anesthesia. For dental procedures on the lower teeth, analgesia is obtained by depositing an anesthetic solution in the proximity of the inferior alveolar nerve as it enters the mandibular foramen. This effectively blocks all sensations carried centrally by this nerve. For anesthesia to the upper teeth, superior alveolar nerve endings are blocked by inserting the needle beneath the mucous membrane and the anesthetic solution is then infiltrated

Table 7-10
Cutaneous Innervation of Face

Nerve	Origin	Course	Distribution
Frontal	Ophthalmic nerve (V_1)	Traverses orbit on superior aspect of levator palpebrae superioris; divides into supraorbital and supratrochlear branches, which traverse foramina or notches of same name	Skin of forehead and scalp to lambdoidal suture
Infratrochlear	Nasociliary nerve	Follows medial wall of orbit to upper eyelid	Skin of upper lid
Dorsal (External) nasal	Anterior ethmoidal nerve	Terminal branch in nasal cavity emerges on face between nasal bone and nasal cartilage	Skin on dorsum of nose
Zygomatic	Maxillary nerve (V_2)	Arises in floor of orbital cavity, divides into zygomaticofacial and zygomaticotemporal, which traverse foramina of same name	Skin over zygomatic arch and anterior temporal region
Infraorbital	Terminal branch of maxillary nerve (V_2)	In floor of orbit runs in orbital groove and canal to emerge at infraorbital foramen	Skin of cheek, lateral side of nose and upper lip
Buccal	Mandibular nerve (V_3)	From anterior division of V_3 in infratemporal fossa courses anteriorly to reach cheek	Skin of cheek
Auriculotemporal	Mandibular nerve (V_3)	From posterior division of V_3 passes between neck of mandible and external auditory meatus to course with superficial temporal artery	Skin in front of ear and temporal region
Mental	Terminal branch of inferior alveolar	Emerges from mandibular canal at mental foramen	Skin of chin and lower lip

slowly throughout the area of the roots of the teeth to be treated.

Muscles (Table 7-11)

The muscles of facial expression lie within the superficial fascia of the face (Fig. 7-14). They gain their origin from fascia or underlying bone and insert into the skin. They are usually described by their action on, or relation to, the orifices of the face; are frequently fused with adjacent muscles; and are innervated by the facial, or seventh, cranial nerve. Muscles associated with or adjacent to the eye include the **frontalis,** which wrinkles the forehead and raises the eyebrows, as in registering surprise, or draws the skin of the scalp forward; the **corrugator** furrows and wrinkles the brow; and the **orbicularis oculi** acts in winking and blinking.

Muscles around the nose are the **procerus** (pyramidalis nasi) over the bridge of the nose, which wrinkles the skin between the eyebrows and draws the latter medially in registering a fierce expression; the **nasalis** (compressor

Table 7-11
Muscles of the Face

Muscle	Origin	Insertion	Action	Nerve
Frontalis	Epicranial aponeurosis	Skin of forehead	Raises the eyebrows, wrinkles forehead	Facial
Corrugator supercilii	Medial portion of supra-orbital margin	Skin of medial half of eyebrow	Draws eyebrows downward and medialward	Facial
Orbicularis oculi	Medial orbital margin, medial palpebral ligament, and lacrimal bone	Skin and rim of orbit and tarsal plate	Sphincter of eyelids	Facial
Procerus	Lower part of nasal bone, upper part of lateral nasal cartilage	Skin between eyebrows	Wrinkles skin over bridge of nose	Facial
Nasalis	Canine eminence lateral to incisive fossa	Nasal cartilages	Draws alae of nostril toward septum	Facial
Depressor septi	Incisive fossa of maxilla	Posterior aspect of ala and nasal septum	Draws nasal septum inferiorly	Facial
Dilator nares	Margins of piriform aperture of maxilla	Side of nostril	Widens nostril	Facial
Orbicularis oris	Surrounds oral orifice, forming intrinsic muscle of lips. Interlaces with other muscles associated with lips.		Acts in compression, and protrusion of lips	Facial
Levator labii superioris	Frontal process of maxilla, infraorbital region, and inner aspect of zygomatic bone	Greater alar cartilage, nasolabial groove, and skin of upper lip	Elevates lip, dilates nostril, and raises angle of mouth	Facial
Zygomaticus	Zygomatic arch	Angle of mouth	Elevates and draws angle of mouth backward	Facial
Depressor labii inferioris	Mandible between symphysis and mental foramen	Into orbicularis oris and skin of lower lip	Depresses and everts lower lip	Facial
Depressor anguli oris	Oblique line of mandible	Angle of mouth	Turns corner of mouth downward	Facial
Risorius	Fascia overlying masseter muscle	Angle of mouth	Retracts angle of mouth	Facial
Platysma	Superficial fascia of pectoral and deltoid regions	Mandible, skin of neck and cheek, angle of mouth, and orbicularis oris	Depresses lower jaw; tenses and ridges skin of neck	Facial
Buccinator	Pterygomandibular raphe, alveolar processes of jaw opposite molar teeth	Angle of mouth	Compresses cheek as accessory muscle of mastication	Facial
Mentalis	Incisive fossa of mandible	Skin of chin	Elevates and protrudes lower lip	Facial
Auricularis anterior, superior, and posterior (all rudimentary)	Temporal fascia, epicranial aponeurosis, and mastoid process	Front of helix, triangular fossa, and convexity of concha	May retract and elevate ear	Facial

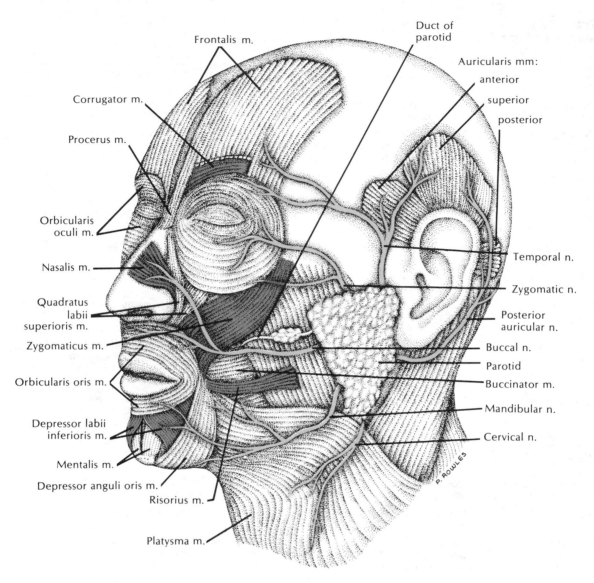

Figure 7-14. *Muscles of the facial expression and the facial nerve.*

nares), which compresses the nostril, drawing its margins toward the septum; the **dilator nares,** which enlarges the nostril; and the **depressor septi,** which draws the cartilaginous portion of the septum inferiorly and constricts the nostril.

The **orbicularis oris** ("kissing" muscle) encircles the mouth, forms the muscular bulk of the lips, interlaces with other muscles of the lips, and acts in compression, contraction, and protrusion of the lips. Two muscles are associated with the upper lip: the **levator labii superioris,** which functions in elevation of the lip, dilation of the nostril, and raising the angle of the mouth; and the **zygomaticus** ("laughing" muscle), which draws the angle of the mouth backward and upward.

Muscles of the lower lip are four in number: the **depressor labii inferioris** expresses terror or grief in depressing and everting the lip; the **depressor anguli oris** draws the angle of the mouth downward; the **risorius** retracts the angle of the mouth in grinning; and the **platysma,** interdigitating with muscles of the lower lip, depresses the lower lip and mandible and tenses or ridges the skin of the neck.

The **buccinator** muscle makes the cheek rigid in sucking or blowing. It acts as an accessory muscle of mastication by eliminating the space of the vestibule between the cheek and the jaws. The **mentalis** muscle of the chin elevates and protrudes the lower lip and, when well developed, may form with its member of the opposite side the dimple of the chin.

Three poorly developed muscles of the external ear, the **auricularis anterior, superior,** and **posterior,** insert into the pinna and act in feeble movements of the ear.

Scalp

The skull is covered by the **scalp** which has three layers. The outer layer is the skin. Next to the skin is a layer of dense connective tissue through which nerves and blood vessels course. The **epicranium** is adjacent to the periosteum. It is a musculoaponeurotic sheet with the **occipitalis muscle** at its posterior end and the **frontalis muscle** attached anteriorly. The strong aponeurosis between these muscles is the **epicranial aponeurosis (galea aponeurotica).** All three layers are bound tightly together and are separated from the periosteum of the skull by a loose connective tissue space called the **danger area of the scalp** when infected.

The blood supply to the scalp comes from **occipital, superficial temporal,** and **posterior auricular** branches of the external carotid. In addition, the internal carotid supplies the scalp by way of **supratrochlear** and **supraorbital** branches of the ophthalmic artery. All of these branches anastomose freely so scalp lacerations tend to bleed profusely.

The nerves to the scalp include the **supratrochlear** and **supraorbital** to the forehead and anterior scalp. The temporal region is supplied by the **auriculotemporal** nerve; the **lesser occipital** innervates the skin behind the ear, and the **greater occipital** nerve is distributed to the back of the head. The latter extends anteriorly to the lambdoidal suture.

The Danger Area of the Scalp

The danger area of the scalp includes the subaponeurotic tissue cleft beneath the epicranial (galea) aponeurosis. Due to the looseness of the connective tissue in this area, blood and pus may spread over the skull, limited only by the attachments of the galea aponeurotica. Infection may spread through the valveless emissary veins to skull bones, which may become infected and necrotic, or to the dural sinuses and result in meningitis. Septic emboli can form and pass into the dural sinuses to be disseminated throughout the venous sytem. An old surgical axiom states that if it were not for emissary veins, wounds of the scalp would lose half their significance.

Arteries and Nerves (Tables 7-12 and 7-13)

The **facial nerve** follows a complicated intracranial course from its origin on the brain stem to its emergence from the cranium at the stylomastoid foramen. Here its branches include the **posterior auricular nerve,** with an **auricular branch** to the posterior auricularis muscle, and an **occipital branch** to the occipitalis muscle. The **digastric branch** supplies the posterior belly of the digastric muscle, and the **stylohyoid branch,** the stylohyoideus muscle. The facial nerve then turns anteriorly to pass lateral to the styloid process and enter the substance of the parotid gland, where it divides into an **upper (temporozygomatic)** and a **lower (cervicomandibular) division.**

The upper division supplies all the superficial

Table 7-12
Terminal Branches of Facial Nerve

Nerve	Origin	Course	Distribution
Posterior auricular	Facial nerve (VII)	From stylomastoid foramen passes posteriorly to occipital region	Posterior and superior auricular muscles and occipitalis
Nerve to stylohyoid	Facial nerve (VII)	From stylomastoid foramen passes to styloid process	Stylohyoid and posterior belly of digastric
Branches to facial muscles	Facial nerve (VII)	From stylomastoid foramen facial turns anteriorly, cleaves parotid, and bifurcates into upper (temporozygomatic) and lower (cervicomandibular) divisions	Five terminal branches (*i.e.*, temporal, zygomatic, buccal mandibular and cervical) pass to and supply facial muscles in respective regions

muscles of the face above the zygomatic arch, with a **temporal branch** innervating the anterior and superior auricularis, frontalis, orbicularis oculi, and corrugator muscles, and a **zygomatic branch** to the orbicularis oculi and zygomaticus muscles. The lower division gives a **buccal branch** to the muscles of expression below the orbit and around the mouth, a **mandibular branch** to the depressor labii inferioris and mentalis muscles, and a **cervical branch** to the platysma (see Fig. 7-14).

Parotidectomy and Facial Paralysis
In partial or complete surgical removal of the parotid gland the facial nerve may be damaged causing facial (Bell's) paralysis. The most important part of removal of the parotid is the identification, dissection, and preservation of the facial nerve. The safest procedure is to identify the nerve as it exits from the stylomastoid foramen and then follow its course through the parotid gland. A malignant parotid tumor may involve the facial nerve and cause unilateral facial palsy. In removing such a tumor it may be necessary to sacrifice the facial nerve or its branches. Nerve grafts may reestablish facial muscle innervation and unilateral motor paralysis of the face can be averted in some cases.

The **facial artery** appears on the face at about the middle of the inferior border of the mandible and passes from the anterior border of the masseter muscle toward the medial angle of the eye. In its course it gives off the **superior** and **inferior labial branches** to the upper and lower lips, respectively, and the **lateral nasal branch** to the ala and dorsum of the nose. It terminates as the **angular artery** at the medial canthus of the eye. The angular artery anastomoses with the dorsal nasal and palpebral branches of the ophthalmic artery to establish a communication between branches of the internal and external carotid arteries. Additional anastomoses are present between the facial artery and the infraorbital branches of the maxillary artery.

The **superficial temporal branch** of the external carotid artery arises in the substance of the parotid gland and courses superficially toward the temporal region. It gives off the **transverse facial artery,** which parallels the course of the parotid duct. It supplies the parotid gland and duct, the masseter and buccinator muscles, the skin over the cheek, and anastomoses with branches of the facial artery. The superficial temporal artery terminates by bifurcating into the **frontal (anterior) branch,** which anastomoses with branches of the ophthalmic artery, and the **parietal (posterior) branch,** which joins with the posterior auricular and occipital arteries.

Table 7-13
Arterial Supply to the Face and Scalp

Artery	Origin	Course	Distribution	Anastomoses
Facial	Anterior aspect of external carotid	Follows tortuous course deep to sub-mandibular gland, at inferior border of mandible angles towards medial can-thus of eye	Anterior aspect of face, submandibular gland, suprahyoid muscles	No direct anasto-moses; branches of facial anastomose across midline
Inferior labial	Facial	Extends towards lower lip	Chin and lower lip	Mental branch of inferior alveolar
Superior labial	Facial	Passes to upper lip	Upper lip and exter-nal nose	Nasopalatine branch of maxillary
Lateral nasal	Facial	Extends towards side of nose	Side and dorsum of nose	External nasal branch of anterior ethmoidal and nasal branch of infraorbital
Angular	Terminal branch of facial	Passes towards me-dial canthus of eye	Anterior aspect of cheek and lower eyelid	Infratrochlear branch of ophthalmic and infraorbital branch of maxillary
Superficial temporal	One of terminal branches of external carotid	Ascends anterior to ear to reach temporal region	Facial muscles and skin over temporal region	Supraorbital and supratrochlear
Transverse facial	Superficial temporal	Courses parallel to zygomatic arch	Muscles and skin of temporal region	Infraorbital, supraor-bital and supratroch-lear
Mental	Terminal branch of inferior alveolar	Emerging at mental foramen, passes to chin	Facial muscles and skin over chin	Inferior labial
Infraorbital	Terminal branch of maxillary	From infraorbital foramen ramifies on cheek	Facial muscles and skin of cheek	Angular, lateral nasal and superior labial
Supraorbital	One of terminal branches of ophthalmic	From supraorbital foramen passes su-peroposteriorly to lambdoidal suture	Muscles and skin of forehead and scalp	Supratrochlear, su-perficial temporal and occipital
Supratroch-lear	One of terminal branches of ophthalmic	From supratrochlear notch courses su-peroposteriorly to midpoint of scalp	Muscles and skin of forehead and scalp	Superficial temporal and supraorbital
Occipital	Posterior aspect of external carotid	Follows inferior bor-der of posterior belly of digastric to reach occipital region	Occipitalis muscle and skin of scalp over occipital bone	Posterior auricular, and supraorbital

Venous Drainage of Face

Under normal conditions the venous return of the face is essentially superficial. The forehead is drained by the **supraorbital** and **supratrochlear veins,** which pass to the medial canthus to unite to form the **angular vein.** The continuation of the angular vein is the **facial vein.** It descends obliquely across the face and is accompanied by the facial artery. Just below the inferior border of the mandible it pierces the external investing layer of deep fascia to join the **internal jugular vein.**

Along its downward course the facial vein is joined by several additional tributaries draining specific regions of the face. These veins include the **palpebral** (from the eyelids), **external nasal** (about the nose), **labial** (from the lips), **deep facial** (draining part of the pterygoid venous plexus), **submental** (from the chin), **ascending palatine** (draining the tonsillar bed), **superficial temporal** (over the temporalis muscle), and the **retromandibular vein.** That part of the facial vein, inferior to its union with the retromandibular vein, is sometimes called the **common facial** vein before it empties into the internal jugular vein.

> #### The Danger Area of the Face and Scalp
> The danger area of the face and scalp extends from the upper lip and the lower nasal regions to the midpoint of the scalp. This region is drained by the facial veins. It is a dangerous area because boils, pustules (pimples) and other skin infections commonly occur here. Manipulation of these infected structures may spread infection into the facial veins. These veins anastomose with the inferior ophthalmic veins, the pterygoid plexus, and the cavernous dural sinus. Clotting of the blood is a specific problem in the cavernous sinus because of the unusually slow movement of the blood. The extensive endothelial surface of this sinus encourages bacterial growth and precipitates the formation of thrombi. Septicemia is a very dangerous condition because of the probability of meningitis. Before antibiotic therapy such conditions had a mortality of over 90%.

Parotid Gland

The **parotid gland,** largest of the three major salivary glands, occupies the depression between the sternocleidomastoideus muscle and the ramus of the mandible. It is roughly quadrilateral in shape and somewhat flattened, with a **deep process** passing to the inner aspect of the mandible.

> #### Mumps
> Mumps, or parotitis, is a common viral inflammatory lesion of the parotid gland that may spread to the testes and cause sterility.

The **superficial facial process** (accessory parotid) extends anteriorly beyond the ramus of the mandible, surrounds the proximal part of the large parotid (Stensen's) duct, and overlies the masseter muscle. The gland is enclosed by the external investing layer of deep cervical fascia. The thick-walled excretory **duct of the parotid** emerges from the facial process to continue across the masseter muscle and turns deeply to pass through the suctorial fat pad and buccinator muscle. It terminates inside the oral vestibule opposite the upper second molar tooth.

> #### Calculus
> A calculus is an abnormal concretion of mineral salts, which usually occurs in excretory ducts of glands. For example, calculi may form in any of the three major salivary glands. Obstruction of the parotid duct by a calculus is often associated with chronic parotitis and is perhaps best treated by surgical removal.

The parotid gland lies in the concavity formed by the sternocleidomastoid muscle and the ramus of the mandible. It is related laterally to the skin and superficial fascia of the face, branches of the greater auricular nerve and parotid lymph nodes, and superiorly to the external auditory meatus. The posteromedial surface of the parotid overlies the mastoid process, the sternocleidomastoideus, posterior belly of the digastric and stylohyoideus muscles, and the styloid process. The anteromedial surface is molded to the posterior border of the ramus of the mandible and the structures attaching to it, that is, the masseter and medial pterygoideus muscles, and the temporomandibular ligament.

The most superficial structures passing through the substance of the gland are the terminal branches of the facial nerve. Anteriorly this nerve divides into upper and lower trunks within the gland. The retromandibular vein, the external carotid artery and some of its branches, and the auriculotemporal and posterior auricular nerves are related to the deep aspect of the parotid.

The above structures radiate from the periphery of the gland to pass to their destinations. The superficial temporal artery and vein, the auriculotemporal nerve, and the temporal branch of the facial nerve pass superiorly; the zygomatic, buccal, and mandibular branches of the facial nerve and the transverse facial artery pass anteriorly. The cervical branch of the facial nerve passes inferiorly, and the posterior auricular artery and nerve pass posteriorly to supply the area behind the ear.

Temporal and Infratemporal Fossae

Objectives

At the completion of the study of the temporal and infratemporal fossae the student should be able to

▶ *Locate the region of the infratemporal fossa, define its boundaries, and delineate the bony processes that circumscribe this region*

▶ *Discuss the emergence of the mandibular nerve (V_3) from the cranial cavity into this region, and give the origin, course, and distribution of its branches*

▶ *Follow the course and give the functional components of the chorda tympani nerve*

▶ *List the branches of the first and second divisions of the maxillary artery; give their course and distribution*

▶ *Name the muscles of mastication; relate their location to their function*

The **temporal fossa** is an oval area on the lateral aspect of the skull. It is continuous inferiorly with the infratemporal fossa and contains the fan-shaped **temporalis muscle** of mastication.

The **infratemporal fossa** is the area bounded laterally by the ramus of the mandible, anteriorly by the body of the maxilla, superiorly by the infraorbital fissure, and inferiorly by the upper second and third molar teeth and their alveolar processes. The **medial wall** consists of the lateral or muscular plate of the pterygoid process, and the **roof** is formed by the inferior surface of the greater wing of the sphenoid and part of the temporal bone. The medial and lateral pterygoid muscles, the pterygoid plexus of veins, the mandibular nerve, and the maxillary artery are the contents of the infratemporal fossa. The foramen ovale penetrates the roof of the fossa at the posterior border of the lateral pterygoid plate. The infraorbital fissure opens into it at right angles to the pterygomaxillary fissure, and the foramen spinosum is situated just posterolateral to the foramen ovale in the roof of the fossa.

Muscles (Table 7-14)

Located in the temporal and infratemporal fossae, the four muscles of mastication act in movements of the mandible and are innervated by the mandibular division of the fifth cranial nerve (Fig. 7-15). The fan-shaped **temporalis muscle** originates from the temporal fossa, inserts into the coronoid process, and elevates and retracts the

Table 7-14
Temporal and Infratemporal Muscles

Muscle	Origin	Insertion	Action	Nerve
Temporalis	Temporal fossa and temporal fascia	Coronoid process and anterior border of ramus of mandible	Raises and retracts mandible	Mandibular
Masseter	Lower border and deep surface of zygomatic arch	Lateral surface of ramus and coronoid process of mandible	Raises and helps protract mandible	Mandibular
Lateral pterygoid	Intratemporal surface of sphenoid and lateral surface of lateral pterygoid plate	Neck of mandible and capsule of temporomandibular joint	Protrudes and depresses mandible; draws it toward opposite side	Mandibular
Medial pterygoid	Maxillary tuberosity and medial surface of lateral pterygoid plate	Mandible between mandibular foramen and angle	Raises and protrudes mandible; draws it toward opposite side	Mandibular

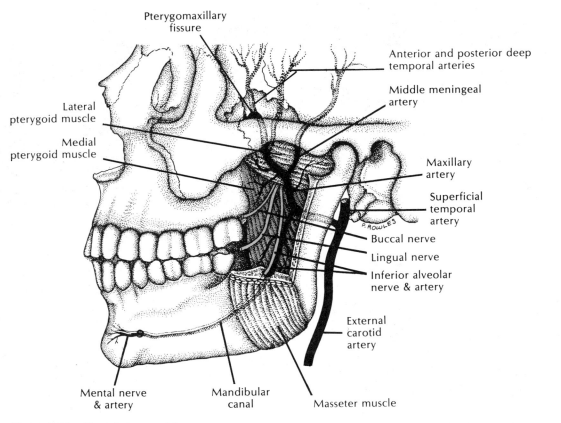

Figure 7-15. *The infratemporal fossa.*

mandible. The thick quadrilateral **masseter muscle** covers the lateral surface of the ramus, the angle, and the coronoid process of the mandible. It raises the lower jaw. In forcible clenching of the jaw this muscle can be felt as it bulges over the angle of the mandible. The **medial** and **lateral pterygoids** are both situated deep to the mandible in the infratemporal fossa. Both have two heads partially originating from respective surfaces of the lateral pterygoid plate and act to protrude and move the mandible from side to side. In addition, the medial pterygoid elevates, and the lateral pterygoid depresses, the mandible as it pulls it over the articular eminence.

Arteries and Nerves (Table 7-15)

The **maxillary artery,** the larger of the two terminal branches of the external carotid, passes deep to the neck of the mandible and courses across the lower border of the lateral pterygoideus muscle to disappear into the pterygopalatine fossa (Fig. 7-15). It is divided into three parts by the lateral pterygoideus. The **first part,** proximal to the muscle, gives four branches: the **deep auricular artery** follows the auriculotemporal nerve to supply the skin of the external auditory meatus and outer aspect of the tympanic membrane; the **anterior tympanic branch** passes

Table 7-15
Arterial Supply to the Infratemporal Fossa

Artery	Origin	Course	Distribution	Anastomoses
Maxillary	One of terminal branches of external carotid at neck of mandible	Traverses the infratemporal fossa; pterygomaxillary fissure; pterygopalatine fossa; to floor of orbit	Supplies midregion of the face	No direct anastomoses
Middle meningeal	Maxillary	Ascends through foramen spinosum to middle cranial fossa	Primary supply to meninges	Anterior and posterior meningeal arteries
Inferior alveolar	Maxillary	Descends to mandibular foramen; traverses mandibular canal; emerges at mental foramen	Mylohyoid, anterior belly of digastric and lower teeth	No direct anastomoses
Anterior tympanic	Maxillary	Courses in external auditory meatus to tympanic membrane	External auditory meatus, and tympanic membrane	Deep auricular
Deep auricular	Maxillary	Passes to external auditory meatus	External auditory meatus	Anterior tympanic
Muscular branches	Maxillary	Pass to each muscle in infratemporal fossa; branches arise from second part of maxillary	Four muscles of mastication	With each other
Posterior superior alveolars	Maxillary	Traverse small foramina in tuberosity of maxilla	Molar teeth	Middle superior alveolar

behind the capsule of the temporomandibular joint to pass into the tympanic cavity by way of the petrotympanic fissure; the **middle meningeal artery** ascends medial to the lateral pterygoideus muscle to pass through the foramen spinosum to supply the meninges; and the **inferior alveolar** accompanies the inferior alveolar nerve as it passes through the mandibular foramen to traverse the mandibular canal and supply the lower teeth. The latter vessel terminates as the **mental branch** as it passes onto the face through the mental foramen. The **second part** of the maxillary artery gives **muscular branches** to the lateral and medial pterygoidei, the masseter, and the temporalis muscles. The terminal, or **third part** of the artery passes into the pterygopalatine fossa and will be described with that area.

The sensory portion of the **mandibular division** of the trigeminal nerve originates from the trigeminal ganglion and enters the infratemporal fossa by passing through the foramen ovale, where it is joined by the small **motor root** of the trigeminal. Two branches originate from the main trunk, one the **recurrent meningeal** (nervus spinosus), which reenters the skull through the foramen spinosum with the middle meningeal artery to supply sensory fibers to the meninges, and the other the **nerve to the medial pterygoideus,** which supplies motor fibers to the muscle and sends sensory twigs to the otic ganglion. The main trunk then divides into an anterior and a posterior division. The smaller **anterior division,** essentially motor, gives a branch to each of the four muscles of mastication: the **masseteric branch** to the masseter, the **deep temporal branch** to the temporalis, and the **nerves to the lateral** and **medial pterygoidei.** The buccal nerve is the only sensory branch from the anterior division.

The **posterior division,** mostly sensory, has three main branches: the auriculotemporal, the lingual, and the inferior alveolar. The **auriculotemporal nerve,** which embraces the middle

meningeal artery near its origin, sends **communicating branches** to the facial nerve and the otic ganglion. In the latter, motor fibers continue through the ganglion without synapsing to innervate the tensor veli palatini and the tensor tympani muscles. An **anterior auricular branch** of the auriculotemporal nerve supplies sensation to the front of the ear. The tympanic membrane and skin of the external auditory meatus are innervated by the **external acoustic branch,** and the small **superficial temporal branch** of the auriculotemporal nerve supplies the parotid gland and skin over the temporal region.

The **lingual nerve** descends anteroinferiorly, medial to the lateral pterygoideus and anterior to the inferior alveolar nerve. Passing between the medial pterygoideus and the mandible, it reaches the submandibular region where it supplies general sensation to the mucous membrane of the floor of the mouth and the anterior two-thirds of the tongue. **Communicating branches** are given to the inferior alveolar and hypoglossal nerves, and the submandibular ganglion. The **chorda tympani branch** from the facial nerve joins the lingual nerve in the infratemporal fossa to supply special sensory (taste) fibers to the anterior two-thirds of the tongue and preganglionic parasympathetic fibers to the submandibular ganglion.

The **inferior alveolar nerve** descends with the inferior alveolar artery to enter the mandibular foramen and traverses the mandibular canal to emerge at the mental foramen as the mental nerve. The **mylohyoid branch** arises as the inferior alveolar nerve enters the mandibular foramen and supplies the anterior belly of the digastricus and the mylohyoideus muscles. Within the mandibular canal the inferior alveolar nerve sends twigs to the lower teeth. Terminal branches include the **mental** and **inferior labial,** supplying the skin of the chin, the skin and mucous membrane of the lower lip, and the anterior incisive branch to the incisors (Table 7-16).

Text continues on page 344.

Table 7-16
Distribution of Mandibular (V₃), Chorda Tympani and Minor Petrosal Nerves

Nerve	Origin	Course	Distribution
Branches to muscles of mastication	Anterior division or main trunk of (V_3)	Pass directly to muscles	Masseter, temporalis, medial and lateral pterygoids
Buccal	Anterior division of (V_3)	Passes anteriorly to cheek	Mucous membrane and skin of cheek
Lingual	Posterior division of (V_3)	Crosses lateral aspect of medial pterygoid to reach tongue	General sensation to anterior ⅔ of tongue
Chorda tympani	Facial nerve (VII)	From vertical portion of facial traverses middle ear cavity to reach infratemporal fossa and join lingual nerve	Taste from anterior ⅔ of tongue; conveys preganglionic parasympathetic fibers to submandibular ganglion. Postganglionics supply submandibular and sublingual glands
Inferior alveolar	Posterior division of (V_3)	Runs on lateral aspect of medial pterygoid, enters mandibular foramen, traverses mandibular canal, and terminates at mental foramen as mental nerve	Sensory to lower teeth, and skin of chin and lower lip
Nerve to mylohyoid	Inferior alveolar	At mandibular foramen, follows mylohyoid groove to reach superficial aspect of mylohyoid	Mylohyoid and anterior belly of digastric
Auriculotemporal	Posterior division of (V_3)	Passes between external auditory meatus and neck of mandible to course with superficial temporal artery	Skin of sideburn area; muscular twigs to tensors tympani and veli palatini
Minor petrosal	Glossopharyngeal (IX) through tympanic nerve and plexus	From tympanic plexus, leaves middle ear cavity via petrosphenoid fissure to reach otic ganglion that is suspended from auriculotemporal nerve	Conveys preganglionic parasympathetic fibers to otic ganglion. Postganglionics supply parotid gland

Skull

The **skull** is the most complex osseous structure of the body (Fig. 7-16). It is adapted to house the brain and special sensory organs, and encloses the openings into the digestive and respiratory tracts. It is composed of twenty-two flattened, irregular bones that, except for the mandible, are joined by immovable, sutural-type articulations. For descriptive purposes, the skull is subdivided into the **cranium,** which houses the brain and special sense organs and is formed by eight bones, and the **facial skeleton,** composed of fourteen bones. There is no special demarcation of this subdivision, but junctional areas contribute to the support of nasal, ocular, and auditory organs. The major **cranial sutures** are the **sagittal,** passing in the midline between the two parietal bones,

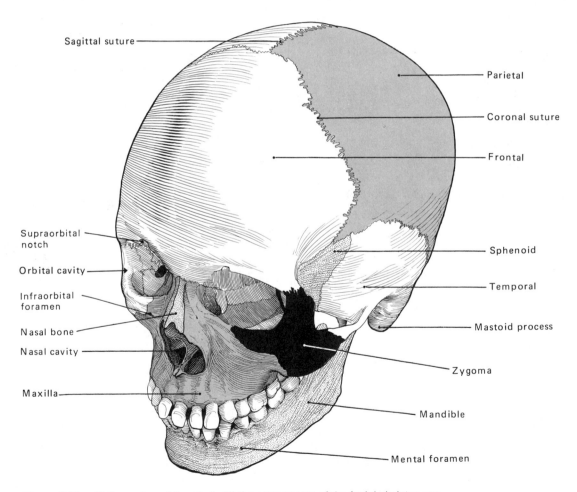

Figure 7-16. *Oblique view of the skull. All the components of the facial skeleton are visible.*

and two transverse sutures, the **coronal** between the frontal and the parietal bones, and the **lambdoidal** between the occipital and parietal bones.

Premature Cranial Synostosis

Premature cranial synostosis is an early union of the bones of the skull before the brain has reached its normal size. Such premature closure reduces the size of the cranial vault substantially, thus preventing normal brain development. Tribasilar synostosis is the fusion in infancy of the three bones at the base of the skull. Mental retardation is a common sequela to any abnormal or early union of cranial bones.

Basic points of reference of the skull include the **nasion,** the midpoint of the nasofrontal sutures; the **bregma,** at the junction of the sagittal and coronal sutures; the **obelion,** that portion of the sagittal suture adjacent to the parietal foramina; the **lambda,** at the junction of the sagittal and lambdoidal sutures; the **inion,** or external occipital protuberance; the **asterion,** at the junction of the occipital, parietal, and temporal bones, and the **pterion,** at the junction of the frontal, sphenoid, parietal, and temporal bones.

The Soft Spot

The "soft spot" of the infant's skull is a diamond-shaped membranous area where the sagittal and coronal sutures eventually will fuse. Usually by the 18th postnatal month the sutures of the two parietal and the frontal bones unite and the anterior fontennelle (soft spot) is no longer palpable. Before closure, the physician can determine, by observation or palpation, increased intracranial pressure by a bulging of the fontennelle. Because the superior sagittal venous sinus lies in the midline of the anterior fontennelle, blood can be obtained by inserting a needle into the sinus. The posterior fontennelle occupies the interval between the sagittal and lamboidal sutures. This

smaller gap closes during the second postnatal month.

Cranial Cavity (Table 7-17)

Objectives

At the completion of the study of the cranial cavity the student should be able to

▶ *List the bones forming the calvaria and the cranial vault*

▶ *Delineate the boundaries of the cranial fossae*

▶ *From both endocranial and ectocranial surfaces, locate the major foramina and the structures they transmit*

▶ *Locate the position of the dural venous sinuses, the bony attachments of the dura and its modifications, the falx cerebri, falx cerebelli, tentorium cerebelli, and diaphragma sellae*

▶ *Describe the ventricular system and the circulation of cerebrospinal fluid*

▶ *Draw and label the constituent arteries that form the cerebral arterial circle of Willis*

Superiorly, in the median plane of the internal aspect of the **calvaria,** or skull cap, a shallow groove lodges the superior sagittal venous sinus. Numerous small pits lie adjacent to this groove, which correspond to the position of the arachnoid granulations. Many grooves on the inner aspect of the calvarium are formed by the meningeal vessels. The largest and most prominent of these grooves, extending laterally over the inner surface from the foramen spinosum, lodges the middle meningeal artery and its branches.

Contrecoup Fracture

Contrecoup fracture is a term applied to a fracture of the skull at some distance from the point of contact of the blow. *Text continues on page 348*

Table 7-17
Foramina of the Skull

Name	Bone	Position on bone	Structures passing through
Foramina Associated with Floor of Skull			
Cecum	Frontal	Between frontal and ethmoid (crista galli) at anterior end of crista galli in midline	Vein from nasal cavity to superior sagittal sinus
Olfactory	Ethmoid	In cribriform plate of ethmoid	Olfactory nerve branches and nasociliary nerve
Pterygoid (Vidian) canal	Sphenoid	Through root of pterygoid process	Nerve, artery, and vein of pterygoid canal
Sphenopalatine	Sphenoid and palatine	Between vertical part of palatine and ventral surface of sphenoid	Superior nasal nerve, nasopalatine nerves, and sphenopalatine vessels
Rotundum	Sphenoid	At junction of anterior and medial parts of sphenoid	Maxillary nerve
Vesalius	Sphenoid	Opposite root of pterygoid process and medial to foramen ovale	Small vein from cavernous sinus
Ovale	Sphenoid	Base of lateral pterygoid plate, in greater wing of sphenoid	Mandibular nerve, motor root of trigeminal, accessory meningeal artery, sometimes minor petrosal nerve
Spinosum	Sphenoid	Posterior angle of sphenoid, medial to spine of sphenoid	Middle meningeal vessels and recurrent branch from mandibular nerve
Lacerum	Sphenoid and petrous part of temporal	Bounded in front by sphenoid, behind by apex of petrous, medially by sphenoid and occipital bones	Internal carotid artery passes along upper part surrounded by nerve plexus, vidian nerve, meningeal branch of ascending pharyngeal artery
Jugular	Temporal and occipital	Behind carotid canal, between petrous part of temporal and occipital	Inferior petrosal sinus and bulb of internal jugular vein; ninth, tenth, and eleventh cranial nerves
Hypoglossal	Occipital	Above base of condyles	Hypoglossal nerve, meningeal branch of ascending pharyngeal artery
Magnum	Occipital	Center of posterior cranial fossa	Medulla oblongata and membranes, spinal accessory nerve, vertebral and spinal arteries
Foramina Associated with Orbit			
Anterior ethmoidal	Frontal and ethmoid	In frontoethmoidal suture on medial wall of orbit	Anterior ethmoidal vessels and nerve
Posterior ethmoidal	Frontal and ethmoid	2 to 3 cm posterior to anterior ethmoidal	Posterior ethmoidal vessels
Optic	Sphenoid	Between upper and lower roots of lesser wing of sphenoid	Optic nerve and ophthalmic artery

(Continued)

TABLE 7-17
(Continued)

Name	Bone	Position on bone	Structures passing through
Superior orbital fissure	Sphenoid	Between greater and lesser wings of sphenoid	Above superior head of lateral rectus: trochlear, frontal, and lacrimal nerves. Between heads of lateral rectus; abducens oculomotor, and nasociliary nerves; ophthalmic veins
Inferior orbital fissure	Sphenoid and maxilla	Between greater wing of sphenoid and maxilla	Maxillary and zygomatic nerves, infraorbital vessels, and veins to pterygoid plexus
Foramina Associated with Mouth			
Mandibular	Mandible	Center of medial surface of ramus of mandible	Inferior alveolar nerve and vessels
Greater palatine	Maxilla and palatine	At either posterior angle of hard palate	Greater palatine nerve and descending palatine vessels
Lesser palatine	Palatine	In pyramidal process of palatine bone (two or more)	Lesser palatine nerves
Incisive	Maxilla	Anterior end of median palatine suture just behind incisor teeth	Terminal branches of descending palatine vessels and nasopalatine nerve
Scarpa's	Maxilla	Incisive foramen in midline	Nasopalatine nerve
Stensen's	Maxilla	Lateral openings in incisive foramen	Terminal branch of descending palatine artery
Foramina Associated with Face			
Supraorbital	Frontal	Supraorbital margin of orbit	Supraorbital nerve and artery
Infraorbital	Maxilla	Above canine fossa and below orbit	Infraorbital nerve and artery
Zygomaticoorbital	Zygomatic	Anteromedial surface of orbital process	Zygomaticotemporal and zygomaticofacial nerves
Zygomaticofacial	Zygomatic	Near center of deep surface of zygomatic	Zygomaticofacial nerve and vessels
Zygomaticotemporal	Zygomatic	Near center of temporal surface of zygomatic	Zygomaticotemporal nerve
Mental	Mandible	Below second premolar tooth	Mental nerve and vessels
Foramina Associated with External Aspects of Skull			
Stylomastoid	Temporal	Between styloid and mastoid processes	Facial nerve and stylomastoid artery
Mastoid	Temporal	Near posterior border of mastoid process of temporal bone	Vein to transverse sinus, small branch of occipital artery to dura
Parietal	Parietal	Posterior aspect of parietal close to midline	Emissary vein to superior sagittal sinus

Since the osseous elements of the skull form a hollow but somewhat elastic shell, the force of a blow to the cranium can be transmitted to the opposite side where a fracture may be sustained. Blows to the top of the head may cause fractures of the bones of the floor of the cranial vault, often involving the sella turcica, one of the weakest parts of the base of the skull. The floor of the middle cranial fossa is especially involved because it is weakened by various foramina and fissures. Further, the strength of the petrous bone is decreased by the cavities of the internal ear, the carotid canal, and the jugular fossa.

The **floor** of the cranial cavity is subdivided into anterior, middle, and posterior cranial fossae

(Fig. 7-17). The **anterior cranial fossa** is formed by portions of the ethmoid, sphenoid, and frontal bones and is adapted for the reception of the frontal lobes of the brain. Posteriorly the fossa is limited by the posterior border of the lesser wings of the sphenoid and the anterior margin of the chiasmatic groove. The **crista galli,** a midline process of the ethmoid bone, affords attachment for a longitudinal fold of dura mater, the **falx cerebri.** The **cribriform plate of the ethmoid,** at either side of the crista galli, admits passage of filaments of the olfactory nerve from the nasal mucosa to synapse in the olfactory bulb.

The floor of the **middle cranial fossa** is composed of the body and great wings of the sphenoid, and the squamosal and petrous portions of the temporal bones. Laterally it contains the temporal lobes of the brain. It is limited posteriorly by

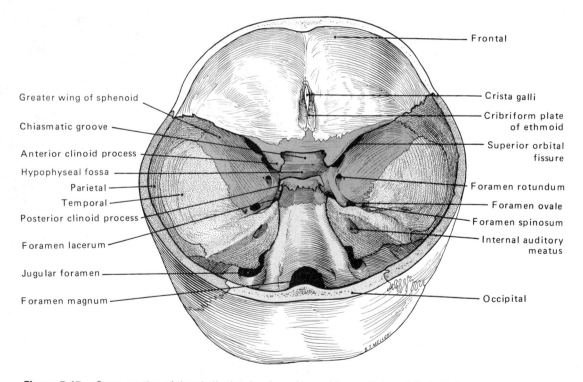

Figure 7-17. *Cross section of the skull, showing foramina and bones that contribute to the floor of the cranial cavity. The shaded, colored area is the sphenoid bone. The stippled, colored area is the temporal bone.*

the superior angle of the petrous portion of the temporal bone and the dorsum sellae centrally. The **sella turcica,** the site of the hypophyseal fossa, is bounded by the tuberculum sellae anteriorly, the dorsum sellae posteriorly, and the anterior and posterior clinoid processes laterally. The anterior and posterior clinoid processes give attachment to a dural fold, the diaphragma sella, which covers the sella. A crescentic arrangement of the **foramina spinosum, ovale,** and **rotundum** and the **superior orbital fissure** is present in the floor of the middle cranial fossa. The **trigeminal impression** on the anterior surface of the petrous portion of the temporal bone lodges the trigeminal (semilunar or gasserian) ganglion of the fifth cranial nerve. The **tegmen tympani,** the lateral part of the anterior and superior surfaces of the petrous bone, forms the roof of the tympanic cavity, the mastoid antrum, and the internal acoustic meatus.

The **posterior cranial fossa** comprises the remainder of the cranial cavity. Its floor is formed by parts of the sphenoid, temporal, and occipital bones. The posterior cranial fossa contains the cerebellum, pons, and medulla oblongata, with a sheet of dura mater, the **tentorium cerebelli,** extending in a transverse plane to separate the cerebellum from the cerebrum. The basilar part of the occipital bone articulates with the sphenoid at the dorsum sellae and is related posteriorly to the pons and medulla of the brain stem. The **internal occipital protuberance** is located at the region of the confluence of the dural sinuses. The inferiormost portion of the posterior cranial fossa presents the large **foramen magnum,** through which the spinal cord passes. At either side of the foramen magnum, the **hypoglossal canal** transmits the hypoglossal nerve. Laterally the deep groove for the **lateral (transverse** and **sigmoid) dural sinus** extends from the internal occipital protuberance to the **jugular foramen.** At the foramen the glossopharyngeal, vagus, and spinal accessory nerves leave the cranial cavity, and the sigmoid sinus becomes the internal jugular vein. On the posterior surface of the petrous portion of the temporal bone, the prominent **internal acoustic meatus** transmits the seventh and eighth cranial nerves as well as branches of the basilar artery to the internal ear.

Cranial Meninges

Three distinct connective tissue membranes **(meninges),** the dura, the arachnoid, and the pia mater cover the brain. The **dura mater** forms a tough outer covering and is composed of two closely adherent layers. The **outer layer** fuses to the periosteum lining the internal surface of the cranial cavity (endosteum) and is continuous through the foramina of the skull with the ectocranial periosteum.

Extradural Hemorrhage

An extradural hemorrhage, so named from the location of the hemorrhage in relation to the meninges, may result from a blow to the side of the head over the inferior part of the parietal bone that causes fracture of the bone and rupture of the middle meningeal artery. The bleeding occurs extradurally, stripping away the periosteum and dura from the internal surface of the cranium.

Within the outer layer **arachnoid granulations,** bulging cauliflowerlike masses, pit the inner surface of the parietal bones. The meningeal vessels, grooving the calvaria, also run in this layer. The **inner layer** forms four inward-projecting reduplicated folds, namely, the falx cerebri, the falx cerebelli, the tentorium cerebelli, and the diaphragma sellae, which partially divide the cranial cavities into compartments. Separations of the outer and inner layers create venous spaces called the **dural sinuses.**

The midline, sickle-shaped **falx cerebri** projects inward between the cerebral hemispheres. Anteriorly it is attached to the crista galli of the ethmoid bone; superiorly the convex upper border extends from the crista galli, along the midline of the inner surface of the calvarium, to the internal occipital protuberance, and is the site of the superior sagittal sinus. The lower concave border is free anteriorly and contains the inferior sagittal sinus. Posteriorly the lower border unites with the tentorium cerebelli where it forms the straight sinus. The **tentorium cerebelli** is situated between the cerebellum and the occipital and posterior portions of the temporal lobes of the cerebrum. The **falx cerebelli** is a slight fold, attached posteriorly to the internal occipital crest and the tentorium. Anteriorly it is free and projects between the cerebellar hemispheres. The occipital sinus is in its posterior attachment. The **diaphragma sellae** bridges over the sella turcica and covers the hypophysis. A central aperture admits passage of the stalk of the hypophysis.

Subdural Hemorrhage

A subdural hemorrhage is a blood collection in the potential space between the dura and the arachnoid. It results from a rupture of the large veins that return blood from the surface of the brain to the superior sagittal sinus. It often results from a blow on the front or back of the head that causes considerable anteroposterior movement of the brain within the cranium.

The **arachnoid** is separated from the dura mater by a capillary (subdural) space that contains just sufficient fluid to keep the adjacent surfaces moist. From the inner surface of the arachnoid, cobweblike trabeculae extend across the subarachnoid space to become continuous with the pia mater. Between the arachnoid and the pia the cerebrospinal fluid is contained within the relatively large **subarachnoid space.** The **arachnoid villi** or **granulations** project into the superior sagittal sinus to permit absorption of the cerebrospinal fluid.

Subarachnoid Hemorrhage

A subarachnoid hemorrhage may come from leakage of blood vessels that traverse the subarachnoid space. Aneurysms of the cerebral arteries often occur in or near the circle of Willis. Rupture of such an aneurysm bleeds into the subarachnoid space and is a common cause of a cerebrovascular accident in a young person. Blood detected in the cerebral spinal fluid aspirated during a lumbar spinal puncture may be due to a hemorrhage of this nature.

At certain areas around the base of the brain the arachnoid and the pia mater are widely separated as the **cerebellomedullary, pontine, interpeduncular, chiasmatic,** and **ambiens cisternae,** which contain large amounts of cerebrospinal fluid. The subarachnoid space communicates with the ventricular system of the brain through small apertures in the roof of the fourth ventricle (the midline **foramen of Magendie** and the lateral **foramina of Luschka**) and is continuous with the perineural space around nerves emerging from the brain and spinal cord.

The **pia mater** is a thin, highly vascular layer intimately adherent to the cortex of the brain and follows closely the contours of the brain.

Dural Sinuses

The **dural sinuses** are venous channels that drain blood from the brain and meninges (Fig. 7-18). They contain no valves and are located between the inner and outer layers of the dura, except for the inferior sagittal and straight sinuses that are between the reduplications of the inner dural layer. The **superior sagittal sinus** is triangular in cross section, occupies the entire length of the attached superior portion of the falx cerebri, and increases in size as it passes posteriorly.

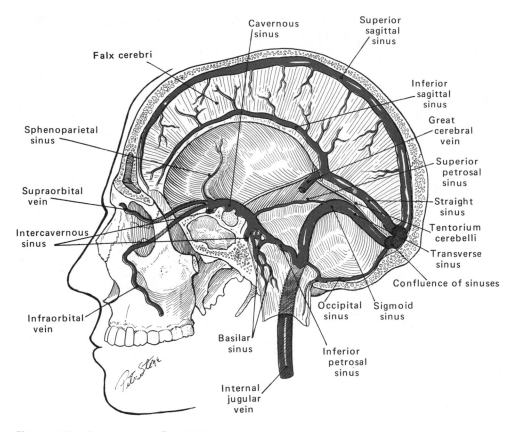

Figure 7-18. *Dural sinuses. Superficial veins of the face empty into the cavernous sinus.*

At the internal occipital protuberance it usually continues as the right lateral sinus. Arachnoid granulations bulge into its lateral expansions, the **lacunae laterales.** The superior sagittal sinus receives several cerebral veins, diploic veins, and some drainage from the meningeal veins. The smaller **inferior sagittal sinus** occupies the free inferior edge of the falx cerebri. It receives adjacent cerebral veins and, at the junction of the falx cerebri with the tentorium cerebelli, receives the **great cerebral vein** (of Galen) to become the **straight sinus.** The latter is situated along the fusion of the falx cerebri to the tentorium cerebelli. At the internal occipital protuberance the straight sinus usually continues as the left lateral sinus. The **lateral sinuses** are continuations of

either the superior sagittal or straight sinuses as noted above, but at their origin may form a common space, the confluence of sinuses. The lateral sinuses are subdivided into transverse and sigmoid portions. The **transverse sinus** occupies the attached portion of the tentorium cerebelli (between inner and outer layers of dura) and receives as tributaries the superior petrosal sinus, diploic veins, and adjacent cerebral and cerebellar veins. The **sigmoid sinus,** a continuation of the transverse, follows an S-shaped course internal to the junction of the petrous and mastoid portions of the temporal bone. It receives the occipital sinus and, at the jugular foramen, becomes the internal jugular vein.

As noted above, the **confluence of sinuses** is

a dilatation commonly at the beginning of the right transverse sinus and forms a wide, shallow depression where the right and left transverse sinuses communicate. It may receive the superior sagittal and straight sinuses, and may give origin to the **occipital sinus.** The latter occupies the attached border of the falx cerebelli and is variable in size. Inferiorly the occipital sinus bifurcates, partially encircles the foramen magnum, and ends in the sigmoid sinus. It receives cerebellar veins and communicates with the vertebral plexus of veins.

The **cavernous sinus** is an expanded, trabeculated dilatation at either side of the hypophyseal fossa. The oculomotor, trochlear, ophthalmic, and maxillary nerves are embedded in its lateral wall, while the internal carotid artery with its sympathetic plexus and the abducens nerve pass through the sinus close to its medial wall. It receives ophthalmic and cerebral veins, and the sphenoparietal sinus, and is drained by the superior and inferior petrosal sinuses. The **intercavernous (circular) sinus** connects the two cavernous sinuses.

The **sphenoparietal sinus** lies beneath the lesser wing of the sphenoid, where it receives the anterior branch of the middle meningeal vein and diploic veins, and drains into the cavernous sinus. Within the attached margin of the tentorium cerebelli, at the superior border of the petrous bone, the **superior petrosal sinus** bridges the trigeminal ganglion and drains the cavernous sinuses into the junction of the sigmoid and transverse sinus. The larger **inferior petrosal sinus** occupies a groove between the petrous and basioccipital bones, where the abducens nerve passes. This sinus drains the cavernous sinus, the internal auditory and adjacent veins, and passes independently through the jugular foramen to empty into the internal jugular vein. The **basilar sinus** is on the posterior aspect of the dorsum sellae and the superior surface of the basioccipital bone. It drains both the cavernous and inferior petrosal sinuses and communicates inferiorly with the vertebral plexus of veins. All dural sinuses ultimately drain into the lateral sinuses and finally into the **internal jugular vein,** except for the inferior petrosal sinus, which empties directly into the internal jugular.

Diploic veins drain the diploë (the marrow-filled space between the inner and outer tables of the cranial bones) into dural sinuses adjacent to the bone. They have no accompanying arteries, the diploë being supplied by meningeal and superficial arteries of the scalp.

Emissary veins pass through foramina of the skull and connect the dural sinuses with the veins on the external surface of the skull. These vessels include a connection between the veins of the nose and the superior sagittal sinus through the foramen cecum; a communication between the veins of the scalp and the superior sagittal sinus through the parietal foramina; a connection (the largest of the emissary veins) between the posterior auricular vein and the sigmoid sinus through the mastoid foramen; and a communication between the suboccipital veins and the sigmoid sinus through the condylar foramen. Additional emissary veins include communications between the pterygoid plexus and the cavernous sinus through the formamen ovale; between supraorbital, ophthalmic, and facial veins and the cavernous sinus through the supraorbital fissure and the optic foramen; and between the pharyngeal venous plexus and the cavernous sinus by way of the carotid canal.

Hypophysis

The **hypophysis cerebri** is an endocrine gland situated in the sella turcica. Superiorly the sella turcica is roofed by the diaphragma sellae, which has a central aperture through which the **infundibulum,** or stalk, of the hypophysis connects the posterior lobe of the gland to the **tuber cinereum** of the hypothalamus. The cavernous sinuses are at either side of the sella turcica, while anteriorly the diaphragma sellae separates the anterior lobe of the hypophysis from the optic chiasma.

Central Nervous System

The central nervous system consists of the brain and the spinal cord. Subdivisions of the brain include the cerebrum, diencephalon, midbrain, pons, medulla oblongata, and the cerebellum.

Cerebrum

The **cerebrum** is the largest subdivision of the brain and fills most of the cranial cavity (Fig. 7-19). It is partially divided by a deep midline cleft, the **longitudinal fissure,** into bilaterally symmetrical hemispheres. At the depth of the fissure a thickened band of transverse fibers, the **corpus callosum,** unites the two hemispheres.

The surface of the cerebrum is corrugated in appearance. The furrows in this corrugation are designated as **sulci,** the ridges as **gyri.** Similar patterns of sulci and gyri are present in all brains, though they may differ in detail. Certain sulci and fissures divide each of the cerebral hemispheres into **lobes.** The prominent central sulcus extends transversely at about the middle of the cerebral hemisphere and separates the frontal lobe from the parietal lobe. The **frontal lobe** lies subjacent to the frontal bone of the skull; the **parietal lobe** is internal to the parietal bone. Posteriorly the **parietooccipital fissure** divides the parietal lobe from the **occipital lobe.** The latter lies in the posterior portion of the cranial cavity, internal to the occipital bone, and rests on the superior surface of the **tentorium cerebelli.**

A deep cleft, the **transverse fissure,** separates the elongated **temporal lobe** from the frontal lobe. The temporal lobe is located in a pocketlike recess of the cranial cavity internal to the temporal bone.

The surface of the cerebrum, the **cerebral cortex,** is composed of gray matter that extends about 1 cm to 2 cm internally. The corrugation of the surface of the cerebrum is more extensive in humans than in any other animal and greatly increases the surface area of the cortex. The cerebral cortex controls all conscious motor activity. Sensory impulses must reach the cortex for conscious interpretation. Distinct functional areas of the cortex have been localized. They include the **precentral gyrus,** which lies immediately in front of the central sulcus, and is the primary motor area; the **postcentral gyrus,** located immediately behind the central sulcus, is the primary area for interpretation of general sensation; the posterior portion of the occipital lobe is the primary area of visual interpretation; while the temporal lobe is associated with interpretation of auditory, gustatory stimuli, and speech activity.

Lesions of the Cerebrum

Lesions of the cerebrum are many and varied, but their symptoms and signs are relatively few: vomiting, dizziness, headache, convulsions, and partial or complete paralysis. The location of the lesion is more important in eliciting specific symptoms than the nature of the pathological disturbance. For example, any lesion that blocks the motor pathway between the motor cortex of the cerebrum and the skeletal muscles where the nerves terminate will cause paralysis whether the obstruction is due to a tumor, blood clot, inflammation, compression, or scar.

The underlying white matter of the cerebral cortex includes numerous **association fibers** that connect portions of the cortex with one another. For example, a great mass of fibers, the **corpus callosum,** which connects the cerebral hemispheres; the **optic radiation** that sweeps posteriorly to the occipital (visual) cortex, and fibers of the **internal capsule** that connect lower centers of the central nervous system to the cortex.

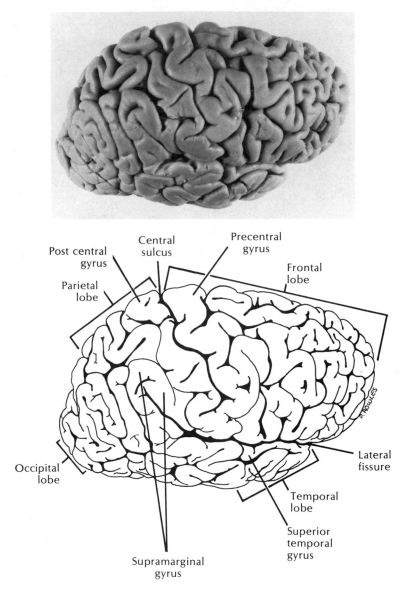

Figure 7-19. *Photo and schematic drawing of cerebrum.*

Diencephalon

The **cerebral hemispheres** cover the centrally located **diencephalon.** It lies to either side of the third ventricle and is subdivided into the **thalamus, hypothalamus** and **epithalamus.** The latter forms the roof of the third ventricle. The **thalamus,** an oval mass of cell bodies, is an important sensory center. It functions as a relay station, receiving impulses from sensory fibers ascending from the spinal cord, and transmitting impulses to the cerebral cortex. The thalamus is under the control of the cerebral cortex, and therefore, some of the impulses from the cerebral cortex to the spinal cord are relayed through this center. It also acts as an integrating center through its connections with **subcortical nuclei** (collections of nerve cell bodies situated deeply in the central nervous system) in the basal ganglia and the nuclei in the hypothalamus.

The **hypothalamus** occupies the floor and a portion of the lateral wall of the third ventricle. It serves as the great integrating center for the autonomic nervous system.

A number of structures are visible on the inferior aspect of the diencephalon. The most prominent of these is the midline **hypophysis** or **pituitary gland,** which, with the hypothalamus, functions to facilitate rapid and effective adjustments to sudden changes in the environment and is the chief regulator of the endocrine glands. Additional structures visible on the ventral aspect of the diencephalon include the **optic chiasma** and the paired, rounded **mammillary bodies.**

Hypophyseal Tumors

Hypophyseal tumors cause erosion and enlargement of the sella turcica, which are evident on routine skull roentgenograms. The pituitary is in close proximity to the optic tracts, chiasma and nerves; thus, any enlargement of the gland may cause pressure on these vital structures, causing partial blindness in one or both eyes. After surgical removal of the tumor, vision is usually improved and may return to normal.

Midbrain

The **midbrain** is the smallest subdivision of the brain stem. It is situated between the pons and the diencephalon as it surrounds the **cerebral aqueduct (of Sylvius),** which connects the third and fourth ventricles. Four rounded eminences, the **corpora quadrigemina,** are visible on the dorsal aspect of the midbrain. The two more superiorly situated bodies are called the **superior colliculi** and serve as synaptic stations in the visual pathway. The two **inferior colliculi** are synaptic areas in the auditory pathway.

Two large ropelike bundles of nerve fibers, the **cerebral peduncles** (crura cerebri), are visible on the ventral aspect of the midbrain. They are composed of motor fibers that extend from the cerebral cortex to the spinal cord. These fibers are cell processes from the initial (upper motor) neuron in the two-neuron pathway for motor impulses from the cerebral cortex to skeletal muscle fibers (Fig. 7-20).

Internally the midbrain includes the **oculomotor** and **trochlear nuclei** containing cell bodies for the fibers that form the third and fourth cranial nerves. The **red nucleus,** a part of the **reticular formation,** which appears pinkish in color in the fresh state, is also present in the midbrain.

Pons

When viewed from the ventral aspect of the brain stem the **pons** is seen as a prominent band of fibers that extends transversely toward the cerebellum. These fibers form the **middle cerebellar peduncle** (brachium pontis). Internally interspersed in white matter are the **pontine nuclei,** which lie ventral to the **reticular formation.** The nuclei of the trigeminal nerve are also present in the pons. Fiber tracts forming the **superior cerebellar peduncle** or **brachium conjunctivum**

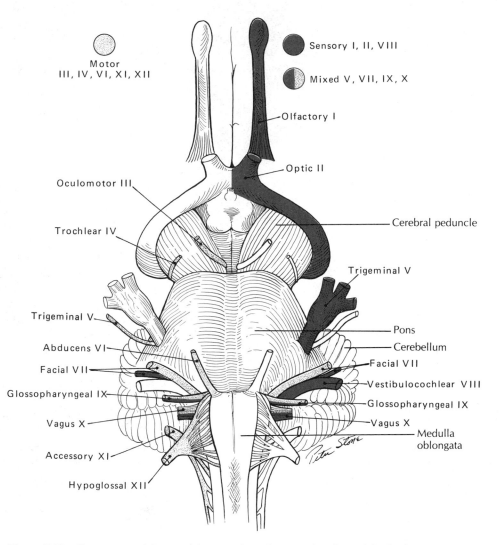

Figure 7-20. *Emergence of the cranial nerves from the ventral surface of the brainstem.*

are also visible on the cross section of the pons. The most anterior part of the fourth ventricle extends into the pons.

Medulla Oblongata

The **medulla oblongata** is the caudalmost extent of the brain stem. It extends from the pons to the **foramen magnum** where it is continuous inferiorly as the spinal cord. Anteriorly the medulla oblongata rests on the basal portion of the occipital bone, while posteriorly it is covered by the cerebellum. The extension of the cerebral aqueduct widens considerably in the midportion of the medulla as the **fourth ventricle.** Fiber tracts of the spinal cord extend into the medulla where they become rearranged and may decussate or cross to the opposite side.

A series of longitudinal ridges or elevations are visible on the ventral aspect of the medulla. The more prominent of these, located adjacent to the midline, are the two **pyramids** that contain fibers that will form the lateral corticospinal tracts of the spinal cord. The crossing of these tracts is visible as the **pyramidal decussation.** Laterally two ridges corresponding to the **nucleus gracilis and nucleus cuneatus** are evident. These nuclei contain cell bodies of **second order sensory neurons.** Two flattened masses, the **olives,** are present at the upper end of the medulla oblongata. Rootlets of origin of the ninth, tenth, and eleventh cranial nerves emerge in the more lateral of two parallel grooves on the ventral aspect of the medulla. Fascicles of origin of the **hypoglossal nerve** emerge along the more medial groove between the olive and the pyramid. At the junctional area between the pons and the medulla the **abducens, facial,** and **vestibulocochlear nerves** are visible as they emerge from the brain stem.

Nuclei of the above nerves are demonstrable within the medulla, as are the nuclei of the reticular substance and the olive. The **vestibular** and **cochlear nuclei** associated with impulses originating in the internal ear are also located in the medulla oblongata.

Cerebellum

The **cerebellum** is situated posteriorly and superiorly to the pons and medulla oblongata. It occupies the **posterior cranial fossa** inferior to the level of the **tentorum cerebelli.** It is divided into two lateral hemispheres that are united in the midline by a median portion, the **vermis.**

The surface of the cerebellum is covered by a superficial layer of gray matter, the **cerebellar cortex.** It appears laminated due to the presence of sulci and delicate gyri that are arranged in a parallel pattern. These laminae are referred to as **folia.** Deep to the cortex a number of nuclei are embedded in the white matter. The cerebellum functions to coordinate muscular activity of the body, particularly gross muscular movements.

Cerebellar peduncles connect the cerebellum to other components of the brain stem. These include the **superior cerebellar peduncle** (brachium conjunctivum), which connects the cerebellum to the midbrain; the **middle cerebellar peduncle** (brachium pontis), connecting the cerebellum to the pons; and the **inferior cerebellar peduncle** (restiform body), connecting the cerebellum to the medulla oblongata.

Lesions of the Cerebellum

Lesions of the cerebellum may be from tumors, abscesses, cysts, or inflammation. The symptoms for all of these lesions are essentially similar: instability of equilibrium and locomotion with severe dizziness. Voluntary movements are impaired by a marked ataxia (incoordination of muscular activity) and nystagmus. Because of the ataxia, the patient tends to lean or fall towards the side of the lesion.

Ventricles

Some parts of the central nervous system are not solid structures but have cavities. In the brain and its subdivisions these cavities are termed **ventricles,** while in the spinal cord the cavity is called the **central canal** (Fig. 7-21). The cavity associated with each cerebral hemisphere is called the **lateral ventricle.** The **third ventricle** is the cavity of the diencephalon, which communicates with the two lateral ventricles through the **interventricular foramina (of Monro).** The **cerebral aqueduct (of Sylvius)** associated with the midbrain, extends between the third and **fourth ventricle.** The latter lies internal to the pons and medulla oblongata and is under cover of the cerebellum. Small openings in the fourth ventricle, the **foramen of Magendie** in the midline, and the two **foramina of Luschka** laterally, communicate with the **subarachnoid space.** The fourth ventricle continues distally as the small **central canal** of the spinal cord.

In the roof of the ventricles, specialized vascular structures, the **choroid plexuses,** elaborate

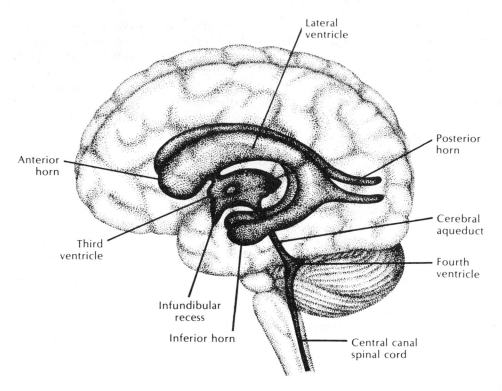

Figure 7-21. *Ventricles of the brain.*

cerebrospinal fluid. The latter circulates from the ventricles through the foramina of Magendie and Luschka into the subarachnoid space where it is reabsorbed into the vascular system at the arachnoid villi. The latter are specializations of the **arachnoid membrane** that project into the **dural venous sinuses.**

Hydrocephalus

Hydrocephalus (water on the brain) is due to an abnormally large accumulation of cerebrospinal fluid within the ventricles of the brain. It may be caused by: (1) obstruction of some part of the ventricular system, (2) excessive production of fluid, or (3) interference with the absorption of the fluid. Obstruction is most commonly caused by scar tissue produced by inflammation of the meninges, which may close the foramina of the fourth ventricle. It may also be due to a tumor of the mesencephalon, which blocks the cerebral aqueduct. The resultant pressure dilates the ventricles and compresses the cerebral cortex against the unyielding skull. If this occurs in infants when the cranial bones are not yet united, the internal pressure greatly enlarges the skull as well as the ventricles of the brain. The cerebral cortex becomes markedly thinned-out and degenerates, causing severe mental retardation.

Blood Vessels (Table 7-18)

The brain is supplied by two pairs of arteries, the internal carotid and the vertebral arteries, which form an anastomosis at the base of the brain called

Table 7-18
Arterial Supply to the Brain and Spinal Cord

Artery	Origin	Course	Distribution	Anastomoses
Vertebral	Subclavian	Traverses foramina transversarii C_6–C_1 to reach foramen magnum	Meninges, spinal cord and cerebellum	With each other to form basilar
Spinal (anterior and posterior)	Vertebral at foramen magnum	Pass on respective aspects of spinal cord; anterior spinals unite to form a single trunk	Spinal cord with assistance from all arteries that course adjacent to vertebral column	Segmental branches to spinal cord
Posterior inferior cerebellar	Vertebral	Follows posteroinferior aspect of cerebellum	Posteroinferior aspect of cerebellum	Anterior inferior cerebellar
Basilar	Formed by junction of right and left vertebrals	Ascends on anterior aspect of brainstem	Brainstem, cerebellum, cerebrum	No direct anastomoses
Pontine	Basilar	Anterior aspect of brainstem	Numerous twigs to brainstem	No direct anastomoses
Labyrinthine	Basilar	Courses in internal auditory meatus	Internal ear	No direct anastomoses
Anterior inferior cerebellar	Basilar	Anteroinferior aspect of cerebellum	Inferior aspect of cerebellum	Inferior and superior cerebellars
Superior cerebellar	Basilar	Follows superior aspect of cerebellum	Superior aspect of cerebellum	Anterior inferior cerebellar
Internal carotid	Common carotid at superior border of thyroid cartilage	Traverses neck to reach carotid canal→cavernous sinus→cranial cavity	Gives branches in cavernous sinus, provides primary supply of brain and orbital cavity	Circle of Willis, formed by internal carotid; anterior and posterior cerebrals; anterior and posterior communicating
Anterior cerebral	Internal carotid	Passes anteroposteriorly over corpus callosum and along opposing surface of the cerebral hemispheres	Medial aspect of the cerebral hemispheres, except occipital lobe	With opposite anterior cerebral via anterior communicating
Middle cerebral	Continuation of internal carotid distal to anterior cerebral	Passes through lateral fissure to reach convex surface of cerebral hemisphere	Convex surface of cerebral hemisphere except occipital lobe	Posterior cerebral via posterior communicating
Posterior cerebral	Terminal branch of basilar	Courses along inferior aspect of cerebral hemisphere	Inferior aspect of cerebral hemisphere and occipital lobe	Middle cerebral via posterior communicating
Anterior communicating	Anterior cerebral	Passes between anterior cerebrals	Circle of Willis	Anterior communicating link of circle of Willis
Posterior communicating	Posterior cerebral	Passes between posterior cerebral and middle cerebral	Circle of Willis	Posterior communicating link of circle of Willis

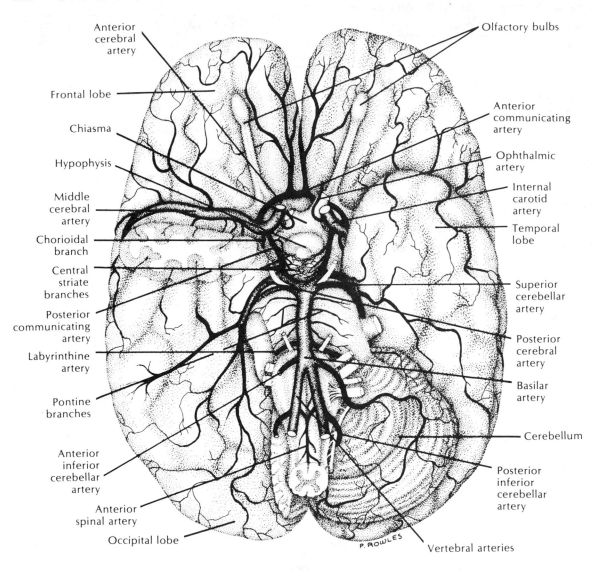

Figure 7-22. *Circle of Willis and arterial supply to brain.*

the **circulus arteriosus cerebri,** or the circle of Willis (Fig. 7-22). The **internal carotid** is subdivided for descriptive purposes into four parts. The **cervical portion** begins at the bifurcation of the common carotid artery opposite the upper border of the thyroid cartilage and passes vertically upward anterior to the transverse processes of the upper three cervical vertebrae. In its course it is crossed by the twelfth cranial nerve, the digastric and stylohyoideus muscles, and the occipital and posterior auricular arteries. At the base of the skull the ninth, tenth, and eleventh cranial nerves pass between the artery and the internal jugular vein. The internal carotid artery has **no branches in the neck.** Its **petrous portion** follows an S-shaped course through the petrous portion of the temporal bone within the carotid canal. It is separated by a thin layer of bone from the tympanic cavity and the trigeminal ganglion and gives the **caroticotympanic branch** to the tympanic cavity. The **cavernous portion,** within the cavernous sinus, follows a second sigmoid course passing upward to the posterior clinoid process, then forward and superiorly again at the anterior clinoid process. Branches within the sinus include twigs to the cavernous sinus, the hypophysis, the trigeminal ganglion, and the meninges. The **ophthalmic artery** arises just as the internal carotid passes through the roof of the cavernous sinus.

Passing between the second and third cranial nerves, the **cerebral portion** of the internal carotid artery extends to the medial end of the lateral cerebral fissure, where it divides into terminal branches, the anterior cerebral and the much larger middle cerebral. Its **posterior communicating branch** courses posteriorly beneath the optic tract to join the posterior cerebral branch of the basilar artery. The internal carotid artery supplies the optic tract and cerebral peduncles and gives deep branches to the internal capsule and thalamus of the brain. Its small **choroidal branch** follows the optic tract and cerebral peduncles to the choroid plexus. The **anterior cerebral artery** passes anteromedially above the optic chiasma, then along the fissure over the corpus callosum to the parietooccipital sulcus, giving cortical branches to the medial surface of the hemisphere and deep central branches. The **anterior communicating branch** connects the anterior cerebral arteries of either side as they course over the corpus callosum. The direct continuation of the internal carotid, the **middle cerebral artery,** passes upward and laterally in the lateral cerebral fissure to spread out over the surface of the brain, giving orbital, frontal, temporal, and parietal cortical branches to the lateral surface of the hemisphere, and **central striate branches** (the arteries of cerebral hemorrhage), which pass deeply into the corpus striatum of the brain.

The **vertebral artery,** a branch of the subclavian, courses superiorly through the foramina transversarii of the cervical vertebrae to enter the cranial cavity by way of the foramen magnum. As it enters the cranium it gives off **anterior** and **posterior spinal arteries** that descend on either surface of the spinal cord. They anastomose with each other and, with segmental spinal branches, form longitudinal vessels along the length of the cord. The **meningeal branch** supplies the meninges of the posterior cranial fossa. The largest branch of the vertebral artery, the **posterior inferior cerebellar,** passes between the tenth and eleventh cranial nerves to supply the undersurface of the cerebellum.

The **basilar artery** is formed by the union of the two vertebral arteries. It passes in the midline of the inferior surface of the pons and gives the following branches: the **anterior inferior cerebellar** to the under surface of the cerebellum; the **pontine branches** to the pons; and the **labyrinthine artery,** which passes through the internal auditory meatus to the internal ear. Its **superior cerebellar branch** courses laterally, just posterior to the third cranial nerve, supplies the superior surface of the cerebellum, and anastomoses with the anterior and posterior inferior cerebellar arteries. The **posterior cerebral branch** passes laterally to join the **posterior communicating vessel** from the internal carotid and contributes to the circle of Willis. The latter branch winds around the cerebral peduncle giving cortical branches to the occipital lobe, to the area adjacent to the parietooccipital fissure, and to the temporal lobe.

The **meningeal arteries** are periosteal arteries that groove the inner surface of the calvaria, lie in the outer layer of the dura mater, and supply the dura, the inner table of the skull, and the diploë. The **middle meningeal** is a branch of the internal maxillary artery and enters the cranial cavity through the foramen spinosum to divide into anterior and posterior branches, which supply most of the meninges. Other arteries to the meninges include branches from the anterior and posterior ethmoidal arteries to the anterior cranial fossa; the accessory meningeal and branches of the internal carotid to the middle cranial fossa; and branches of the vertebral, the ascending pharyngeal, and the occipital artery to the posterior cranial fossa.

A Stroke or Cerebrovascular Accident (CVA)

A stroke or CVA is the rupture or occlusion of certain cerebral arteries, especially the striate arteries leading to the internal capsule of the brain. Such a lesion usually produces hemiplegia on the opposite side of the body. If the hemorrhage is massive, the area of the brain supplied by the ruptured artery will degenerate, and the neurologic deficits will be permanent. Fortunately, in many cases the motor and sensory functions partially or completely return with an early dissolution of the blood clot and resultant release of pressure on the cerebral cortex.

Cranial Nerves

Ten of the **twelve cranial nerves** originate directly from the brain stem (Table 7-19). Each has a superficial attachment as well as a deeply located nucleus of origin or termination. The olfactory (I) and optic (II) nerves are unusual in that they do not arise from the brain stem. The optic nerve is actually a brain tract rather than a nerve. The olfactory nerve begins in the olfactory mucosa and terminates in the olfactory bulb. Five of the twelve cranial nerves attach to the ventral aspect of the brainstem: the olfactory (I) at the olfactory bulb on the cribriform plate; the optic (II) at the anterolateral angle of the optic chiasma; the oculomotor (III) in a groove, the oculomotor sulcus, between the cerebral peduncle and the interpeduncular fossa; the abducens (VI) in a groove between the pons and the lateral aspect of the pyramid; and the hypoglossal (XII) as a row of rootlets

Table 7-19
Cranial Nerves and Their Components

Number	Name	Component	Cell bodies	Distribution
I	Olfactory	Sensory (smell)	Olfactory mucosa	Nasal mucosa of upper part of septum and superior concha
II	Optic	Sensory (vision)	Ganglionic layer of retina	Retina
III	Oculomotor	Somatic motor	Brain stem	Superior, inferior, and medial rectus; inferior oblique and levator palpebrae superioris muscles
		Parasympathetic	Preganglionic, brain stem; postganglionic, ciliary ganglion	Sphincter muscle of iris; ciliary muscle
IV	Trochlear	Somatic motor	Brain stem	Superior oblique muscle
V	Trigeminal	Sensory (general)	Semilunar ganglion	Skin and mucosa of the head; meninges
		Motor	Brain stem	Muscles of mastication (masseter, temporalis, external and internal pterygoids), mylohyoid, anterior belly of digastric, tensor tympani, and tensor veli palatini

(Continued)

Table 7-19 **(Continued)**

Number	Name	Component	Cell bodies	Distribution
VI	Abducens	Somatic motor	Brain stem	Lateral rectus muscle
VII	Facial	Motor	Brain stem	Facial muscles, posterior belly of digastric, stylo-hyoid, and stapedius
		Parasympathetic	Preganglionic, brain stem; postganglionic, pterygo-palatine ganglion	Nasal, palatine, and lacrimal glands
		Parasympathetic	Preganglionic, brain stem; postganglionic, subman-dibular ganglion	Submandibular and sublin-gual glands
		Sensory (general)	Geniculate ganglion	Skin of mastoid region and external acoustic meatus
		Sensory (taste)	Geniculate ganglion	Anterior two-thirds of tongue by way of chorda tympani; soft palate by way of major petrosal
VIII	Vestibulocochlear	Sensory (hearing)	Spiral ganglion	Organ of Corti
		Sensory (equilib-rium)	Vestibular ganglion	Semicircular canals, utricle, and saccule
IX	Glossopharyngeal	Motor	Brain stem	Stylopharyngeus muscle
		Parasympathetic	Preganglionic, brain stem; postganglionic, otic gan-glion	Parotid gland
		Sensory (general)	Superior ganglion	Mucosa of pharynx, tym-panic cavity, posterior one-third of tongue, and carotid sinus
		Sensory (taste)	Inferior ganglion	Posterior one-third of tongue
X	Vagus	Motor	Brain stem	Muscles of pharynx, larynx, levator veli palatini, pala-toglossus
		Parasympathetic	Preganglionic, brain stem; postganglionic, on, in, or near viscera	Thoracic and abdominal viscera
		Sensory (general)	Superior ganglion	Skin of external acoustic meatus
		Sensory (general)	Inferior ganglion	Pharynx, larynx, and tho-racic and abdominal viscera
		Sensory (taste)	Inferior ganglion	Epiglottis and base of tongue
XI	Spinal accessory Cranial root	With motor branches of vagus	Brainstem	
	Spinal root	Somatic motor	Spinal cord (C_1–C_5)	Sternocleidomastoid and trapezius
XII	Hypoglossal	Somatic motor	Brain stem	Intrinsic and extrinsic mus-cles of tongue

in a groove between the pyramid and the olive. Six cranial nerves attach to the lateral aspect of the brain stem: the trigeminal (V) by two roots, a large sensory and a small motor at the side of the pons; the facial and vestibulocochlear (VII and VIII) in line with the trigeminal, at the border of the pons and the inferior cerebellar peduncle; and the glossopharyngeal, vagus, and spinal accessory (IX, X, and XI) as a row of rootlets in a narrow groove along the entire lateral side of the medulla. This origin of the spinal accessory is the cranial or accessory portion of that nerve; the spinal portion is derived from the upper five cervical segments of the cord. Only one cranial nerve, the trochlear

(IV), attaches to the dorsum of the brain stem, emerging immediately posterior to the inferior colliculus of the midbrain at the superior medullary velum (Fig. 7-23).

The **olfactory nerve (I)** is entirely sensory. It arises in the olfactory mucosa of the nasal cavity as bipolar neurons and is limited in origin to the mucous membrane covering the superior nasal concha and the adjacent nasal septum. Numerous filaments from this distribution area pierce the cribriform plate to synapse with secondary neurons in the **olfactory bulb.** The **olfactory tract** passes posteriorly from the olfactory bulb to the olfactory trigone of the brain.

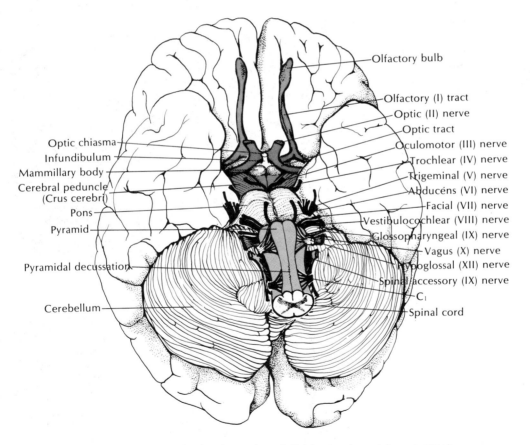

Figure 7-23. *Inferior view of brain showing major subdivisions and cranial nerve attachments.*

Lesion: Olfactory (I)
If the lesion is unilateral, anosmia (loss of smell) results on the affected side.

The **optic nerve (II)** is also entirely sensory and originates in the **ganglionic cells** of the retina, which are tertiary neurons in the visual pathway. The orbital portion of the nerve is about 5 cm long, invested by meninges, and passes posteriorly through the optic foramen to the **optic chiasma,** which rests on the tuberculum sellae. Fibers partially decussate at the optic chiasma and continue posteriorly as the **optic tract,** which winds around the cerebral peduncle and terminates in the lateral geniculate body of the mesencephalon.

Lesion: Optic (II)
Depending on the location of the lesion, blindness may be total in both eyes (at the chiasma), in one eye only (along an optic nerve), or may affect both visual fields (along the optic tract).

The **oculomotor nerve (III),** from its emergence at the oculomotor sulcus, passes between the posterior cerebral and superior cerebellar arteries. Lateral to the clinoid processes it pierces the dura to traverse the cavernous sinus and enter the orbital cavity through the supraorbital fissure. The oculomotor nerve is accompanied by the abducens and nasociliary nerves, all of which pass between the heads of the lateral rectus muscle. Its terminal distribution to the extraocular muscles will be described with the section on the orbit.

Lesion: Oculomotor (III)
A lesion results in pupil dilatation, inability to focus, eyeball directed downwards and outwards, and ptosis (drooping of upper eyelid) on the damaged side.

The **trochlear nerve (IV)** is motor in function. It is the most slender of the cranial nerves, yet has the longest intracranial course. Originating from the dorsum of the brain stem, it winds around the midbrain, enters the edge of the tentorium cerebelli, and passes with the oculomotor nerve between the posterior cerebral and the superior cerebellar arteries. It continues forward around the cerebral peduncle to penetrate the dura and enter the cavernous sinus between the third and sixth nerves. It passes along the lateral wall of the sinus to enter the orbit through the supraorbital fissure above the origin of the ocular muscles. Its orbital course will be described with the orbital cavity.

Lesion: Trochlear (IV)
No obvious dysfunction occurs if this nerve is damaged. However, if patient looks downward and outward, he has double vision.

The large **trigeminal nerve (V)** is both motor and sensory. It is formed by a large **sensory root from the trigeminal ganglion** and a smaller **motor root.** The trigeminal ganglion occupies a cavity (cavum trigeminale) in the dura at the sutural area between the petrous portion of the temporal and the greater wing of the sphenoid bones. From the ganglion the **ophthalmic division (V_1)** courses anteriorly to enter the orbit by way of the supraorbital fissure. The **maxillary division (V_2)** passes through the foramen rotundum to the pterygopalatine fossa, and the **mandibular division (V_3)** traverses the foramen ovale to reach the infratemporal fossa. The motor root passes independently through the foramen ovale to join the mandibular nerve in the infratemporal fossa. Attached by sensory roots to branches of the trigeminal nerve are the four small parasympathetic ganglia of the head. The trigeminal nerve, however, has no parasympathetic components in its brain stem nuclei. The **ophthalmic division** will be discussed with the orbital cavity; the **maxillary division** will be described with the pterygopalatine fossa; and the **mandibular division** has been covered with the infratemporal fossa.

Lesion: Trigeminal (V)
If the lesion involves the entire nerve, extensive anesthesia is present on the affected side over the face, scalp, conjunc-

tiva, gingivae, anterior two-thirds of the tongue, mucous membrane of the nose, hard and soft palate, cheek, and lips. The four muscles of mastication on the injured side are also paralyzed. If only the first or second divisions are damaged, only sensation over appropriate distribution is lost. If the third division is divided, there is sensory loss to the lower facial region and motor loss to the muscles of mastication.

The **abducens nerve (VI)** is motor in function and pierces the dura at the dorsum sellae to pass below the posterior clinoid process and enter the cavernous sinus at the lateral side of the internal carotid artery. Traversing the sinus, it passes through the supraorbital fissure to enter the orbital cavity above the ophthalmic artery. Its distribution will be discussed with the orbital cavity.

Lesion: Abducens (VI)
With damage to this nerve the eye cannot move laterally.

The **facial nerve (VII)** contains both motor and sensory fibers. The large **motor** and smaller **sensory (nervus intermedius) roots** traverse the internal acoustic meatus (in company with the eighth nerve) to unite at the **geniculate ganglion** located on the sharp posterior bend of the seventh nerve within the facial canal. From the geniculate ganglion the **major petrosal nerve,** transmitting preganglionic parasympathetic fibers, traverses the petrous portion of the temporal bone to enter the middle cranial fossa through the hiatus of the facial canal. This branch then courses forward between the dura and the trigeminal ganglion, passing deep to the latter, to unite with the **deep petrosal nerve.** The latter carries postganglionic sympathetic fibers from the internal carotid plexus with their cell bodies located in the superior cervical sympathetic ganglion. These two nerves unite to form the **nerve of the pterygoid canal** (Vidian nerve), which passes through the

pterygoid (Vidian) canal in the sphenoid bone to terminate in the pterygopalatine ganglion. Distal to the geniculate ganglion the facial nerve passes through the facial canal to emerge from the skull at the stylomastoid foramen. Additional branches within the facial canal include the **nerve to the stapedius muscle, communicating twigs** to the auricular branch of the vagus, and the **chorda tympani nerve.** The latter turns upward through a separate small canal to enter the tympanic cavity through the posterior wall and courses forward on the internal surface of the tympanic membrane. Arching across the handle of the malleus, it leaves the tympanic cavity by passing through the anterior wall. It then traverses the petrotympanic fissure and joins the lingual branch of the trigeminal nerve in the infratemporal fossa. The **chorda tympani** carries preganglionic parasympathetic fibers to the submandibular ganglion and special taste fibers to the anterior two-thirds of the tongue. The **terminal branches** of the facial nerve have been described with the face.

Lesion: Facial (VII)
Symptoms vary in extent and severity, depending on the site of the lesion. However, paralysis of the facial muscles always occurs. The taste fibers to the anterior two-thirds of the tongue and the corneal reflex may or may not be affected. Spontaneous paralysis of the facial nerve is often called Bell's palsy.

The **vestibulocochlear nerve (VIII),** entirely sensory in function, consists of two parts, the cochlear and vestibular portions, which differ in peripheral endings, central connections, and function. The eighth nerve courses with the facial nerve through the internal acoustic meatus, where it divides at the termination of the canal into its respective parts. The **cochlear portion,** the nerve of hearing, consists of bipolar neurons associated with the **spiral ganglion** of the cochlea. The peripheral processes pass to the spiral organ **(of Corti),** while the central processes pass from the modiolus to the lateral end of the internal acoustic meatus. The **vestibular portion,** the nerve of equilibrium, consists of the bipolar cells of the **vestibular ganglion** located in the superior part of the lateral end of the internal acoustic meatus. Peripheral processes of these bipolar neurons pass to the utricle, the saccule, and the ampullae of the semicircular ducts. The central processes join with those of the cochlear division to form the eighth nerve.

Lesion: Vestibulocochlear (VIII)

The eighth cranial nerve has two functional components, auditory and vestibular. Partial or complete deafness may result from damage of any of its auditory nerve components, such as the organ of Corti, spiral ganglion, or cochlear division. Lesions of the vestibular division will cause disturbances of equilibrium, for example, dizziness, loss of balance, or nystagmus. Meniere's syndrome is a fairly common disturbance of the vestibular apparatus causing dizziness (vertigo), nausea, ringing in the ears (tinnitus), and often deafness. Its etiology is obscure and its cure still eludes us.

The **glossopharyngeal nerve (IX)** is both motor and sensory and exits through the jugular foramen in company with the tenth and eleventh nerves. Two **sensory ganglia,** the **superior** and the **inferior (petrosal),** are associated with the glossopharyngeal nerve as it passes through the foramen. The **tympanic branch** (nerve of Jacobson), transmitting preganglionic parasympathetic fibers to the otic ganglion, passes through a small canal (of Jacobson) within the temporal bone to the tympanic cavity. Here it joins with branches of the facial and caroticotympanic (sympathetics) nerves to form the **tympanic (promontory) plexus,** which supplies the mucous membrane of the tympanic cavity. From this plexus, the **minor petrosal nerve** is reconstituted and courses through the petrous portion of the temporal bone to run forward in the middle cranial fossa, traverses the foramen ovale, and synapses in the **otic**

ganglion. At its exit from the jugular foramen, the glossopharyngeal nerve passes deep to the styloid process, gives a **branch to the stylopharyngeus muscle,** then joins with branches of the vagus and sympathetic fibers to form the **pharyngeal plexus.** Terminal branches of the glossopharyngeal nerve supply the posterior third of the tongue with general and special (taste) sensation.

Lesion: Glossopharyngeal (IX)

Severing this nerve results in loss of sensation of the posterior one-third of the tongue and pharynx. The patient may also have some difficulty in swallowing.

The **vagus nerve (X),** containing both motor and sensory fibers, has the most extensive course and distribution of any of the cranial nerves. It leaves the cranial cavity through the jugular foramen in company with the ninth and eleventh nerves. The superior and inferior sensory ganglia of the vagus are located within the jugular foramen or just below it. Passing to the cervical region in the carotid sheath, in company with the internal jugular vein and carotid artery, the vagus nerve gives a **recurrent branch** to the meninges, **auricular branches** to the ear, **pharyngeal branches** to the pharyngeal plexus and soft palate, the **nerve to the carotid sinus, cervical cardiac branches,** and a **superior** and an **inferior (recurrent) laryngeal branch** to the larynx. At the root of the neck, the left vagus passes between the common carotid and subclavian arteries to pass anterior to the arch of the aorta and posterior to the root of the lung, where it joins with the right vagus to form the **esophageal plexus.** The right vagus passes anterior to the subclavian artery, then descends along the trachea to the posterior aspect of the root of the lung and joins the left vagus as above. The **inferior laryngeal nerves** loop under the arch of the aorta on the left side and the subclavian artery on the right to reach the tracheoesophageal groove and ascend to supply all the intrinsic musculature of the larynx, except the cricothyroideus. In the thorax, the vagus gives off **cardiac** and **pulmonary branches** and then forms the **esophageal plexus.** After passing through the esophageal hiatus, the right vagus reforms as the **posterior vagal trunk,** the left as the **anterior vagal trunk,** to aid in the formation of the **celiac plexus.** The vagi contribute to the innervation of all abdominal viscera, except those portions of the gastrointestinal tract distal to the splenic flexure of the colon.

Lesion: Vagus (X)
Trauma to the vagus nerve will result in unilateral paralysis of larynx and palate, with hoarseness due to loss of inferior (recurrent) laryngeal nerve. Unilateral anesthesia of the larynx on the same side will also be present. If both vagi are cut in the head or neck, tachycardia and decreased respiration result, and the patient is often unable to breathe or speak because of loss of innervation of both vocal cords.

The **spinal accessory nerve (XI)** is a motor nerve formed from both cranial and spinal components. The smaller **cranial portion** unites with the spinal part to pass through the jugular foramen, then separates to join with the vagus for distribution to the pharynx and larynx. It contributes to the supply of the musculus uvulae, the levator veli palatini, the pharyngeal constrictor muscles, and muscles of the larynx and esophagus. The **spinal portion,** originating in the motor cells of the ventral horns of the first through the fifth cervical nerves, passes superiorly along the side of the spinal cord and through the foramen magnum, where it joins with the cranial portion. It continues a short distance with the latter as it exits through the jugular foramen, then separates to pass inferiorly behind the jugular vein, the stylohyoideus, the digastric, and the upper part of the sternocleidomastoideus. It crosses the posterior triangle of the neck, picking up communicating twigs of the cervical plexus (C_2 and C_3), and passes to the deep surface of the trapezius to supply this muscle and the sternocleidomastoideus.

Lesion: Spinal Accessory (XI)
If damaged, the sternocleidomastoideus and trapezius muscles are paralyzed. The patient will have a "drooped" shoulder.

The **hypoglossal nerve (XII),** the motor nerve to the tongue musculature, passes through the hypoglossal canal to descend almost vertically to a point opposite the angle of the mandible. It courses deep to the internal carotid artery and internal jugular vein, then lies between the artery and the vein, deep to the stylohyoideus and digastric muscles. At the intermediate tendon of the digastric it loops around the occipital artery and passes anteriorly between the hyoglossus and mylohyoideus muscles to **supply the intrinsic and extrinsic muscles of the tongue.** Communicating twigs from the first loop of the cervical plexus join the hypoglossal nerve and run with it for a short distance before most of the fibers leave the hypoglossal nerve as the superior limb (descendens hypoglossi) of the ansa cervicalis to supply infrahyoid muscles of the neck. Some of the fibers from this communication with the cervical plexus continue with the hypoglossal nerve to branch from the latter as individual twigs to supply the thyrohyoideus and geniohyoideus muscles.

Lesion: Hypoglossal (XII)
The affected side of the tongue becomes wrinkled and atrophied. When the tongue is protruded, the tip deviates towards the side of the lesion.

Cranial Parasympathetic Ganglia (Table 7-20)

Located within the orbital cavity, the small **ciliary ganglion** receives preganglionic parasympathetic fibers from the **ciliary branch of the oculomotor nerve.** From cell bodies within the ganglion, the postganglionic parasympathetic fibers leave through the **short ciliary nerves** to penetrate

Table 7-20
Parasympathetics of Head

Nerve	Nucleus of Preganglionic Cell Bodies	Ganglion of Postganglionic Cell Bodies	Transmission Pathway
Oculomotor (III)	Edinger-Westphal	Ciliary	Oculomotor → motor twigs → ciliary ganglion → short ciliaries → middle layer of eyeball → ciliary muscle and sphincter muscle of iris
Facial (VII)	Superior salivatory	Pterygopalatine	Facial → major petrosal → nerve of pterygoid canal → pterygopalatine ganglion → maxillary → zygomatic → communicating to lacrimal → lacrimal gland; from ganglion other branches (sphenopalatine, greater and lesser palatine) → pass to midregion of face
		Submandibular	Facial → chorda tympani → lingual → submandibular ganglion → submandibular and sublingual glands
Glossopharyngeal (IX)	Inferior salivatory	Otic	Glossopharyngeal → tympanic nerve → tympanic plexus → minor petrosal → otic ganglion → parotid gland
Vagus (X)	Motor nucleus of vagus	Diffuse	Vagus → cardiac nerves → cardiac plexus → heart; pulmonary nerves → pulmonary plexus → lungs; esophageal plexus → gastric trunks → vascular plexuses (e.g., celiac) → postganglionic cell bodies in myenteric and submucosal plexuses in digestive organs and in associated glands

the sclera at the area cribrosa and supply the ciliary and sphincter pupillae muscles. Sensory fibers from the **nasociliary nerve** traverse the ganglion without synapsing and become components of the short ciliary nerves, which supply general sensation to the eyeball. Sympathetic fibers from the ophthalmic plexus also pass through the ganglion and are distributed with the short ciliary nerves to innervate the dilator pupillae muscle and the smooth muscle of orbital blood vessels.

Within the pterygopalatine fossa, the **pterygopalatine ganglion** is attached to the maxillary division of the trigeminal nerve by two **short sensory roots.** Preganglionic parasympathetic fibers pass from the **nervus intermedius of the seventh nerve** to the ganglion through the **major petrosal nerve.** Postganglionic parasympathetic fibers are distributed with branches of the

maxillary division of the trigeminal nerve to the lacrimal gland. Branches from the ganglion also supply the nasopharynx, nasal cavity, palate, upper lip, and gingiva. Postganglionic sympathetic fibers, transmitted along the internal carotid plexus, pass as the **deep petrosal nerve** to join the **major petrosal nerve** in the middle cranial cavity and form the **nerve of the pterygoid canal.** The sympathetic fibers pass through the ganglion to be distributed with its branches. **Sensory fibers,** by way of the two roots to the ganglion from the maxillary division of the fifth cranial nerve, also pass through the ganglion and are distributed as sensory components of the ganglionic branches.

The **otic ganglion,** located in the infratemporal fossa, is attached to the mandibular division of the trigeminal immediately distal to the foramen ovale. Its preganglionic parasympathetic fibers are derived from a branch of the glossopharyngeal nerve, **the tympanic nerve,** which passes to the promontory plexus. From the latter, the **minor petrosal nerve** is reconstituted and transmits the preganglionic parasympathetic fibers to the otic ganglion. Postganglionic fibers are distributed to the parotid gland by the auriculotemporal branch of the trigeminal. Sympathetic fibers passing through the ganglion are derived from the plexus surrounding the middle meningeal artery and are distributed to the blood vessels of the parotid, along with sensory fibers derived from the mandibular nerve.

The **submandibular ganglion** is adjacent to the submandibular gland. Its preganglionic parasympathetic fibers are derived from the seventh cranial nerve through the **chorda tympani.** The latter arises from the facial nerve within the facial canal to course through the middle ear cavity and traverses the petrotympanic fissure to reach the infratemporal fossa where it joins the lingual nerve. The preganglionic fibers pass to the submandibular ganglion by way of two short roots that suspend the ganglion from the lingual nerve. Postganglionic parasympathetic fibers are distributed to the submandibular and sublingual glands. Fibers from the sympathetic plexus around the external maxillary artery may pass through the ganglion and continue with its branches. Sensory fibers from the lingual nerve are distributed with the branches of the ganglion to the submandibular and sublingual glands, and to the mucous membrane of the oral cavity.

Orbital Cavity

Objectives

At the completion of the study of the orbit the student should be able to

▶ *Describe the bony components and foramina or fissures associated with the orbital cavity; list the structures that pass through them*

▶ *Delineate the periorbital fascia; list the seven extraocular muscles of the orbit and give the action of each muscle*

▶ *Give the course and distribution of the third, fourth, and sixth cranial nerves, and the ophthalmic division of the fifth cranial nerve*

▶ *Follow the course and distribution of the branches of the ophthalmic artery*

▶ *Describe the autonomic innervation to structures of the orbit and follow the pathways of preganglionic and*

postganglionic neurons providing this innervation

▶ *Label on a diagram the component parts of the eyeball*

▶ *Describe the optic nerve and its coverings*

▶ *Locate the elements of the lacrimal apparatus and follow the flow of lacrimal fluid from the lacrimal gland to the nasal cavity*

The pyramidal-shaped **orbital cavity** presents a base, an apex, and four walls (Fig. 7-24). The optic foramen is located at the **apex,** while the quadrangular **base** opens onto the face and is formed about equally by the frontal, maxillary, and zygomatic bones. Each bone transmits cutaneous nerves: the supraorbital in the frontal, the infraorbital in the maxillary, and the zygomaticofacial and zygomaticotemporal in the zygomatic

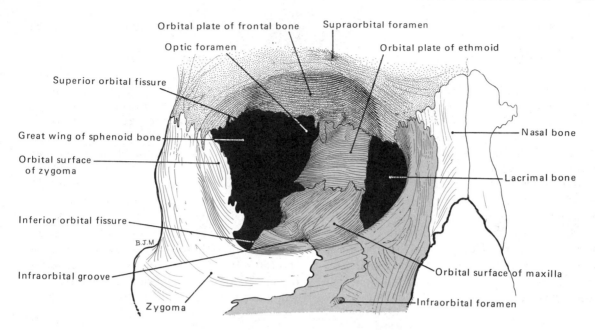

Orbital plate of frontal bone
Supraorbital foramen
Optic foramen
Orbital plate of ethmoid
Superior orbital fissure
Great wing of sphenoid bone
Nasal bone
Orbital surface of zygoma
Lacrimal bone
Inferior orbital fissure
B.J.M.
Infraorbital groove
Orbital surface of maxilla
Zygoma
Infraorbital foramen

Figure 7-24. *The orbital cavity, frontal view. Note that the zygoma, maxilla* (shaded light color), *and frontal bone* (stippled area) *contribute the margin of the orbital cavity.*

bone. The **medial walls** of the cavities are parallel and about 2 cm to 3 cm apart. They are separated by the nasal cavities and are formed by the fragile lacrimal bone and the orbital plates of the ethmoid and palatine bones. The strong **lateral walls,** at right angles to each other, are formed by processes of the zygomatic and the greater wings of the sphenoid bones. The fossa for the lacrimal gland is in the superolateral portion of the lateral wall. The **superior wall,** or roof of the cavity, is formed by the orbital plate of the frontal bone. The orbital plate of the maxillary and a small part of the zygomatic bone forms the **inferior wall,** or floor. In the floor of the orbital cavity, the **infraorbital groove** continues anteriorly as the **infraorbital canal** to open onto the face at the **infraorbital foramen.**

The **orbital periosteum** is a funnel-shaped sheath attached to the bony walls of the orbital cavity, continuous posteriorly with the outer layer of the dura and anteriorly with the periosteum covering the external surface of the skull. It en-

closes the contents of the orbit, except for the zygomatic nerves and the infraorbital nerves and vessels that lie between the periosteum and the bone. The thin, membranous **fascia bulbi (Tenon's capsule)** encloses the eyeball, forming a socket in which it moves, and separates the eyeball from the orbital fat pad. Its smooth inner surface forms, with the sclera, the **periscleral space.** Expansions from the fascial sheath pass to the lateral and medial recti muscles as the **check ligaments.** The sheath reflects onto and encloses the eye muscles as they attach to the sclera. Posteriorly the bulbar fascia is perforated by ciliary vessels and nerves at the **lamina cribosa sclerae.**

Exophthalmos
Exophthalmos is an abnormal protrusion of the eyeball. It is associated with hyperthyroidism (toxic goiter), or an aneurysm of one of the arteries in the orbit, which pushes the eye forward (pulsating exophthalmos).

Muscles (Fig. 7-21)

The **extrinsic muscles** of the eye consist of four straight muscles, the superior, inferior, medial, and lateral recti; two oblique muscles, the superior and inferior; and the levator palpebrae superioris (Table 7-21). The **recti muscles** originate from the margin of a **fibrous cuff,** which is fixed posteriorly to the periosteum and the dural sheath of the optic nerve at the optic foramen, and laterally to the margins of the supraorbital fissure. The lateral rectus is split into upper and lower heads at its origin, with vessels and nerves entering the orbital cavity between the two heads. The recti muscles spread out like the staves of a barrel to insert into a bandlike aponeurosis encircling the sclera just behind the corneoscleral junction. Each muscle has a fascial sheath, with adjacent sheaths

joining to form a **fibromuscular cone** (Fig. 7-25).

The **levator palpebrae superioris** is separated superiorly from the superior rectus and inserts into the tarsal plate of the upper eyelid and the superior fornix of the conjunctivum. The tendon of the **superior oblique** passes through a fascial pulley, the **trochlea,** attached to the medial wall of the orbit and reverses direction before inserting into the sclera. The **inferior oblique,** located in the anterior part of the orbital cavity, has an origin on the floor of the orbital cavity, apart from the other extrinsic ocular muscles. The lateral rectus is supplied by the **abducens nerve,** the superior oblique by the **trochlear nerve,** and the remaining muscles by the **oculomotor nerve.** The medial, superior, and inferior recti, acting together, move the eye medially, while the

Table 7-21
Extrinsic Muscles of the Eye

Muscle	Origin	Insertion	Action	Nerve
Rectus superior, inferior, medial, and lateral	All originate from fibrous cuff fixed posteriorly to optic foramen and anteriorly to dural sheath of optic nerve, and insert superiorly, inferiorly, medially, and laterally by bandlike aponeuroses into sclera just behind corneoscleral junction	See below	See below	Lateral rectus, abducens; others by oculomotor
Levator palpebrae superioris	Orbital roof anterior to optic foramen	Upper tarsal plate and superior fornix of conjunctivum	Elevates upper lid	Oculomotor
Superior oblique	Roof of orbital cavity between superior and medial recti and anterior to optic foramen	Slender tendon passes through fibrous ring (trochlea), reverses direction to insert deep to superior rectus	See below	Trochlear
Inferior oblique	Floor of orbital cavity lateral to lacrimal canal	To sclera between superior and lateral recti	See below	Oculomotor

Action of Eye Muscles

	Action	Muscle
	Adduction	**Medial,** superior, and inferior recti
	Abduction	Inferior and superior oblique; **lateral rectus**
	Elevation	Inferior oblique; **superior rectus**
	Depression	Superior oblique; **inferior rectus**
	Medial rotation	Superior rectus; superior oblique
	Lateral rotation	Inferior oblique; inferior rectus

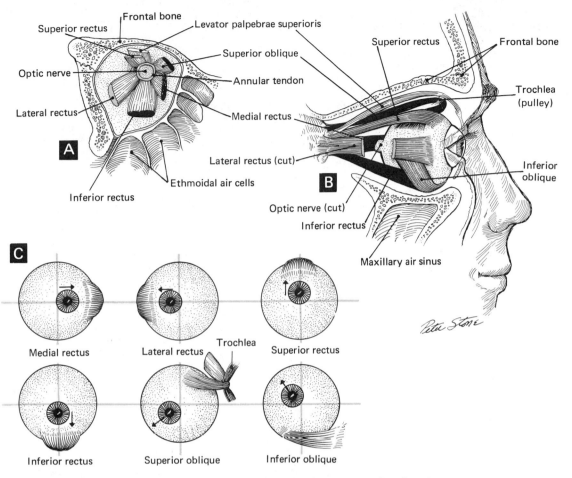

Figure 7-25. *The extrinsic eye muscles.* **A.** *The eyeball has been removed to show the origin of the muscles in frontal section.* **B.** *Lateral view.* **C.** *Action of the extrinsic eye muscles. Arrow indicates direction of movement of the eyeball.*

inferior and superior oblique and the lateral rectus muscles, acting in concert, shift it laterally. In conjugate eye movement, the medial and lateral recti act as "yolk" muscles; that is, to look to the left, the left lateral rectus will abduct the left eye and the right medial rectus will adduct the right eye. The inferior oblique and superior rectus direct the eye upward; the superior oblique and inferior rectus move it downward. The superior rectus and superior oblique medially rotate the eye, and the inferior oblique and inferior rectus

are lateral rotators. Additional actions include the recti muscles acting as retractors and the oblique muscles as protractors to keep the eyeball in balance.

Blood Vessels and Nerves (Table 7-22)

As the internal carotid artery leaves the cavernous sinus, it gives off the **ophthalmic artery** that passes through the optic foramen. Inferior to the

Table 7-22
Arterial Supply to the Orbital Cavity

Artery	Origin	Course	Distribution	Anastomoses
Ophthalmic	Internal carotid	Traverses optic formen to reach orbital cavity	Orbital cavity and muscles and skin of scalp	No direct anastomoses
Supraorbital	Ophthalmic	From supraorbital foramen passes superoposteriorly to lambdoidal suture	Muscles and skin of forehead and scalp to lambdoidal suture	Occipital, superficial temporal and supratrochlear
Supratrochlear	Ophthalmic	From supratrochlear notch passes superoposteriorly to midpoint of scalp	Muscles and skin of forehead	Supraorbital and superficial temporal
Lacrimal	Ophthalmic	Passes along superior border of lateral rectus to reach lacrimal gland	Lacrimal gland	Supratrochlear, supraorbital, middle meningeal
Dorsal nasal	Ophthalmic	Courses along dorsum aspect of nose	Supplies outer surface of nose	Angular and anterior ethmoidal
Long posterior ciliaries (2)	Ophthalmic	Pierces posterior aspect of sclera to reach middle layer of eyeball	Wall of eyeball, ciliary muscle, and iris	Anterior ciliaries
Posterior ethmoidal	Ophthalmic	Traverses posterior ethmoidal foramina	Posterior ethmoidal air cells and posterosuperior aspect of nasal cavity	Sphenopalatine branches
Anterior ethmoidal	Ophthalmic	Passes through anterior ethmoidal foramen; to anterior cranial fossa; along cribriform plate; nasal slit; to anterosuperior aspect of nasal cavity; to dorsum of nose	Anterior ethmoidal air cells, anterosuperior aspect of the nasal cavity, and skin of dorsum of nose	Lateral and dorsal nasal, angular
Infraorbital	Internal maxillary	Passes along infraorbital groove and foramen onto face	Lower eyelid, upper lip, lateral nose	Angular, dorsal nasal, lacrimal
Central artery to retina	Ophthalmic	Pierces optic nerve; courses in center of optic nerve; appears at center of optic disc	Sole supply to retina	No anastomoses
Anterior ciliaries	Muscular branches of ophthalmic	Pierce sclera at periphery of iris	Forms arterial network in iris	Posterior ciliaries

optic nerve, it pierces the dural sheath and lies free within the fibromuscular cone of the orbital cavity. It courses above the optic nerve to give several branches to the structures within the cavity (Figs. 7-26 and 7-27). A very important small branch, the **central artery to the retina,** pierces the optic nerve sheath about 1 cm to 2 cm posterior to the eyeball. It courses in the center of the optic nerve to reach and supply the retina.

Central Artery of Retina

The central artery of the retina is an end artery. Therefore, if it is blocked, there are no anastomoses with other arteries to sustain circulation to the retina. The result is sudden, total blindness of the affected eye. An occlusion may result from several causes, such as a tumor, a thrombus, or massive edema.

Several **short posterior ciliary arteries** arise from the ophthalmic artery to pierce the sclera and form a plexus in the choroid. Additional branches, the **long posterior ciliary arteries,** extend anteriorly to anastomose with the anterior ciliary artery at the margin of the iris. The **anterior ciliary arteries** arise from muscular branches to the recti muscles. They pierce the sclera just behind the corneoscleral junction to supply the ciliary body and iris, and give twigs to the deep conjunctival plexus. The relatively large **lacrimal branch** of the ophthalmic artery begins near the optic foramen and courses, with the lacrimal nerve, along the lateral wall above the lateral rectus muscle. It supplies the latter, the superior oblique muscle, the lacrimal gland, upper eyelid, and the conjunctivum. Terminal branches of the ophthalmic artery include the **supraorbital, supratrochlear, dorsal nasal, anterior ethmoidal,** and **posterior ethmoidal,** which pass from the orbital cavity to anastomose freely with branches of the external carotid in the upper face.

Venous drainage of the orbital cavity is accomplished by the **superior** and **inferior ophthalmic veins.** The former, formed by the junction of the supraorbital and supratrochlear veins, drains into the cavernous sinus. The inferior ophthalmic vein, originating in the floor of the cavity, communicates through the infraorbital fissure with the pterygoid plexus or through the supraorbital fissure with the cavernous sinus.

The **optic nerve** enters the orbital cavity through the **optic foramen,** while all the other nerves to the orbit traverse the supraorbital fissure. The optic nerve passes anterolaterally and slightly inferiorly from the optic foramen to penetrate the posterior aspect of the eyeball, slightly medial to the posterior pole of the eye. Its meningeal coverings fuse with the sclera of the eyeball. The optic nerve is slightly longer than its course and, therefore, does not interfere with movements of the eyeball. Within the retina, the optic nerve fibers spread out as the third-order neurons in the visual pathway.

The **oculomotor nerve** is motor to the superior, medial and inferior recti, the inferior oblique, and the levator palpebrae superioris muscles. It also carries preganglionic parasympathetic fibers to the ciliary ganglion. Passing between the heads of the lateral rectus muscle, it divides into an **upper division,** supplying the superior rectus and the levator palpebrae superioris, and a **lower division** to the medial and inferior recti and the inferior oblique muscles. The **trochlear nerve** innervates the superior oblique muscle as it passes along the superior border of that muscle. The **abducens nerve** passes between the heads of the lateral rectus to course on the ocular surface of this muscle, which it innervates.

The smallest division of the trigeminal nerve, the **ophthalmic nerve,** is sensory and terminates as frontal, lacrimal, and nasociliary branches. The **frontal nerve** passes through the supraorbital fissure above the lateral rectus muscle. It courses between the levator palpebrae superioris and the orbital plate to terminate as **supraorbital** and **supratrochlear branches,** which supply the eyelids, forehead, and scalp. The **lacrimal nerve,** passing just below the frontal branch, follows the upper border of the lateral

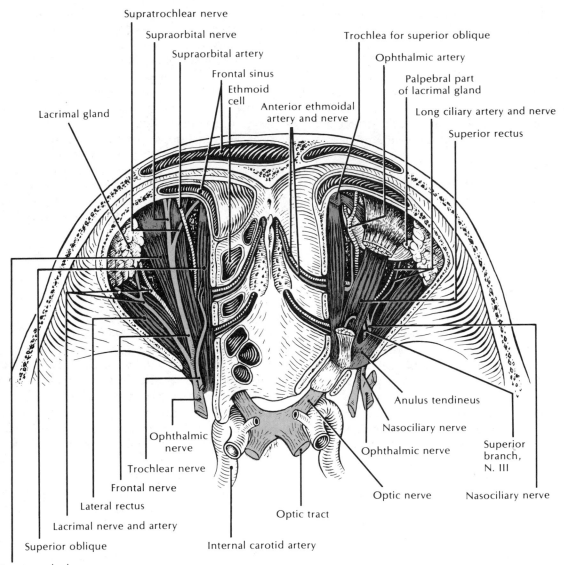

Figure 7-26. *Superior view of the interior of the orbit. (Hollinshead WH: Textbook of Anatomy, 3rd ed. New York, Harper & Row, 1974)*

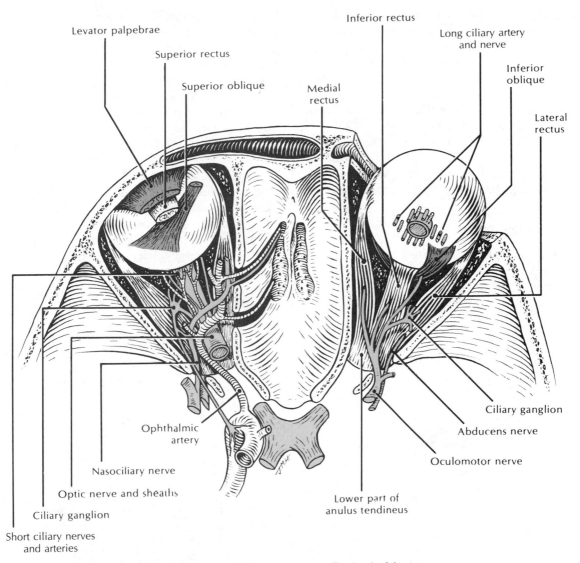

Figure 7-27. *Deep aspect of the orbital cavity. (Hollinshead WH: Textbook of Anatomy, 3rd ed. New York, Harper & Row, 1974)*

rectus muscle. It supplies the lacrimal gland and terminates as twigs to the conjunctivum, skin of the eyelids, and the skin over the zygomatic process. Communications between the lacrimal nerve and the zygomatic branch of the maxillary division of the trigeminal form the pathway for postganglionic parasympathetic fibers to pass from the pterygopalatine ganglion to the lacrimal gland.

The **nasociliary nerve** courses between the heads of the lateral rectus muscle to cross the medial wall of the orbital cavity above the optic nerve. Its branches include the **long ciliary nerves,** carrying sensory and postganglionic sympathetic fibers to the iris; **posterior ethmoidal nerves,** supplying the ethmoidal and sphenoidal air sinuses; the **infratrochlear nerve** innervating the lacrimal sac, conjunctivum, eyelids, and upper nose; and the **anterior ethmoidal nerve.** The latter passes through the anterior ethmoidal foramen to enter the anterior cranial fossa, courses along the cribriform plate, and enters the nasal cavity by traversing the nasal slit. It terminates by dividing into the **internal** and **external nasal branches** (Table 7-23).

The **ciliary ganglion,** located in the posterior third of the orbital cavity, is a synaptic station for parasympathetic neurons and has been described in this chapter in the section on parasympathetics.

Lacrimal Apparatus

The **lacrimal gland** is located in the superolateral aspect of the orbital cavity, partly within the **lacrimal fossa** and partially embedded in the upper eyelid (Fig. 7-28). It drains through six to ten

lacrimal ducts that pierce the superior fornix of the conjunctivum to empty onto the opposing palpebral and ocular surfaces of the conjunctivum. The action of blinking spreads a uniform layer of lacrimal fluid over the conjunctivum. At the medial canthus, a **punctum lacrimale** opens at the summit of a **lacrimal papilla** on the free margin of each lid as the beginning of the **lacrimal canaliculi.** The latter drain the lacrimal fluid into an expansion of the proximal end of the nasolacrimal duct, the **lacrimal sac.** The **nasolacrimal duct** passes through the nasolacrimal canal of the maxilla and lacrimal bones to open into the inferior meatus of the nasal cavity.

The lacrimal gland is innervated by parasympathetic fibers originating from the facial nerve. Postganglionic fibers from the pterygopalatine ganglion travel with a communicating branch from the zygomatic nerve to the lacrimal branch

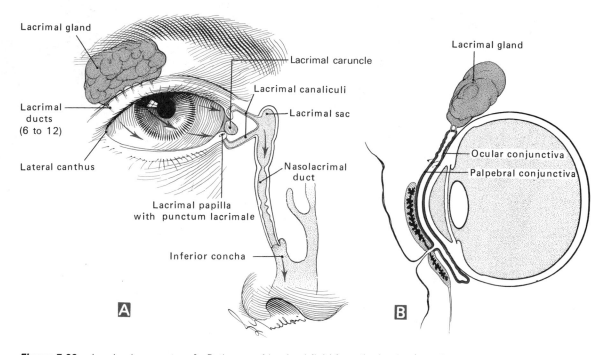

Figure 7-28. *Lacrimal apparatus.* **A.** *Pathways of lacrimal fluid from the lacrimal sac to the nasal cavity.* **B.** *Relation of the lacrimal gland to the conjunctival surfaces with the eye closed. Parasagittal section.*

Table 7-23
Nerve Distribution to Orbit

Nerve	Origin	Course	Distribution
Optic (II) —Actually a brain tract. Optic "nerve" formed by nerve cell processes of tertiary neurons in visual pathway, neuronal elements in retina include (1) rods and cones, (2) bipolar neurons, and (3) neurons of ganglion layer.			
	Posterior pole of eyeball conveys nerve cell processes of neurons in ganglion layer of retina	Traverse orbital cavity to pass through optic foramen and joins nerve of other eye at chiasma; extends to brain as optic tract	Nerve of vision
Motor nerves (III, IV, VI)—All arise from brain stem, traverse cavernous sinus, and enter orbit via supraorbital fissure to extrinsic eye muscles			
Oculomotor (III)		Divides into superior and inferior divisions	Superior division to superior rectus and levator palpebrae superioris; inferior division to medial and inferior rectus and inferior oblique, and conveys preganglionic parasympathetic fibers to ciliary ganglion; postganglionics to ciliary muscle and circular (sphincter) muscle of iris
Trochlear (IV)		Along superior border of superior oblique	Superior oblique
Abducens (VI)		Along medial aspect of lateral rectus	Lateral rectus
Ophthalmic (V₁)—From trigeminal ganglion traverses cavernous sinus to enter orbit via supraorbital fissure and trifurcates into lacrimal, frontal, and nasociliary nerves			
Lacrimal		Follows lateral wall of orbit to reach lacrimal gland	Sensory to lacrimal gland
Supraorbital and supratrochlear	Terminal branches of frontal	Run on superior aspect of levator palpebrae superioris to reach foramen or notch of same name	Supply skin of forehead and scalp to lambdoidal suture
Posterior ethmoidal	Nasociliary	Passes through posterior ethmoidal foramen	Sensory to ethmoidal air cells
Anterior ethmoidal	Nasociliary	Passes through anterior ethmoid foramen, reenters cranial cavity, runs along cribriform plate and, via nasal slit, traverses nasal cavity to emerge on dorsum of nose	Internal nasal—to mucosa of anterosuperior portion of nasal cavity; external nasal—to skin on dorsum of nose
Infratrochlear	Nasociliary	Along medial wall of orbit to upper lid	Skin of upper lid
Long ciliary	Nasociliary	Pierces area cribrosa of sclera to enter middle tunic of eyeball	Sensory to eyeball

of the ophthalmic division of the trigeminal nerve. The lacrimal gland receives its blood supply by way of the lacrimal branch of the opthalmic artery.

Conjunctivum

The **conjunctivum** is the mucous membrane of the eye that lines the inner surface of the eyelids, and covers the outer surfaces of the sclera and cornea. It forms a partial sac where it reflects from the deep surface of the lid on to the eye as the **conjunctival fornix.** There is a fornix for each lid, that is, a superior and an inferior fornix.

The conjunctivum is divided into a vascular **palpebral part** that lines the back of the eyelids, and a thin, transparent, bulbar portion covering the sclera and cornea. A vertical fold of bulbar conjunctiva at the medial angle of the eye forms the **plica semilunaris. The lacrimal caruncle** is a small mound of modified skin at the inner canthus. The conjunctivum is innervated by a rich plexus from **ciliary, lacrimal,** and **infratrochlear nerves,** whose fibers largely terminate as free nerve endings.

Eyeball

Myopia

Myopia is a condition in which the eyeball is too long and the focus of objects lies anterior to the retina, rather than on it. Hyperopia is the opposite condition; the focus lies posterior to the retina. Obviously either condition causes faulty vision, which is correctable by lenses that will focus the image on the retina.

The **eyeball,** approximately 2 cm to 3 cm in diameter, is spherical with a slight anterior bulge (Fig. 7-29). Its wall consists of three concentric coats: an outer fibrous, a middle vascular, and an inner nervous tunic. The sclera and cornea compose the outer **fibrous tunic,** with the **sclera** forming a firm cup covering the posterior five-sixths of the eyeball. Its outer surface is smooth, separated from the bulbar fascia by loose connective tissue, and perforated posteriorly by the optic nerve, and the central artery and vein at the lamina cribrosa. At the corneoscleral junction, it is continuous with the transparent **cornea,** which bulges slightly over the anterior one-sixth of the eyeball. The cornea is avascular and receives its nutrients by diffusion from a capillary network at its margin. The cornea is richly supplied with free sensory nerve endings derived from the ciliary nerves.

The **vascular tunic** lies internal to the sclera and consists of the choroid, the ciliary body, and the iris. The **choroid** is a thin, highly vascular membrane, and is brown in color from pigmented cells. It covers the posterior two-thirds of the eyeball and extends anteriorly to the ora serrata. The choroid consists of a dense capillary network of small arteries and veins held together with connective tissue. It is loosely connected to the sclera, except at the entrance of the optic nerve, where it is firmly fixed. It is intimately attached to the inner pigmented layer of the retina. The **ciliary body,** consisting of a thickening of the vascular tunic as the ciliary ring, ciliary processes, and ciliary muscle, connects the choroid at the **ora serrata** to the peripheral circumference of the iris. (Fig. 7-30). The **ciliary ring** extends from the ora serrata to the **ciliary processes,** while the latter, sixty to eighty small projections, are continuous peripherally with the ciliary ring and give attachment anteriorly to the **suspensory ligament of the lens.** The nonstriated **ciliary muscle** originates at the posterior margin of the scleral spur and inserts into the ciliary ring and ciliary processes. It contracts in accommodation to draw the ciliary processes forward, which relaxes the suspensory ligament permitting the natural elasticity of the lens to become more rounded and so effects focus. The **iris** is a thin, pigmented, contractile diaphragm with a central aperture, the **pupil.** The opening varies in size. It constricts by the action of the circularly arranged **sphincteric muscle** that is innervated by para-

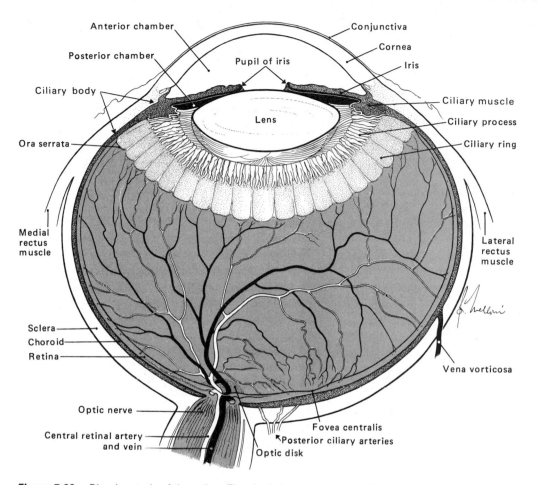

Anterior chamber

Conjunctiva

Posterior chamber

Cornea

Pupil of iris

Iris

Ciliary body

Ciliary muscle

Lens

Ciliary process

Ciliary ring

Ora serrata

Medial rectus muscle

Lateral rectus muscle

Sclera

Choroid

Retina

Vena vorticosa

Optic nerve

Central retinal artery and vein

Fovea centralis

Posterior ciliary arteries

Optic disk

Figure 7-29. *Blood vessels of the retina. The shaded area represents the extent of visual receptors in the retina of the intact eye.*

sympathetic fibers from the ciliary ganglion and third cranial nerve, or by dilating through the action of the radially arranged **dilator muscle**— innervated by sympathetic fibers from the superior cervical ganglion. The iris, the conspicuous colored portion of the eye, separates the **anterior chamber** (posterior to the cornea and anterior to the iris) from the **posterior chamber** (posterior to the iris and anterior to the suspensory ligament and ciliary processes). Both chambers are continuous with each other through the pupil and are filled with a clear, refractile fluid, the **aqueous humor.**

Glaucoma

Glaucoma is a condition of the eye that is characterized by increased intraocular pressure. It occurs when the drainage of the aqueous humor fails to keep up with the production of the fluid by the ciliary processes. The buildup of pressure may cause severe pain and damage to the

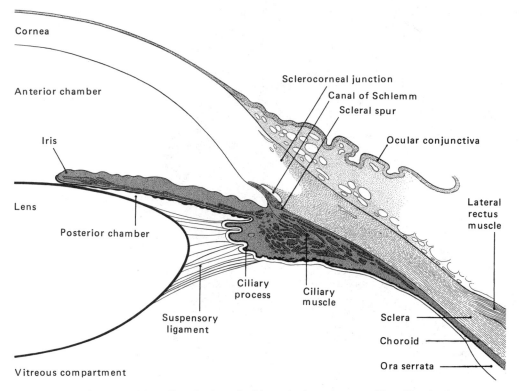

Figure 7-30. *Relation of the ciliary body to the iris and other structures. The ciliary body consists of the ciliary muscle and ciliary processes.*

nerve fibers in the retina, especially at the optic disc. If the pressure continues, blindness results. Drugs that reduce the rate of aqueous humor production, and thus lower the intraocular pressure, will prevent permanent retinal and optic nerve injury.

The **nervous tunic** or **retina** is developmentally an evagination of the brain. It has an outer pigmented layer and an inner light-receptor portion. Three regions of the retina are differentiated: the optic, ciliary, and iridial portions.

Detachment of the Retina

Detachment of the retina may occur in trauma, such as a blow to the head. The tear in the retina causes blindness in the corresponding field of vision. The actual detachment occurs between the sensory part of the retina and the underlying pigmented layer. Fluid accumulates between these layers, forcing the thin, friable retina to billow out toward the vitreous humor. The retina may be reattached by various surgical procedures, such as photocoagulation by laser beam, cryosurgery, or scleral resection.

The optic part is light-sensitive and contains a three-neuron pathway. The first neuron constitutes the light-receptor **rods** and **cones;** the second, **bipolar cells;** and the third, the **ganglion cells,** whose axons form the fibers of the optic nerve. The optic portion occupies the posterior part of the bulb and ends at the ora serrata. It is firmly attached at the ora serrata anteriorly and at the entrance of the optic nerve posteriorly. The **ciliary portion** of the retina begins at the ora serrata and continues anteriorly to line the internal surface of the ciliary body. The pigmented **iridial (iris) portion** of the retina covers the posterior aspect of the iris.

Color Blindness
Color blindness is the inability to distinguish colors correctly. To have normal color vision, there must be three types of cones in the retina. Each type responds only to a particular color: red, green, or violet. When all three cones are equally stimulated, sensation of white results. Very few people are truly color blind, but many are color weak because they have difficulty in distinguishing hues. Color blindness is controlled by sex-linked recessive genes and manifests itself much more frequently in the male than in the female.

The **macula lutea** is a yellowish, oval area at the posterior pole of the eye that presents a slight central depression, the **fovea centralis,** the area of greatest visual acuity. Light rays are focused at the fovea if the eyes are correctly accommodating. The **optic disc,** the site of emergence of the optic nerve, is about 3 mm medial to the macula lutea. Axons of the third-order neuron of the retina converge at this point in an area devoid of light-receptor cells, known as the **blind spot** of the eye. At the center of the optic disc, the central artery of the retina emerges and distributes to the retina.

Choked Disc or Papilledema
A choked disc or papilledema is the swelling of the optic disc, so that it protrudes into the vitreous body. Changes in the disc, as viewed through an ophthalmoscope, reflect evidence of the condition of the brain proper since the optic nerve and disc are an extension of the brain. Edema of the papilla suggests increased intracranial pressure that compresses the thin veins from the optic nerve, causing congestion and edema of the retina's optic disc. Several pathologic conditions may cause the increase in intracranial pressure, for example, brain abscess, aneurysm, brain tumor, or hydrocephalus.

The refractile **lens,** a transparent biconvex body more flattened anteriorly than posteriorly, is composed of laminated transparent lens fibers. It is enclosed by a transparent **capsule** and held in place by the **suspensory ligament.** Its shape is modified in focusing by the action of the ciliary muscle. Posterior to the lens, the refractile **vitreous humor,** or **body,** occupies the central portion of the eyeball.

Cataract
A cataract is a condition in which the lens becomes opaque or milk-white in appearance. When the lens has a cataract, little light is transmitted to the retina and blurred images and poor vision result. To remove this obstruction of light, the lens with the cataract is removed and light can freely pass through the eye to the retina. Loss of the lens requires the patient to wear thick convex lenses that focus a clear sharp image on the retina. A recently developed alternative is a lens implant.

Mouth and Pharynx

Objectives

At the completion of the study of the oral cavity and the pharynx the student should be able to

▶ *Delineate the major subdivisions of the oral cavity*

▶ *Describe the gross anatomy of the floor of the oral cavity*

▶ *Describe the general morphology of the tongue; identify the extrinsic and intrinsic musculature of the tongue*

▶ *Delineate the subdivisions of the pharynx*

▶ *Describe the musculature of the pharynx*

▶ *List and give the origin, course, and distribution of the nerve and arteries that supply the oral cavity and pharynx*

Oral Cavity

The **oral cavity** is subdivided for descriptive purposes into the **vestibule** and the **oral cavity proper** (Fig. 7-31). The former is the cleft separating the lips and cheeks from the teeth and gingivae, or gums. With the mouth closed the vestibule communicates with the **oral cavity proper** through the interval between the last molar teeth and the rami of the mandible. At the lips the skin of the face is continuous with the mucous membrane of the oral cavity. The bulk of the **lips** is formed by the orbicularis oris muscle and contains a vascular arch arising from labial branches of the facial artery. The **cheek,** consisting for the most part of the buccinator muscle and the buccal fat pads, is pierced by the parotid duct, which opens into the vestibule opposite the upper second molar tooth. The **gingiva** is composed of dense fibrous tissue covered by a smooth vascular mucosa. It attaches to the alveolar margins of the jaws where it embraces the necks of the teeth as the **periodontal membrane.** Thirty-two **teeth** are normally present in the adult; two incisors, one canine, two premolars, and three molars are found in each half of the upper and lower jaws.

Pyorrhea

Pyorrhea is a discharge of pus around the teeth with progressive necrosis of the alveolar bone and looseness of the teeth. Recession of the gums continues until the teeth lose their bony and gingival attachments and fall out. When the teeth are lost, the inflammatory symptoms subside. Good oral hygiene with semi-annual professional cleaning of the teeth will usually prevent pyorrhea, the principal cause of loss of teeth in the adult.

Posteriorly the mouth cavity communicates with the pharynx through the **fauces.** The **hard** and **soft palates** make up the roof of the mouth, and the floor is formed by the tongue and mucous membrane. Anteriorly the tongue lies more or less free in the mouth, with a median fold of mucous membrane, the **frenulum linguae,** passing from the floor of the mouth to the under surface of the tongue. Two transverse **sublingual folds** overlie the sublingual glands and the minute orifices of these glands open along the summit of the fold. Near the midline on each sublingual fold, the sublingual papillae surround the openings of the submandibular ducts.

The tongue, a mobile mass of muscle and mucous membrane, functions in taste, chewing, swallowing, and speech. It is shaped like an upside-down high-topped shoe, with the sole of the shoe being the dorsum of the tongue and the upper portion of the shoe the root of the tongue. At the back of the tongue, a V-shaped groove, the **sulcus terminalis,** is flanked anteriorly by a ridge of

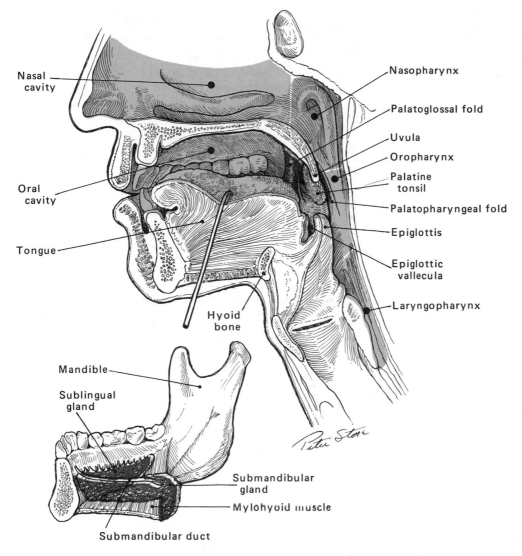

Nasal cavity

Oral cavity

Tongue

Hyoid bone

Nasopharynx

Palatoglossal fold

Uvula

Oropharynx

Palatine tonsil

Palatopharyngeal fold

Epiglottis

Epiglottic vallecula

Laryngopharynx

Mandible

Sublingual gland

Submandibular gland

Mylohyoid muscle

Submandibular duct

Figure 7-31. *Sagittal view of oral, nasal, and pharyngeal cavities. The lower drawing shows the relation of the sublingual and submandibular glands to the floor of the mouth.*

large **circumvallate papillae,** and divides the dorsum of the tongue into two parts, an **anterior** two-thirds, **horizontal** or **palatine portion** and a **posterior** one-third, **vertical or pharyngeal portion.** A small pit, the **foramen cecum,** is present at the vertex of the sulcus terminalis. **Fungiform** and **filiform papillae** are distributed over the palatine, or horizontal portion of

the tongue. With the mouth open only the palatine portion of the tongue is visible; the less apparent vertical pharyngeal portion forms the anterior wall of the orpharynx and is related inferiorly to the epiglottis. On the ventral surface of the tongue, the midline **frenulum linguae,** flanked by the **deep lingual veins,** attaches the tongue to the floor of the mouth. A delicate fringed ridge

Table 7-24
Arterial Supply to the Oral Cavity and Pharynx

Artery	Origin	Course	Distribution	Anastomoses
Lingual	Anterior aspect of external carotid	At superior cornua of hyoid bone pierces hyoglossus muscle to pass to deep region of tongue	Primary blood supply to tongue; within substance of tongue gives off dorsal lingual and deep lingual branches	Across the midline
Sublingual	Lingual	From lingual artery passes between the mylohyoid muscle and mucous membrane on floor of mouth	Mylohyoid, geniohyoid and mucous membrane of mouth; duct of submandibular, sublingual gland	Submental
Submental	Facial	Passes anteriorly between mylohyoid and anterior belly of digastric	Muscular floor of mouth	Sublingual
Descending palatine	Maxillary	From pterygopalatine fossa descends in palatine canal to reach palate; bifurcates into greater and lesser palatine	Greater palatine to hard palate; lesser palatine to soft palate	Greater with long septal; lesser with tonsillar arteries
Ascending pharyngeal	External carotid	Ascends along lateral aspect of pharynx	Muscular wall of pharynx	Tonsillar branches
Buccal branches	Maxillary and infraorbital	From maxillary within infratemporal fossa passes anteriorly to orbit; from infraorbital descends to cheek	Skin, muscle and mucous membrane of cheek	With each other
Tonsillar branches	Facial, ascending pharyngeal, lingual, descending palatine	Arise as parent arteries course adjacent to bed of tonsil	Tonsil, soft palate and fauces	With each other
Inferior alveolar	Maxillary	From infratemporal fossa enters mandibular foramen; traverses mandibular canal; emerges at mental foramen as terminal mental branch	Lower teeth and gingiva, mental supplies muscles and skin of chin	Mental with inferior labial
Superior alveolars (anterior, middle, and posterior)	Anterior and middle from infraorbital; posterior from maxillary	Anterior and middle from floor of orbit pass through tiny canals in bony wall of maxillary sinus; posterior from infratemporal fossa through foramina in tuberosity of maxilla	Upper teeth, gingiva and maxillary sinus	No direct anastomoses

of mucous membrane, the **fimbriated fold,** is present on the lateral aspect of the ventral surface of the tongue. Numerous lymph follicles constituting the **lingual tonsil** are present on the pharyngeal portion of the tongue. As the mucous membrane reflects onto the epiglottis, it forms **median** and **lateral glossoepiglottic folds.** The arterial supply to the oral cavity and pharynx are presented in Table 7-24.

Tongue-Tie

Tongue-tie (ankyloglossia) results from a congenital shortening of the frenulum, which attaches the tongue to the floor of the mouth. Such a condition often causes a severe speech impediment. Recognizing the condition early and simply cutting the frenulum frees the tongue and cures the problem.

Muscles of the Tongue

The muscles of the tongue consist of four pairs of intrinsic and four pairs of extrinsic muscles (Table 7-25; Fig. 7-32). The muscles of either side are separated by a midline fibrous septum. The **superior longitudinal muscle** extends from the tip to the root of the tongue, and the **inferior longitudinal muscle** is present in the interval between the extrinsic genioglossus and the hyoglossus muscles. The **transversus linguae** extend from the septum to the sides of the tongue, while the **verticalis linguae** originate from the dorsum of the tongue and sweep inferiorly and laterally to interdigitate with the other tongue musculature.

Extrinsic muscles include the fan-shaped **genioglossus,** radiating from the genial tubercle of the mandible into the tongue, with its lowermost fibers inserting into the hyoid bone. The quadrilateral **hyoglossus,** under cover of the mylohyoideus, passes from the body and greater horn of the hyoid to the sides of the tongue. The slip-like **styloglossus** sweeps forward from the tip of the styloid process and stylomandibular ligament to blend with, and insert into, the hyoglossus and

palatoglossus with some of its fibers extending along the side of the tongue as far as the tip. The **palatoglossus** originates from the palate, passes to the side of the tongue, and forms the anterior pillar of the palatine fossa.

All the muscles of the tongue, except the palatoglossus (pharyngeal plexus), are innervated by the hypoglossal nerve (Table 7-26). The intrinsic muscles act to alter the shape of the tongue, and the extrinsic group functions to change the position and, to a limited extent, the shape of the tongue. The anterior two-thirds of the tongue is supplied with special (taste) sensation by the chorda tympani branch of the facial nerve and with general sensation by the lingual branch of the mandibular division of the trigeminal nerve. The glossopharyngeal nerve supplies both general and special sensation to the posterior one-third of the tongue.

Injury to Hypoglossal Nerve

If cranial nerve XII is damaged or divided, the half of the tongue on the same side of the lesion becomes atrophic and wrinkled. Asking a patient to stick out his tongue tests the function of the hypoglossal nerve. When the tongue is protruded, the muscles of the intact side push the tip of the tongue towards the paralyzed side and hence the tongue points to the side of the lesion.

The **roof of the mouth** is a vaulted dome formed by the hard and soft palates. The **hard palate** is composed of the palatine processes of the maxillary bone anteriorly, and the horizontal laminae of the palatine bones posteriorly. The **soft palate** consists of muscles, glands, and the palatine aponeurosis. The soft palate is attached anteriorly to the posterior margin of the hard palate and laterally to the wall of the pharynx. Posteriorly in the midpoint of its free margin, it forms the conical **uvula** directed inferiorly into the fauces. During deglutition the soft palate elevates to help close the nasopharynx.

Table 7-25
Muscles of the Tongue

Muscle	Origin	Insertion	Action	Nerve
Genioglossus	Genial tubercle of mandible	Ventral surface of tongue and body of hyoid bone	Aids in protrusion, retraction, and depression of tongue	Hypoglossal
Hyoglossus	Body and greater cornu of hyoid bone	Sides of tongue	Depresses and draws tongue laterally	Hypoglossal
Styloglossus	Styloid process	Sides of tongue	Aids in retraction and elevation of tongue	Hypoglossal
Palatoglossus	Soft palate	Dorsum and sides of tongue	Elevates tongue and narrows fauces	Pharyngeal plexus
Longitudinalis linguae (superior and inferior); transversus and verticalis linguae	From intrinsic musculature of tongue; named according to their relationship		Alter shape of tongue	Hypoglossal

Table 7-26
Innervation to Mouth and Pharynx

Nerve	Origin	Course	Distribution
Hypoglossal (XII)	Brain stem	Exits cranial cavity via hypoglossal canal, deep to posterior belly of digastric picks up and conveys fibers of ansa cervicalis; in neck, lies on hyoglossus and passes deep to intermediate tendon of digastric and mylohyoid to reach floor of mouth	Supplies all muscles of tongue except palatoglossus
Lingual (V)	Posterior division of mandibular (V_3)	Passes on lateral surface of medial pterygoid to reach tongue; receives and conveys fibers of chorda tympani	General sensation to anterior $2/3$ of tongue; chorda tympani supplies taste to anterior $2/3$ of tongue, conveys preganglionic fibers to submandibular ganglion, postganglionics to submandibular and sublingual glands
Lingual (IX)	Glossopharyngeal (IX) near styloid process	Passes anteriorly to tongue	General and taste sensation to posterior $1/3$ of tongue
Lingual (X)	Vagus (X) near thyroid cartilage	Passes directly to tongue	General and taste sensation to root of tongue
Greater and lesser palatine	Maxillary (V_2)	From V_2 (pterygopalatine ganglion) pass through palatine canal to reach hard and soft palate	Sensory to hard (greater) and soft (lesser) palate; convey postganglionics to glands of palate
Pharyngeal plexus	Glossopharyngeal (IX) and vagus (X)	Twiglets form plexus as the nerves traverse the neck	Vagus motor, glossopharyngeal–sensory; glossopharyngeal also supplies motor innervation to stylopharyngeus

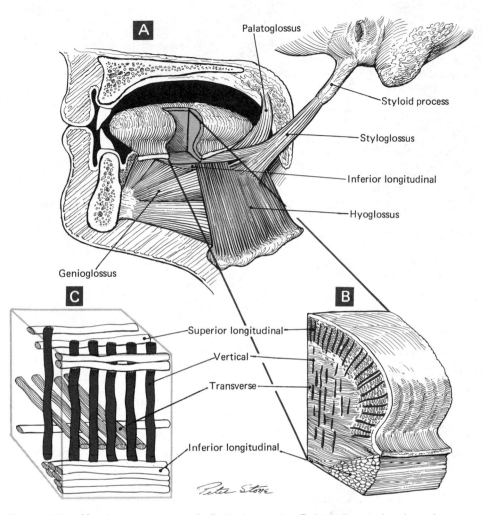

Figure 7-32. *Muscles of the tongue.* **A.** *Extrinsic muscles.* **B.** *Intrinsic muscles shown in a section of the tongue.* **C.** *The arrangement of intrinsic muscles in planes.*

Cleft Palate and Hare Lip

Cleft plate and hare lip may occur separately or together, and in varying degress or extent. Developmentally during the 8th to 9th week of gestation, the palate is formed by fusion of the two lateral palatine processes and the single central premaxilla. Failure of union of these structures leaves a gap or cleft at the fusion site. A hare (cleft) lip is the failure of the medial nasal processes to merge with the maxillary processes. Although they sometimes are bilateral, single cleft palate and cleft lip are the most common anomalies.

Muscles of the Palate

Five pairs of muscles are associated with the soft palate (Table 7-27). The **palatoglossus** and **palatopharyngeus** originate from the palatine apo-

Table 7-27
Muscles of the Palate

Muscle	Origin	Insertion	Action	Nerve
Tensor veli palatini	Scaphoid fossa, spine of sphenoid, and cartilaginous portion of pharyngotympanic tube	Tendon passes around hamulus of pterygoid to insert into soft palate	Tenses soft palate	Mandibular division of trigeminal
Levator veli palatini	Petrous portion of temporal bone and cartilaginous portion of pharyngotympanic tube	Midline of soft palate	Elevates soft palate	Pharyngeal plexus
Palatoglossus	Soft palate	Dorsum and sides of tongue	Narrows fauces and elevates tongue	Pharyngeal plexus
Palatopharyngeus	Soft palate	Posterior border of thyroid cartilage and musculature of pharynx	Elevates pharynx and helps to close nasopharynx	Pharyngeal plexus
Uvulae	Palatine aponeurosis	Mucous membrane of uvula	Elevates uvula	Pharyngeal plexus

neurosis and the posterior part of the hard palate and are contained within the palatoglossal and palatopharyngeal folds, respectively. The palatoglossus inserts into the dorsum and side of the tongue. The palatopharyngeus splits to pass to either side of the levator veli palatini and musculus uvulae, reunites, and then blends with the salpingopharyngeus before inserting into the thyroid cartilage and pharynx. Both **musculi uvulae** arise from the posterior nasal spine and unite as they pass backward to insert into the mucous membrane of the uvula. The rounded **levator veli palatini** originates from the temporal bone adjacent to the opening of the carotid canal. It passes obliquely downward and inserts into the palatine aponeurosis, interdigitating with fibers of the muscle from the opposite side. The flat, triangular **tensor veli palatini muscle** arises from the scaphoid fossa, spine of the sphenoid, and cartilaginous portion of the pharyngotympanic tube. It tapers to form a rounded tendon that hooks around the pterygoid hamulus, turns medially, and then spreads out to form the palatine

aponeurosis attached to the posterior border of the hard palate.

The region of communication between the oral cavity and the pharynx, the **fauces,** which enclose the palatine tonsils, is bounded superiorly by the soft palate, inferiorly by the dorsum of the tongue, and laterally on either side by the palatoglossal and palatopharyngeal arches.

Pharynx

The **pharynx** is a wide muscular tube, about 12 cm to 14 cm long, lined with mucous membrane. It extends from the base of the skull to the level of the sixth cervical vertebra, where it becomes the esophagus. The attachments of the pharynx, from superior to inferior, are: the pharyngeal tubercle at the base of the skull, medial pterygoid lamina, pterygomandibular raphe, inner aspect of the ramus of the mandible, hyoid bone, and thyroid and cricoid cartilages. The muscular wall of the pharynx consists of the three overlapping constrictor muscles (superior, middle, and inferior), the sty-lopharyngeus and palatopharyngeus muscles. Various fascial layers and a mucous membrane also contribute to the wall of the pharynx.

The pharynx is subdivided into the nasal, oral, and largyngeal portions. Anteriorly the wall is interrupted by, and related to, structures associated with openings into these portions. The **nasal pharynx** is situated above the level of the soft palate, posterior to the nasal cavity, and is related superiorly to the sphenoid bone and basilar portions of the occipital bone. It is the widest part of the pharynx and normally remains patent for the passage of air. Anteriorly it is bounded by the **choanae** (internal nares), which open into the nasal cavity. Laterally it receives the opening of the **auditory** or **pharynogotympanic tube,** around which the mucous membrane is raised as the **torus tubarius.** A mucous membrane covering the salpingopharyngeal muscle, the **salpingopharyngeal fold,** descends vertically from the torus tubarius. The roof and the posterior wall of the nasopharynx form a continuous curve, and contain an aggregate of lymphoid tissue between the roof and pharyngeal recesses, the **pharyngeal tonsils.** When enlarged these are called **adenoids.**

Adenoids

Adenoids is a pathologic condition of the pharyngeal tonsil characterized by infection and hypertrophy of its lymphoid tissue. The marked enlargement of the gland blocks the internal nares (choana), which necessitates the person to breathe through the mouth. If the condition persists, the child develops a characteristic facial expression called adenoid facies defined as a ''dull expression, with open mouth seen in children with adenoid growth.'' Infection may spread to the lymphoid tissue surrounding the pharyngotympanic tube (tubal tonsil), causing swelling and closure of the tube. Recur-

ring attacks of middle ear infection usually follow, which may result in temporary or permanent hearing loss.

The pharyngeal isthmus, located between the nasal and oral parts of the pharynx, is bounded laterally by the palatopharyngeal arch and the mucous membrane covering the palatopharyngeal muscle. It is closed during swallowing by the elevation of the soft palate and contraction of the superior constrictor muscle of the pharynx.

Pharyngotympanic Tube Infections

The pharyngotympanic (Eustachian) tube connects the middle ear cavity with the nasopharynx. Such a communication provides a pathway for nasal and oral infections to spread to the middle ear and in time to the mastoid antrum and mastoid air cells. Prior to the advent of antibiotics, mastoid abscesses were very common sequelae of respiratory infections. Now they are rare because penicillin and other antibiotics help eliminate the offending pathogenic organisms, and cure the middle ear infections. Since the mastoid is separated only by a thin plate of bone from the temporal lobe of the brain and the cerebellum, a mastoid abscess or mastoiditis may erode the bone and cause a brain abscess and meningitis.

The **oropharynx** is located posterior to the oral cavity. The lower portion of the anterior wall is formed by the root of the tongue and by the epiglottic cartilage. Three mucous membrane folds, a median glossoepiglottic fold between the tongue and the epiglottic cartilage, and two lateral glossoepiglottic folds between the epiglottis and the junction of the tongue and pharynx, bound depressions, the **epiglottic valleculae.** Each lateral wall houses a mass of lymphoid tissue, the **palatine,** or **true tonsil,** located between the **palatine arches.** The latter are formed anteriorly by the palatoglossus muscle and posteriorly by the palatopharyngeus muscle. The lingual tonsil is a diffuse collection of lymphoid tissue at the root of the tongue.

Tonsillectomy

Tonsillectomy is a common operation, especially in children. Frequent bouts of tonsillitis originate in the many deep crypts that are embedded in the tonsillar mucosal surface. Because the palatine tonsils are very close to the internal cartoid artery, severe hemorrhage may follow a careless operation. The highly vascular tonsillar tissue may also bleed profusely after the operation.

The **laryngopharynx** lies behind the larynx, extending from the inlet of the larynx to the cricoid cartilage, where it becomes the esophagus. The posterior and lateral walls of the laryngopharynx have no characteristic features. Superi-

orly, the anterior wall presents the **inlet** of the larynx with the epiglottis anteriorly, the aryepiglottic folds laterally, and the **piriform recesses** to either side of the folds. These recesses lie between the aryepiglottic membrane medially, and the thyroid cartilage and thyrohyoid membrane laterally. Inferiorly, the muscles and mucous membrane on the posterior aspect of the arytenoid and cricoid cartilages form the anterior wall of the laryngopharynx.

Muscles of the Pharynx

Most of the wall of the pharynx is formed by the three paired superior, middle, and inferior constrictor muscles, which overlap or telescope into one another (Table 7-28; Fig. 7-33). All the constrictors insert posteriorly into the median raphe, with the **inferior constrictor** originating from

Table 7-28
Muscles of the Pharynx

Muscle	Origin	Insertion	Action	Nerve
Inferior constrictor	Side of cricoid and oblique line of thyroid cartilages	Median raphe of pharynx	Constricts pharynx in swallowing	Vagus
Middle constrictor	Greater and lesser cornua of hyoid and stylohyoid ligament	Median raphe	Constricts pharynx in swallowing	Vagus
Superior constrictor	Continuous line from medial pterygoid plate, pterygoid hamulus, pterygomandibular ligament, and side of tongue	Median raphe; superiormost fibers reach pharyngeal tubercle of skull	Constricts pharynx in swallowing	Vagus
Stylopharyngeus	Styloid process	Superior and posterior borders of thyroid cartilage and musculature of pharynx	Raises pharynx	Glossopharyngeal
Palatopharyngeus	Soft palate	Posterior border of thyroid cartilage and musculature of pharynx	Elevates pharynx and helps to close nasopharynx	Vagus
Salpingopharyngeus	Cartilaginous portion of pharyngotympanic tube	Musculature of pharynx	Opens pharyngotympanic tube during swallowing	Vagus

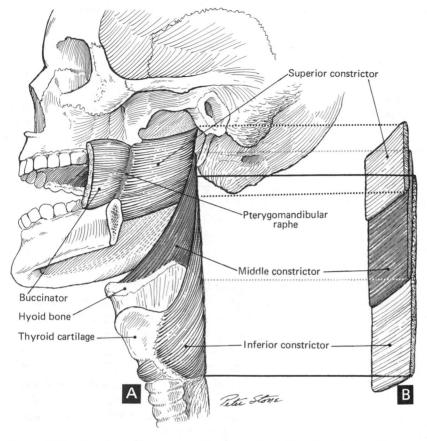

Figure 7-33. *Muscles of the pharynx. These muscles initiate the voluntary phase of swallowing.*

the oblique line of the thyroid and arch of the cricoid cartilages; the **middle constrictor** from the greater and lesser cornua of the hyoid bone and the stylohyoid ligament, and the **superior constrictor** from the medial pterygoid plate and hamulus, pterygomandibular raphe, and mylohyoid line of the mandible. Fibers from the inferior constrictor ascend obliquely toward the median raphe to overlap the middle constrictor. Fibers from the middle constrictor fan out to descend internal to the inferior, and ascend to overlap the superior constrictor. Fibers of the superior constrictor form a gap inferiorly through which the stylopharyngeus muscle passes, and are deficient superiorly at the pharyngeal recess. The lat-

ter is filled in by the levator and tensor veli palatini.

The thin conical **stylopharyngeus** passes from the tip of the styloid process anteroinferiorly to interdigitate and insert between the superior and middle constrictors. The **salpingopharyngeus** descends vertically from the pharyngotympanic tube to insert with the **palatopharyngeus,** which passes from the palatine aponeurosis into the wall of the pharynx internal to the constrictors. All the muscles are innervated by the vagus nerve through the pharyngeal plexus, except the stylopharyngeus, which is the only muscle supplied by the glossopharyngeal nerve (see Table 7-20).

Nasal Cavity

Objectives

At the completion of the study of the nasal cavity and pterygopalatine fossa the student should be able to

▶ *Delineate the boundaries of the nasal cavity*

▶ *List the osseous and cartilaginous components of the external nose and nasal septum*

▶ *Label on a diagram the subdivisions and component elements of the lateral wall of the nasal cavity*

▶ *Locate the paranasal sinuses and the position of their openings into the nasal cavity*

▶ *Locate the pterygopalatine fossa on a skull; list the fissures, foramina and canals related to this region*

▶ *Give the origin, course, and distribution of branches of the maxillary (V_2) nerve and pterygopalatine ganglion*

▶ *List the functional components of the nerve of the pterygoid (Vidian) canal and follow their origins and distributions*

▶ *Follow the course and distribution of branches of the maxillary artery in the pterygopalatine fossa*

Situated above the hard palate and divided by the nasal septum, the **nasal cavity** opens anteriorly at the external nares (nostrils) and posteriorly into the nasopharynx at the internal nares (choanae). The **external nares** are kept patent by the presence of the U-shaped greater alar cartilages. The oblong **internal nares** are rigid, being bounded by bone.

The horizontal **floor** of the nasal cavity is formed by the superior surface of the hard palate,

the palatine process of the maxilla, and the horizontal plate of the palatine bone. It is approximately 7 cm to 8 cm long and 1 cm to 2 cm wide. The long, very narrow **roof** is formed anteriorly by the upper nasal cartilage and nasal bones, and posteriorly by the cribriform plate of the ethmoid. The latter is pierced by twelve to twenty filaments of the olfactory nerve. The osseous portion of the nasal septum is formed by the thin vertical (perpendicular) plate of the ethmoid superiorly and the vomer inferiorly. The septal cartilage is situated anteriorly between the bones of the septum.

Deviated Septum

A deviated septum is deflected laterally from the midline of the nose. The deviation usually occurs at the junction of the osseous with the cartilaginous portion of the septum. If the deformity is severe, it may entirely block the nasal passageway. Even though the blockage may not be complete, infection and inflammation develop, causing nasal congestion, occlusion of the paranasal orifices, and chronic sinusitis.

The **lateral wall** of the nasal cavity presents bony projections, the **conchae** (turbinate bones), which shelter a number of openings (Fig. 7-34). The conchae are three curled bony plates projecting from the lateral wall into the nasal cavity that are covered by thick mucous membrane. The **superior concha,** a process of the ethmoid, is very short; the **middle concha,** also a process of the ethmoid, is larger; the **inferior concha,** longer than the middle, is an individual bone located midway between the middle concha and the floor of the nasal cavity. The **meatuses** are air-flow tracts lying deep to, or under cover of, their respective conchae. The area above the superior concha, into which the sphenoidal air sinuses

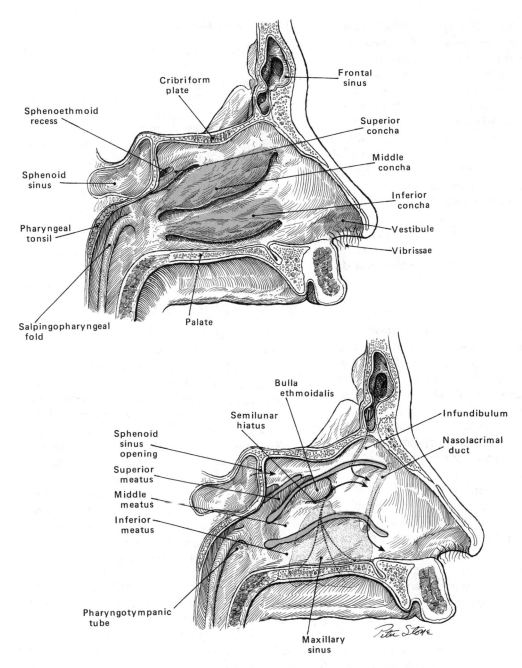

Figure 7-34. *Nasal cavity. At the top, structures of the lateral wall. At the bottom, drainage pathways of the paranasal sinuses and the nasolacrimal duct, with conchae removed.*

open, is designated as the **sphenoethmoidal recess.** The short, narrow **superior meatus** between the superior and middle conchae receives drainage from the posterior and middle ethmoidal air sinuses. Between the middle and inferior conchae, the rather extensive **middle meatus** receives the **infundibulum,** and anterosuperior funnel-shaped opening of the frontal air sinus. Located posterior to the infundibulum a curved slit, the **hiatus semilunaris,** drains the anterior and middle ethmoidal and the maxillary air sinuses. The **bulla ethmoidalis** is a prominent bulging of the ethmoidal air sinuses forming the upper margin of the hiatus semilunaris. The **inferior meatus** is the horizontal passage deep to the inferior concha, into which the **nasolacrimal duct** opens.

Nasal Polyps
Nasal polyps are protruding growths of the mucous membrane that usually hang down from the posterior wall of the nasal septum. In the rhinoscopic mirror, the polyps appear as bluish white tumors, which may fill the nasopharynx. If untreated, nasal polyps usually undergo cystic degeneration, but they are easily removed with a nasal snare and cautery.

Mucoperiosteum, consisting of mucous membrane closely adherent to periosteum, lines the nasal cavity, except for the area of the vestibule. The latter is lined with skin, while part of the roof, the superior concha, and adjacent septum are lined with olfactory epithelium. The mucoperiosteum is continuous through the nasolacrimal duct with the conjunctivum, through various apertures with the mucous membrane lining the several air sinuses, and through the choanae with the mucous membrane of the pharynx. It is thick and spongy, owing to the presence of rich sinusoidal venous plexuses and numerous mucous cells. It functions to moisten and warm the incoming air.

Arteries and Nerves

All the vessels to the nasal cavity form a rich irregular anastomosis deep to the mucous membrane, with the **sphenopalatine artery,** a branch of the maxillary, as the principal supply. Branches of the sphenopalatine artery include the **posterior lateral nasal branch,** supplying the conchae and the ethmoidal, frontal, and maxillary air sinuses; a **posterior septal branch** to the upper part of the septum, and the **nasopalatine artery** that continues anteriorly on the septum to pass through the incisive foramen and anastomose with the great palatine artery supplying the hard palate. The ophthalmic artery gives **anterior** and **posterior ethmoidal branches,** which supply the anterior portion of the superior and middle conchae, adjacent septal areas, and give twigs to the frontal and ethmoidal air sinuses.

Nosebleed
Nosebleed, or epistaxis, is frequent because of the exposure of the nose to trauma, the rich venous plexus deep to the conchal mucosa, and the extensive vascularity of the nose. Bleeding, either arterial or venous, usually occurs on the anterior part of the septum and can be arrested by firm packing of the anterior nares. If the point of bleeding is in the posterior region, plugging of both the anterior and posterior nares may be necessary. In extreme emergency, ligation of the external carotid artery may be needed to control the hemorrhage.

Nerves to the nasal cavity include bipolar neurons of the olfactory nerve, nerves of general sensation from branches of both the ophthalmic and the maxillary divisions of the trigeminal nerve, and the autonomic nerves from the pterygopalatine ganglion.

The cells of origin of the **olfactory nerve** are limited to the small area of **olfactory epithelium** lining a portion of the roof and adjacent surfaces of the septum and superior nasal concha. Peripheral processes begin as osmoreceptors, and central processes pass through the **cribriform plate** to synapse in the **olfactory bulb,** which gives rise to the **olfactory tract** leading to the brain. Anterior and posterior ethmoidal nerves are branches of the nasociliary nerve from the ophthalmic division of the trigeminal. The **posterior ethmoidal nerve** passes through the posterior ethmoidal foramen to the posterior ethmoidal and sphenoidal air sinuses, while the **anterior ethmoidal nerve** passes through the anterior ethmoidal foramen to reenter the anterior cranial fossa. It then crosses the cribriform plate and enters the nasal cavity by way of the nasal slit (fissure) at the side of the crista galli. In the nasal cavity this nerve divides into an **external nasal branch** to pass down the nasal bone to supply the skin on the dorsum of the nose, and an **internal nasal branch.** The latter sends a medial branch to the superoanterior part of the nasal septum and a lateral branch to the anterior portion of the superior and middle conchae. From the maxillary

division of the trigeminal nerve, **sensory twigs** pass to the **pterygopalatine ganglion** and distribute with its branches. These include the **lateral posterior superior nasal branch** to the posterior part of the superior and middle conchae; the **nasopalatine branch,** which crosses the roof of the nasal cavity to supply the septum and then follows the nasopalatine artery through the incisive foramen to supply the anterior portion of the hard palate. The **greater palatine nerve** passes through the palatine canal to give branches to the inferior concha and terminate in the mucous membrane of the hard and soft palates.

Paranasal Air Sinuses

The bilateral **paranasal air sinuses** are located in bones adjacent to the nasal cavity (Fig. 7-35). The **sphenoidal air cavities,** occupying the body of the sphenoid, are rarely symmetrical. Their ostia usually open into the middle or superior part of the sphenoethmoidal recesses. The

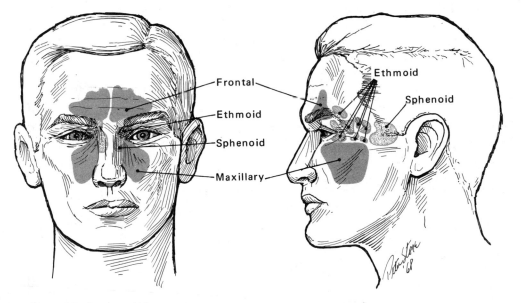

Figure 7-35. Paranasal sinuses.

large **maxillary sinuses** are four-sided, hollow pyramids between the lateral walls of the nasal cavities and the infratemporal fossae. The slitlike ostium of each sinus opens into the posterior part of the hiatus semilunaris. The sockets of the upper molar or premolar teeth may project into these sinuses. The numerous **ethmoidal air cells** are usually limited to the lamina papyracea and the orbital portion of the ethmoid bone. The posterior air cells drain into the superior meatus, and the middle and the anterior into the hiatus semilunaris. **Frontal sinuses,** forming the brow ridges, are developmentally anterior ethmoidal air cells that migrate or extend into the frontal bone. Each drains inferiorly into a middle meatus by way of an infundibulum.

Sinusitis

Sinusitis is a frequent complaint resulting from infection of the paranasal sinuses. This condition occurs most commonly in the frontal and maxillary sinuses. In health the sinuses contain air, but when infected they collect fluid, which can be demonstrated radiographically. Since all paranasal sinuses drain into the nasal cavity, they are easily infected by nasal secretions. Maxillary sinusitis may be confused with toothache since only a thin layer of bone separates the roots of the teeth from the sinus cavity. Ethmoidal and sphenoidal sinusitis may infect the cranial meninges because only a thin bony plate intervenes between these sinuses and the subarachnoid space.

Headaches

Headaches have many causes. Sustained, tense, involuntary contraction of the epicranius muscles causes the common tension headache. A migraine or sick headache is periodic, usually affecting only one side of the head and is accompanied by nausea, vomiting, and visual disturbances. It is probably due to vascular agitation, such as vigorous vasoconstriction followed by vasodilation of the cerebral vessels.

Sinus headache involves severe pain over the sinuses, especially in the frontal and maxillary regions and is caused by pressure in the paranasal sinuses when the drainage is blocked by inflammatory or allergic edema. If the sinusitis is chronic, it is often relieved by surgically enlarging the openings of the sinus into the nasal cavity.

Pterygopalatine Fossa

The **pterygopalatine fossa** is an elongated, triangular area between the posterior aspect of the maxillary bone and the pterygoid processes of the sphenoid bone (Fig. 7-36). The **medial wall** opens into the nasal cavity through the sphenopalatine foramen. The **roof** is formed by the greater wings of the sphenoid. The **lateral wall** is relatively open as the pterygomaxillary fissure. Openings into the pterygopalatine fossa include the **sphenopalatine foramen** at the junction of the roof and the medial wall, for the passage of vessels and nerves to the nasal cavity; the **greater** and **lesser palatine canals** inferiorly, for the passage of the greater and lesser palatine nerves and arteries; posteriorly the **foramen rotundum,** for the maxillary division of the trigeminal nerve; and the **pterygoid** canal, for the passage of its nerve and artery. The fossa communicates with

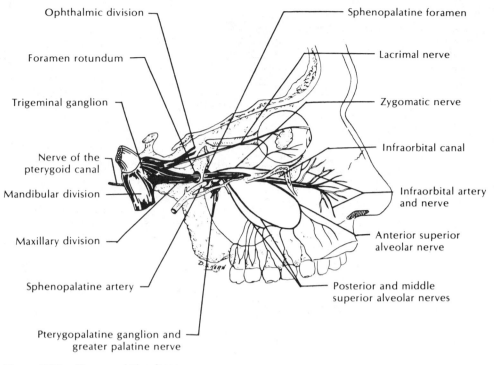

Ophthalmic division

Foramen rotundum

Trigeminal ganglion

Nerve of the
pterygoid canal

Mandibular division

Maxillary division

Sphenopalatine artery

Pterygopalatine ganglion and
greater palatine nerve

Sphenopalatine foramen

Lacrimal nerve

Zygomatic nerve

Infraorbital canal

Infraorbital artery
and nerve

Anterior superior
alveolar nerve

Posterior and middle
superior alveolar nerves

Figure 7-36. *Pterygopalatine fossa.*

the orbital cavity by way of the **inferior orbital fissure** and with the infratemporal fossa by way of the **pterygomaxillary fissure.**

Arteries and Nerves

Contents of the pterygopalatine fossa include the third portion of the **maxillary artery** and its companion **veins,** the **maxillary division of the trigeminal nerve,** and the **pterygopalatine ganglion.** The pterygopalatine, or third portion of the maxillary artery, lies in the fossa lateral to the pterygopalatine ganglion and gives branches to the nasal and orbital cavities, palate, upper teeth, and face (Table 7-29). As the artery enters the fossa, its **posterior superior alveolar branch** descends on the maxillary tuberosity to enter the superior alveolar canal and supply the gingivae and upper molar and premolar teeth. The **descending palatine artery** passes through

the palatine canal to divide into the greater and lesser palatine arteries. The greater palatine emerges at the greater palatine foramen. It passes forward on the hard palate supplying glands and mucous membrane of the palate and gingivae, and anastomoses with the nasopalatine branch through the incisive foramen. The **lesser palatine artery** emerges at the lesser palatine foramen to supply the soft palate and palatine tonsil. The **artery to the pterygoid canal** passes posteriorly to supply twigs to the pharynx, pharyngotympanic tube, and tympanic cavity. **Pharyngeal branches** of the maxillary are distributed to the upper pharynx and pharyngotympanic tube by way of the pharyngeal canal. The major blood supply of the nasal cavity is the **sphenopalatine artery,** which passes through the sphenopalatine foramen to give posterior lateral nasal branches to the conchae, meatuses, and sinuses. It then descends on the nasal septum to anastomose with

Table 7-29
Arterial Supply to Pterygopalatine Fossa and Nasal Cavity

Artery	Origin	Course	Distribution	Anastomoses
Descending palatine	Maxillary	Leaves fossa to descend in pterygo-palatine canal; bifurcates into greater and lesser branches	Greater supplies hard palate, lesser soft palate and fauces	Greater with septal branch, sphenopal-atine; lesser with tonsillar branches
Pharyngeal	Maxillary	Leaves fossa through pharyngeal canal to naso-pharynx	Nasopharynx and pharyngeal tonsil	Ascending pharyn-geal
Infraorbital	Maxillary	Leaves fossa through infraorbital fissure; traverses infraorbital fissure and canal; emerges at infraorbital fora-men	Cheek and temporal region	Dorsal nasal, lateral nasal, superior labial, angular
Sphenopalatine	Maxillary	Leaves fossa through sphenopal-atine foramen, gives lateral and septal branches	Nasal cavity	No direct anasto-moses
Septal	Sphenopalatine	Courses along nasal septum	Nasal septum	Superior labial and greater palatine
Posterior lateral nasal (superior, middle, and inferior)	Sphenopalatine	Pass to superior, middle, and inferior nasal conchae and meatuses	Lateral wall and floor of nasal cavity	Anterior and poste-rior ethmoidals
Posterior ethmoidal	Ophthalmic	Leaves orbital cavity through posterior ethmoidal foramen to reach nasal cavity	Posterior middle ethmoidal air cells; posterosuperior aspect of nasal cavity	Anterior ethmoidal, posterior superior lateral nasal, septal
Anterior ethmoidal	Ophthalmic	Leaves orbital through anterior ethmoidal foramen → anterior cranial fossa → along crib-riform plate → nasal slit → nasal cavity → dorsum of nose	Anterior ethmoidal air cells, anterosu-perior aspect of nasal cavity, dor-sum of nose	Posterior ethmoidal, septal, lateral nasal, angular, infraorbital

the greater palatine and the superior labial arteries as it passes through the incisive foramen.

The **maxillary artery** leaves the pterygopalatine fossa to enter the orbital cavity by way of the infraorbital fissure. Within the orbit it gives rise to the **zygomatic artery.** This branch traverses the zygomatico-orbital foramen to terminate on the face as **zygomaticofacial** and **zygomaticotemporal branches.** Distal to the zygomatic branch, the maxillary becomes the **infraorbital artery.** It traverses the infraorbital groove and canal and emerges on the face at the infraorbital foramen. In the canal it gives branches to the inferior oblique and inferior rectus muscles, the lacrimal sac, and mucous membranes of the maxillary air sinuses. Its **anterior superior alveolar branch** supplies the incisor and canine teeth and its terminal branches on the face ramify and anastomose with branches of the facial artery.

From the midportion of the trigeminal ganglion the **maxillary division** of the **trigeminal nerve** passes through the foramen rotundum, traverses the pterygopalatine fossa and enters the orbit by way of the infraorbital fissure. It continues in the infraorbital groove and canal to emerge on the face at the infraorbital foramen as the **infraorbital nerve.** The branches of the maxillary nerve are entirely sensory and supply the skin and the mucous membrane of the lower eyelid and upper lip.

Within the cranial cavity the **maxillary division** of the fifth cranial nerve gives branches that follow the middle meningeal artery to supply the meninges. The zygomatic, sphenopalatine, and posterior superior alveolar branches arise from the pterygopalatine portion of the maxillary nerve. The **zygomatic branch** enters the orbital cavity through the infraorbital fissure. It sends a communicating branch to the lacrimal nerve. This provides a pathway for postganglionic fibers to travel from the pterygopalatine ganglion to the lacrimal gland. Distally the **zygomatic nerve** passes through the zygomatico-orbital foramen to terminate as **zygomaticofacial** and **zygomaticotemporal branches.** The terminal branches

emerge at their respective foramina to supply skin of the face and temporal regions. The **maxillary nerve** sends two short sensory roots to the pterygopalatine ganglion to be distributed with branches from the ganglion; an orbital branch to the periosteum of the orbit and the posterior ethmoidal sinuses; a **greater palatine branch,** which passes through the palatine canal and out the greater palatine foramen to supply the hard and soft palates, the middle and inferior meatuses, and the inferior concha, and a **lesser palatine branch,** which follows the same course as the greater palatine but emerges through the lesser palatine foramen to supply the soft palate, uvula, and palatine tonsil. The **sphenopalatine nerve** is the primary innervation to the nasal cavity. Branches of the sphenopalatine nerve include the **posterior superior nasal branch** that traverses the sphenopalatine foramen and distributes to the superior and middle conchae and posterior ethmoidal sinus; direct branches to the superior and middle conchae, posterior ethmoidal sinus, and posterior part of the septum; a **pharyngeal branch** passes through the pharyngeal canal to the nasopharynx, and the **nasopalatine (long sphenopalatine) branch** supplies the roof of the nasal cavity and the nasal septum. The latter nerve courses downward on the septum to pass through the incisive canal and supply the anterior portion of the palate. The **posterior superior alveolar nerve** passes through the pterygomaxillary fissure to reach minute foramina on the tuberosity of the maxilla and enters the posterior alveolar canals to supply upper molar teeth, gingivae, and mucous membrane of the cheek.

Within the infraorbital canal, the infraorbital nerve gives off a **middle superior alveolar branch** that supplies the maxillary sinus and the premolar teeth, an **anterior superior alveolar branch** to the maxillary sinus and the canine and incisor teeth, and a **nasal branch** that innervates the anterior part of the inferior meatus and floor of the nasal cavity (Table 7-30).

The terminal (facial) portion of the maxillary

Table 7-30
Distribution of Olfactory (I), Maxillary (V₂) Nerves and Nerve of Pterygoid Canal

Nerve	Origin	Course	Distribution
Olfactory (I)	Bipolar neuronal cell bodies in olfactory mucosa; peripheral processes have osmoreceptors; central processes form filaments of olfactory nerve	Filaments pass through cribriform plate; processes synapse in olfactory bulb	Olfactory mucosa on superior concha and adjacent lateral wall of nasal cavity for sense of smell
Maxillary (V₂)	Trigeminal ganglion	Exits cranial cavity via foramen rotundum, traverses pterygopalatine fossa, and leaves through infraorbital fissure to reach floor of orbital cavity	Sensation to midportion of face; postganglionic autonomic fibers conveyed by most of its branches

Within pterygopalatine fossa the pterygopalatine ganglion is suspended from V₂; most distributing branches are branches of the maxillary nerve or pterygopalatine ganglion. They convey both sensory and postganglionic autonomic fibers (see nerve of pterygoid canal).

Nerve	Origin	Course	Distribution
Nerve of pterygoid canal	Formed by junction of major petrosal transmitting preganglionic parasympathetic fibers (from facial nerve), and deep petrosal from carotid plexus transmitting postganglionic sympathetic fibers (cell bodies in superior cervical ganglion)	Traverses pterygoid canal to terminate in pterygopalatine ganglion	Parasympathetics synapse in ganglion; distributing branches convey both sympathetic and parasympathetic postganglionic processes and sensory fibers to midportion of face
Nasopalatine	Pterygopalatine ganglion	Nasal septum	Sensory and autonomics to nasal mucosa
Sphenopalatine (Lateral nasal branches)	Pterygopalatine ganglion	Lateral wall of nasal cavity	Sensory and autonomics to nasal mucosa
Greater and lesser palatines	Pterygopalatine ganglion	Palatine canal	Mucosa of hard and soft palate
Posterior superior alveolar	Maxillary (V₂)	Through pterygomaxillary fissure to reach small foramina on tuberosity of maxilla	Upper molar teeth
Zygomatic	Maxillary (V₂)	Arises from maxillary in infraorbital groove, courses along lateral aspect of orbit to reach the zygomatico-orbital foramen	Terminal zygomaticofacial and zygomaticotemporal branches supply skin of face; communicating branch to lacrimal nerve conveys postganglionic parasympathetics to lacrimal gland
Infraorbital	Terminal branch of maxillary (V₂)	Traverses infraorbital groove, canal, and foramen	Skin of cheek, lower eyelid, side of nose, upper lip; in infraorbital canal, gives middle and anterior superior alveolar branches to upper teeth

nerve emerges at the infraorbital foramen as the **infraorbital nerve.** It divides into the **inferior palpebral branch** to the lower lid, the **lateral** **nasal branch** to the skin at the side of the nose, and the **superior labial branch** to the skin, mucous membrane, and glands of the upper lip.

Ear

Objectives

At the completion of the study of the ear the student should be able to

▶ Delineate the subdivisions of the ear

▶ Locate the ear ossicles in situ and list their component parts

▶ Define the boundaries of the middle ear cavity (tympanum); identify major landmarks and related structures

▶ Follow the course of the facial nerve through the temporal bone and give its branches in the facial canal

▶ Describe the osseous and membranous labyrinth

▶ Discuss the transmission of a sound wave

External Ear

For descriptive purposes the ear is divided into external, middle, and internal portions. The **external ear** comprises the pinna (auricula) and the external auditory meatus. (Fig. 7-37). The **pinna** collects sound waves and directs them into the

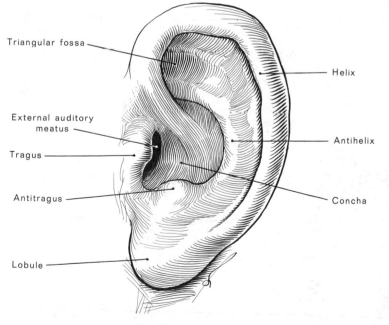

Figure 7-37. *The auricle.*

external auditory canal where they strike the tympanic membrane. Parts of the pinna include the **concha,** the well of the ear leading into the external auditory meatus; the **helix,** the outer rim of the external ear beginning at the concha and ending at the **lobule;** the **antihelix,** rimming the concha opposite the helix; the **tragus,** the small lip overlapping the concha; the **fossa triangularis,** the triangular depression above the concha, and the **scapha,** which forms a depressed groove in front of the helix. The arterial supply to the ear is presented in Table 7-31.

Ear Wax

Ear wax deposits may block the external auditory meatus, causing temporary deafness. Forceful efforts to remove the wax can result in impaction and infection. The wax can usually be dislodged by vigorous irrigation of the canal with warm water.

The lateral third of the **external auditory meatus** is cartilaginous; the remainder is formed by the tympanic part of the temporal bone (Fig. 7-38).

Table 7-31
Arterial Supply to the Ear

Artery	Origin	Course	Distribution	Anastomoses
Anterior tympanic	Maxillary	From infratemporal fossa passes to external auditory meatus	External auditory meatus; tympanic membrane	Anterior auricular, deep auricular
Deep auricular	Maxillary	From infratemporal fossa passes to external auditory meatus	External auditory meatus; tympanic membrane	Anterior tympanic, anterior auricular
Anterior auricular	Superficial temporal	Enters external auditory meatus as superficial temporal passes tragus	External auditory meatus; tympanic membrane	Anterior tympanic, deep auricular
Stylomastoid	Posterior auricular	Enters stylomastoid foramen to course in facial canal	Posterior aspect of tympanic cavity; mastoid air cells	Tympanic branch of internal carotid and petrosal branch of middle meningeal
Tympanic	Internal carotid	From carotid canal pierces anterior wall; tympanic cavity	Anterior aspect of tympanic cavity; pharyngotympanic tube	Stylomastoid, petrosal
Petrosal	Middle meningeal	Traverses hiatus of facial canal	Roof and medial wall of tympanic cavity	Stylomastoid, tympanic
Inferior tympanic	Ascending pharyngeal	Follows pharyngotympanic tube accompanied by a twig from artery of pterygoid canal	Pharyngotympanic tube	Tympanic
Internal auditory	Basilar	Traverses internal auditory meatus	Labyrinth (osseous and membranous)	Stylomastoid
Stylomastoid	Posterior auricular	Traverses internal auditory meatus	Labyrinth (osseous and membranous)	Internal auditory

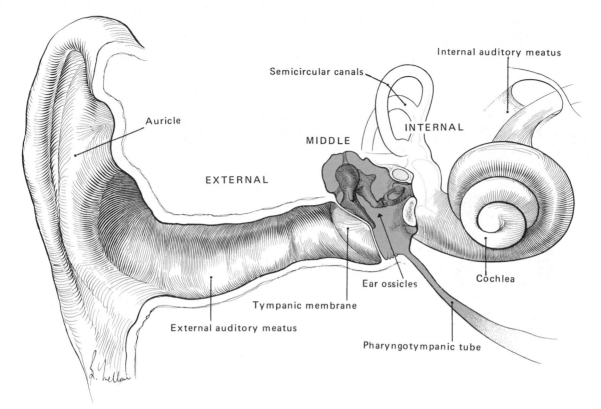

Figure 7-38. *The ear and its subdivisions. The middle ear communicates with the naso-pharynx by means of the pharyngotympanic tube.*

Infections of the External Auditory Canal
Infections of the external auditory canal often result from infected hair follicles. These small abscesses or boils are extremely painful because of the tightness of the skin lining the canal. The pain may be severe enough to cause nausea and vomiting because the nervous impulses of pain, reflexly, stimulate the vagus nerve, which supplies parasympathetic innervation to the upper gastrointestinal tract.

The **tympanic membrane** slopes obliquely, inferomedially, with the lateral surface slightly concave; the maximal point of the concavity is designated as the **umbo.** The handle of the mal-

leus can be seen through the membrane extending inferiorly to the umbo. A flaccid, less tense portion of the tympanic membrane, the **pars flaccida,** lies above the lateral process of the malleus. The whole of the peripheral margin, except for the flaccid part, is lodged in the **tympanic groove.**

Middle Ear (Figs. 7-39 and 7-40)

The **middle ear,** or **tympanic cavity,** is filled with air and communicates anteromedially with the pharynx through the **pharyngotympanic (auditory) tube** and posterosuperiorly with the tympanic antrum and mastoid air cells through the aditus. The area above the tympanic cavity proper is the epitympanic recess. It is the location

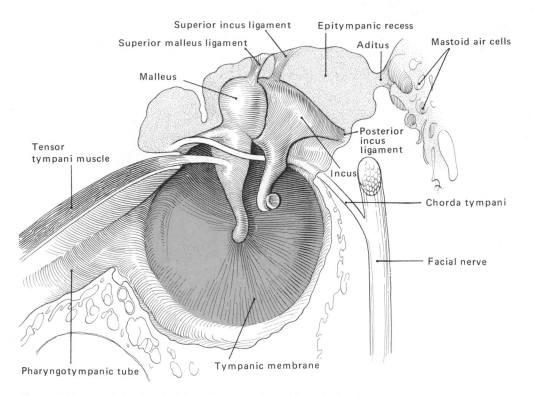

Figure 7-39. *The lateral wall of the middle ear viewed from the interior. The stapes, omitted here, articulates with the lenticular process of the incus.*

of the head of the malleus as it articulates with the body of the incus.

The tympanic cavity contains the three auditory ossicles, stapedius and tensor tympani muscles, tympanic plexus of nerves, and chorda tympani nerve. The tympanic cavity has a roof, floor, and four walls. The **roof** of the cavity, the **tegmen tympani,** is a thin plate of bone separating the epitympanic recess from the middle cranial fossa. The **floor** of the cavity, or **jugulum,** is a thin plate of bone separating the cavity from the jugular fossa and the bulb of the jugular vein. The **posterior (mastoid) wall** contains an opening, the **aditus,** leading from the **epitympanic recess** into the **tympanic antrum.** In the lower part of the posterior wall a small conical projection, the **pyramid,** lodges the **stapedius muscle,** and lateral to the pyramid an opening admits

the chorda tympani nerve. The vertical portion of the facial canal is also related to the posterior wall. Owing to the convergence of the medial and lateral walls, the **anterior (carotid) wall** is narrow, and an opening in the upper portion leads to the canal that houses the tensor tympani muscle. An opening in the midportion leads into the pharyngotympanic tube. Between the canal for the tensor tympani and the pharyngotympanic tube, a septum is prolonged posteriorly on the medial wall as the **processus cochleariformis,** which affords a pulley around which the tensor tympani tendon turns laterally to its insertion. In the lower portion of the anterior wall a thin lamina of bone separates the cavity from the carotid canal. The **medial (labyrinthinc) wall** forms the boundary between the middle and internal parts of the ear, where anteriorly the rounded **promontory**

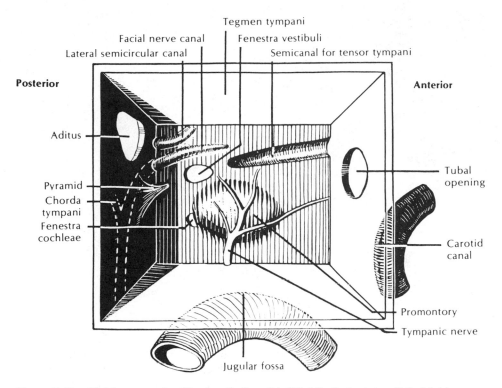

Figure 7-40. *Middle ear cavity. (Gardner E, Gray DJ, O'Rahilly R: Anatomy: Philadelphia, Saunders, 1975)*

is formed by the underlying basal turn of the cochlea. Above the posterior part of the promontory is a depression in which the **fenestra vestibuli (oval window)** is closed by the foot plate of the stapes. Above the oval window, an anteroposterior ridge demarcates the position of the canal containing the facial nerve. Immediately above the ridge of the facial canal a horizontal projection indicates the site of the lateral semicircular canal. Below and behind the promontory is a fossa that contains the **fenestra cochleae (round window),** an opening into the scala tympani, which is closed by the **secondary tympanic membrane.**

The auditory ossicles, the malleus (hammer), incus (anvil), and stapes (stirrup), extend in a chain from the lateral to the medial wall of the middle ear cavity (Fig. 7-41). The **malleus** is described as having a head, neck, handle, and ante-

rior and lateral processes. The **head** of the malleus articulates with the body of the incus. The **handle,** or **manubrium,** of the malleus is attached along its length to the tympanic membrane. It extends to the umbo and receives into its medial surface the insertion of the tendon of the tensor tympani muscle. Just above this insertion the chorda tympani nerve passes between the manubrium and the long crus of the incus. The **incus** consists of a body, and short (horizontal) and long (vertical) crura. The **lenticular process** is a small knob on the end of the long crus, which articulates with the stapes. The **short crus** is attached to the posterior aspect of the epitympanic recess by way of the ligament of the incus and the **body** receives the articulating head of the malleus. The **stapes** presents a head, neck, foot plate, and anterior and posterior limbs. The incus articu-

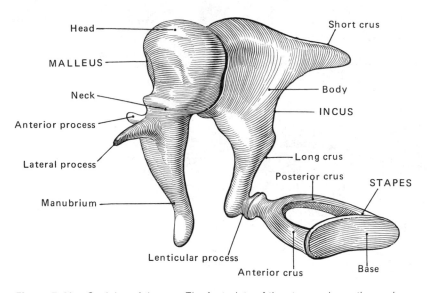

Head — MALLEUS — Neck — Anterior process — Lateral process — Manubrium — Lenticular process — Anterior crus — Short crus — Body — INCUS — Long crus — Posterior crus — STAPES — Base

Figure 7-41. *Ossicles of the ear. The foot plate of the stapes closes the oval window. The manubrium of the malleus is embedded in the tympanic membrane.*

lates at the concave socket on the head of the stapes, and the **foot plate** is attached by the **anular ligament** to the margin of the fenestra vestibuli. The stapedius muscle inserts into the posterior surface of the **neck** of the stapes.

Otitis Media

Otitis media is an acute infection of the middle ear cavity with a reddening and outward bulging of the eardrum, which may rupture. The infection may spread to the mastoid air cells, causing mastoiditis, a very serious complication prior to the advent of antibiotics. Acute otitis media is usually caused by the spread of infection from the nasopharynx along the pharyngotympanic (Eustachian) tube, and into the middle ear.

Internal Ear (Figs. 7-42 and 7-43)

The **internal ear** is located within the petrous portion of the temporal bone and consists of the osseous labyrinth, which contains the membranous labyrinth. The **osseous labyrinth** is composed of bony cavities: the vestibule, three semi-circular canals, and the cochlea, all filled with perilymph, in which the membranous labyrinth is suspended. The **vestibule** is centrally located. The superior, posterior, and lateral semicircular canals open into it posteriorly by way of five openings, with the adjoining ends of the superior and posterior canals forming a common terminal canal. Laterally, the **fenestra vestibuli** is closed by the foot plate of the stapes, while the medial wall of the vestibule presents depressions with small openings for the emergence of filaments of the eighth cranial nerve.

Otosclerosis

Otosclerosis is a pathologic process that deposits new bone around the oval window, which may immobilize the stapes. The resulting deafness may be cured by making a new opening into the internal ear (fenestration). The ankylosed stapes is then freed and attached to the new membranous window. Many attachment techniques are available, but they all have the same objective—to reestablish the functional movement of ear ossicles.

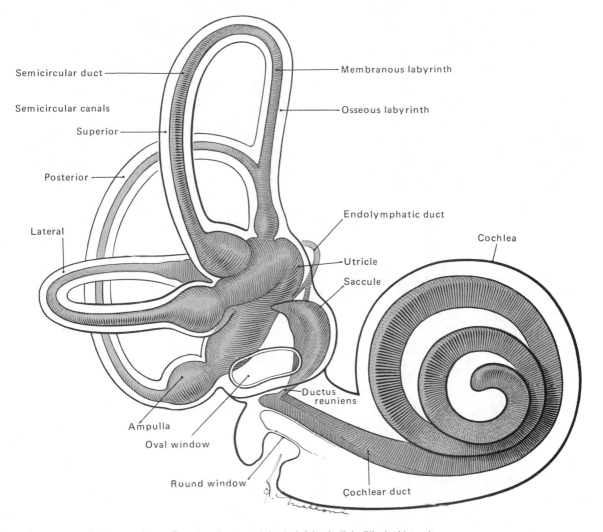

Figure 7-42. *The internal ear. The membranous labyrinth* (shaded) *is filled with endo-lymph; the osseous labyrinth, with perilymph.*

The **semicircular canals** are each approximately 0.8 mm in diameter, and each presents a dilatation, the ampulla, at one end. The vertical **superior canal** is at right angles to the similarly vertical **posterior canal,** which parallels the posterior surface of the petrous portion of the temporal bone. The horizontal **lateral canal** is in the angle between the superior and posterior canals and bulges into the medial wall of the tympanic cavity.

Located anterior to the vestibule, the bony **cochlea** resembles a snail's shell. It is a tapering tube that spirals about 2½ turns around a central core, the **modiolus.** The osseous tube opens into the tympanic cavity through the **fenestra cochlea** (round window) and is closed in the fresh state by the secondary tympanic membrane. The basal, or first, turn around the modiolus bulges as the promontory on the medial wall of the tympanic cavity. The modiolus is thick at the base and

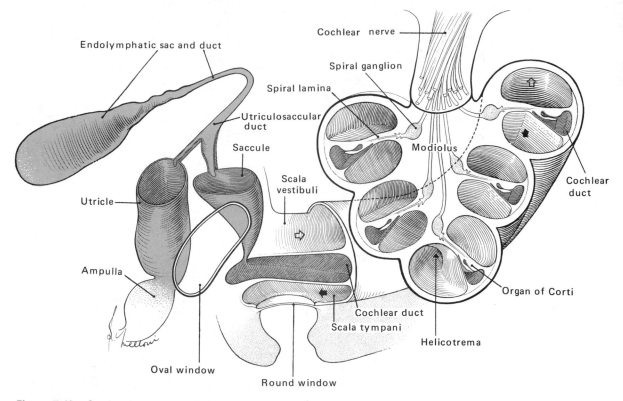

Figure 7-43. *Section through the cochlea and vestibule. Sound waves pass through the perilymph of the cochlea in the direction indicated by arrows.*

tapers rapidly to the apex, with a thin narrow shelf of bone, the **spiral lamina,** turning around the modiolus like the threads of a screw.

Two membranous sacs, the utricle and the saccule, the three semicircular ducts, and the cochlear duct compose the **membranous labyrinth,** which lies within the bony labyrinth but does not completely fill it. The membranous labyrinth contains the fluid endolymph, while the fluid perilymph occupies the space between the membranous and osseous labyrinths. Lying in the posterosuperior part of the vestibule, the **utricle** receives the openings of the semicircular ducts. The small **saccule** lies in the anteroinferior part of the vestibule. A short canal, the **ductus reuniens,** extends from the lower part of the saccule to the cochlear duct. The blind **endolymphatic duct** leaves the posterior part of the saccule, is

joined by the short **utriculosaccular duct** from the utricle, then traverses the aqueduct of the vestibule. It ends under the dura mater on the posterior surface of the petrous portion of the temporal bone as a dilatation, the **endolymphatic sac.** The **semicircular ducts** and their terminal **ampullae** are attached to the convex sides of the semicircular canals and open into the utricle.

Lesions of the Vestibular Nerve

Lesions of the vestibular portion of the vestibulocochlear nerve causes vertigo (dizziness) and nystagmus. The latter is an involuntary rapid movement of the eyeball, and is recognized by a rhythmic oscillation, a slow movement in one direction followed by a rapid jerk back, which is repeated in rapid succession in both eyes.

Within the bony canal of the cochlea the **membranous labyrinth** consists of a closed spiral tube, the **cochlear duct,** which is separated from the internally located **scala vestibuli** by the **vestibular membrane,** and from the externally placed **scala tympani** by the **basilar membrane.** The basilar membrane supports the **spiral organ (of Corti),** which contains the peripheral nerve endings associated with sound reception. The scala vestibuli and scala tympani are continuous with each other at the apex of the cochlea through a small opening, the **helicotrema** (Table 7-32).

Nerve Deafness

Nerve deafness may result from a lesion of the cochlear division of the eighth nerve, its ganglion, the organ of Corti, or the cochlear duct. Tone deafness is a dysfunction of the hair cells in the organ of Corti, so that musical sounds cannot be perceived. Conduction deafness is a defect in the sound conducting apparatus, such as the external auditory meatus, eardrum, or otic ossicles. Word deafness is a lesion of the auditory center of the brain (superior temporal gyrus), where the sounds are heard but convey no meaning to the individual.

Table 7-32
Nerves in Ear

Nerve	Origin	Course	Distribution
Vestibulocochlear	Cochlear division—bipolar neuronal cell bodies in spiral ganglion of modiolus; peripheral processes pass to organ of Corti; central processes join vestibulocochlear nerve	Traverses internal auditory meatus with facial nerve to reach brain stem	Hearing
	Vestibular division—bipolar neuronal cell bodies in vestibular ganglia; peripheral processes in equilibratory receptors; central processes join vestibulocochlear nerve	Traverses internal auditory meatus with facial nerve to reach brain stem	Equilibrium
Nerve to stapedius	Facial (VII) in vertical portion of facial canal	Passes directly to stapedius	Stapedius
Chorda tympani	Facial (VII) in vertical portion of facial canal	From iter chorda posterioris crosses tympanic cavity to pass through iter chorda anterioris to reach infratemporal fossa and join lingual nerve	Taste to anterior ⅔ tongue; preganglionic fibers to submandibular ganglion; postganglionics to submandibular and sublingual glands
Tympanic plexus	From tympanic branch of glossopharyngeal (IX) and twigs of facial (VII); gives rise to minor petrosal	Tympanic branch arises in jugular foramen, passes through tympanic canaliculus, crosses floor of middle ear cavity to reach promontory; minor petrosal traverses petrosphenoidal fissure to reach otic ganglion in infratemporal fossa	Sensory to middle ear; postganglionics from otic ganglion to parotid gland

Larynx

Objectives

At the completion of the study of the larynx the student should be able to

▶ *Describe the relationship of the larynx to structures in the root of the neck*

▶ *Delineate the subdivisions of the larynx*

▶ *Identify the laryngeal cartilages*

▶ *List the laryngeal muscles and their main action on the laryngeal cartilages*

▶ *Follow the course of the sensory and motor nerve supply to the larynx*

▶ *Describe the blood supply of the larynx and thyroid gland*

▶ *Label on a drawing the parts of the thyroid gland*

The **larynx** is specially modified for vocalizations. Situated anteriorly in the neck, below the hyoid bone and the tongue, it has a marked anterior projection, the laryngeal prominence (Adam's apple). The larynx is related anteriorly to skin and fasciae, and laterally to the thin strap muscles of the neck, the thyroid gland, the great vessels of the neck, and the vagus nerves. Posteriorly, it is separated from the vertebral column and prevertebral muscles by the laryngopharynx.

The **skeleton of the larynx** is formed by three single cartilages: the thyroid, cricoid, and epiglottic; and three paired cartilages, the arytenoids, corniculates, and cuneiforms (Fig. 7-44). The thin, leaf-shaped, **epiglottic cartilage** forms the anterior boundary of the inlet (aditus) and the vestibule of the larynx. The superior end of the cartilage is broad and free, with the lateral margins enclosed in the aryepiglottic folds, while the lower end is pointed and connected to the thyroid cartilage by the thyroepiglottic ligament. The large

thyroid cartilage has two broad quadrilateral **laminae,** which are fused anteriorly, open posteriorly, and separated anterosuperiorly by the V-shaped **thyroid notch.** The junction of the notch and the fused laminae forms the **laryngeal prominence.** The posterior border of the thyroid cartilage is thick and rounded, and is prolonged upward as the **superior cornua** and downward as the **inferior cornua.** The superior border gives attachment to the **thyrohyoid membrane,** which is pierced by the internal laryngeal nerve and the superior laryngeal vessels. The thyrohyoid membrane, between the greater horn and the hyoid bone, is free posteriorly and encloses the small **cartilago triticea.** The relatively flat lateral surface of the thyroid cartilage gives attachment to the sternothyroideus, thyrohyoideus, and inferior pharyngeal constrictor muscles. The medial (internal) surface is smooth and gives attachment to the thyroepiglottic, vestibular, and vocal ligaments, and to the thyroarytenoideus and vocalis muscles. The short, thick, inferior horn articulates with the cricoid cartilage (Fig. 7-45).

The signet ring-shaped **cricoid cartilage** presents posteriorly a broad, quadrilateral **lamina** with two convex facets on the upper border that articulate with the base of the arytenoid cartilages. The anterior arch gives attachment to the cricothyroid muscles and the cricovocal ligament. The superior border of the cricovocal ligament attaches to the vocal process of the arytenoid cartilage and presents a free edge, which attaches to the vocal ligament.

The two **arytenoid cartilages** are three-sided pyramidal structures with their bases articulating on the upper border of the cricoid lamina and their apices curving posteromedially. The posterolateral angle of the base presents the thick, projecting **muscular process** that gives attachment to laryngeal muscles, and a spinelike **ante-**

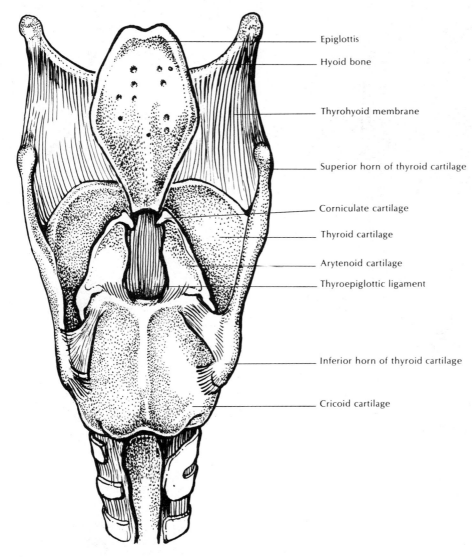

— Epiglottis

— Hyoid bone

— Thyrohyoid membrane

— Superior horn of thyroid cartilage

— Corniculate cartilage

— Thyroid cartilage

— Arytenoid cartilage

— Thyroepiglottic ligament

— Inferior horn of thyroid cartilage

— Cricoid cartilage

Figure 7-44. *Posterior aspect of larynx. (After Romanes GJ, ed: Cunningham's Textbook of Anatomy, 11th ed. New York, Oxford University Press, 1972)*

rior **(vocal) process** for attachment to the vocal ligament. The **corniculate cartilages** are small conical bodies at the apex of the arytenoid cartilages located in the posterior edge of the aryepiglottic folds. The small, rod-shaped **cuneiform cartilages** are embedded in the aryepiglottic folds above the level of the corniculate cartilages.

Interior of the Larynx (Fig. 7-46)

The interior of the larynx is smaller than might be expected. It is subdivided into three portions by the ventricular (true vocal cords) and vestibular (false vocal cords) folds, which extend anteroposteriorly and project inwardly from the sides of the cavity. The upper subdivision, the **vestibule,** ex-

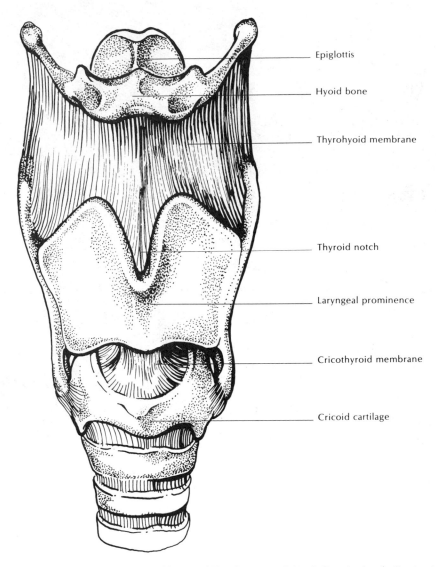

Epiglottis

Hyoid bone

Thyrohyoid membrane

Thyroid notch

Laryngeal prominence

Cricothyroid membrane

Cricoid cartilage

Figure 7-45. *Anterior view of larynx. (After Romanes GJ, ed: Cunningham's Textbook of Anatomy, 11th ed. New York, Oxford University Press, 1972)*

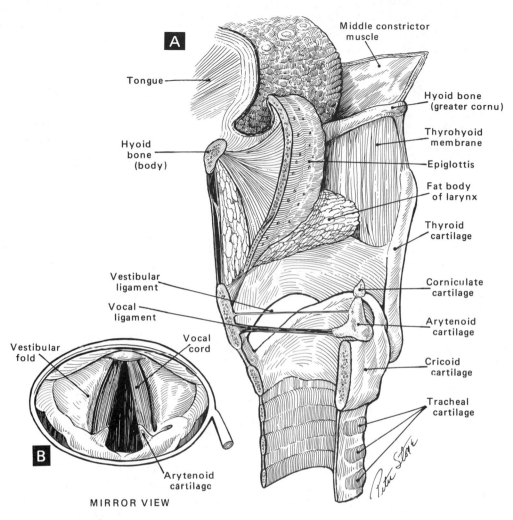

A

Tongue

Hyoid
bone
(body)

Middle constrictor
muscle

Hyoid bone
(greater cornu)

Thyrohyoid
membrane

Epiglottis

Fat body
of larynx

Thyroid
cartilage

Vestibular
ligament

Vocal
ligament

Corniculate
cartilage

Arytenoid
cartilage

Cricoid
cartilage

Tracheal
cartilage

Vestibular
fold

Vocal
cord

Arytenoid
cartilage

B

MIRROR VIEW

Figure 7-46. **A.** *Sagittal section of the larynx in relation to the tongue and trachea.* **B.** *The vocal folds, relaxed as in quiet breathing, as seen by means of a laryngoscope.*

tends from the aditus to the vestibular folds. It diminishes in width from superior to inferior, and its anterior wall is longer than the posterior. The anterior wall is formed, in part, by the epiglottic cartilage, and thyroid lamina, and the thyroepiglottic ligament. The lateral walls are formed by the **aryepiglottic folds,** which cover the aryepiglottic muscle, and the posterior wall is the interarytenoid membrane.

The middle portion, the **ventricle,** is the

smallest of the three regions and is bounded by the vestibular folds above and the vocal folds below. The soft, flaccid **vestibular folds** stretch anteroposteriorly across the side of the cavity, and the interval between them, the **rima vestibuli,** is wider than the interval between the **vocal folds.** The latter are sharp, prominent bands that enclose the **vocal ligament.** They are prismatic in cross section and appear pearly white in the fresh state. The elongated interval between the vocal folds, the **rima glottidis,** forms the narrowest part of the laryngeal cavity, and the shape of the opening varies with respiration and vocalization. The vocal folds and the interval between them form the glottis.

The remainder of the laryngeal cavity, the **infraglottic portion,** extends from the rima glottidis to the trachea. Superiorly it is narrow and compressed from side to side, then it gradually widens to become circular as it becomes the trachea.

Functions of the Larynx

The larynx has three functions. The inlet acts as a sphincter in swallowing to prevent food or drink from entering the trachea. The rima glottidis acts also as a sphincter to close the larynx during coughing, sneezing, or whenever an increased intrathoracic pressure is needed, for example, micturition, defecation, or parturition. The third function is vocalization. In the latter, short bursts of expired air cause vibration of the closely approximated vocal cords that produce the sound in vocalization. Direct visualization of the larynx, including the vocal cords, is possible through a laryngoscope.

Muscles

The suprahyoid and infrahyoid musculature of the neck aids in phonation (Table 7-33). The former muscles elevate the larynx in the production of high notes, while the latter depress the larynx in the formation of low notes. The intrinsic muscles

Table 7-33
Muscles of the Larynx

Muscle	Origin	Insertion	Action	Nerve
Cricothyroideus	Arch of cricoid	Inferior horn and lower border of thyroid cartilage	Chief tensor of vocal ligament	External laryngeal
Posterior cricoarytenoideus	Posterior surface of lamina of cricoid cartilage	Muscular process of arytenoid cartilage	Abductor of vocal folds	Inferior laryngeal
Transverse arytenoideus (only unpaired muscle of larynx)	Passes from posterior aspect of one arytenoid cartilage to other		Closes rima glottidis	Inferior laryngeal
Oblique arytenoideus	Muscular process of arytenoid cartilage	Some fibers into apex of arytenoid, most prolonged as aryepiglottic muscle	Closes rima glottidis	Inferior laryngeal
Lateral cricoarytenoideus	Upper border of cricoid arch	Muscular process of arytenoid cartilage	Adducts vocal folds	Inferior laryngeal
Thyroarytenoideus	Inner surface of lamina of thyroid	Anterolateral surface of arytenoid cartilage	Slackens vocal folds and closes rima glottidis	Inferior laryngeal
Thyroepiglottis	Anteromedial surface of lamina of thyroid cartilage	Lateral margin of epiglottic cartilage	Aids in closure of laryngeal inlet	Inferior laryngeal
Vocalis	Anteromedial surface of lamina of thyroid cartilage	Vocal process	Adjusts tension of vocal ligament	Inferior laryngeal

control the airway through the larynx. The cricothyroideus, lateral cricoarytenoideus, transverse arytenoideus, and thyroarytenoideus **adduct,** or close, the vocal folds, while the important posterior cricoarytenoideus, "the safety muscle of the larynx," **abducts** the vocal folds and opens the glottis. In phonation the thyroarytenoideus relaxes the vocal folds, while the vocalis muscle acts to tense portions of the folds (Fig. 7-47).

The **cricothyroideus** bridges the lateral portion of the interval between the cricoid and thyroid cartilages. The **posterior cricoarytenoideus** passes from the lamina of the cricoid cartilage to the muscular process of the arytenoid cartilage. The only unpaired muscle in the larynx, the **transverse arytenoideus,** extends from the posterior aspect of one arytenoid cartilage to the other. The **oblique arytenoidei** lie on the posterior aspect of the arytenoid cartilage, where they cross, like the limbs of an X, superficial to the

transverse arytenoideus. Some of the fibers of the oblique arytenoideus insert into the apex of the arytenoid cartilage, but most of the fibers are prolonged anteriorly as the **aryepiglottic muscle,** which inserts into the margin of the epiglottis. The **lateral cricoarytenoideus,** applied to the upper border and side of the cricoid arch, passes posterosuperiorly to insert into the muscular process of the arytenoid cartilage. The **thyroarytenoideus muscle** extends as a sheet between the thyroid and arytenoid cartilages, where the uppermost fibers continue superiorly as the **thyroepiglottic muscle,** and the deepest fibers stretch from the thyroid cartilage to the lateral side of the vocal process as the **vocalis muscle.** All the intrinsic muscles of the larynx are supplied by the inferior (recurrent) laryngeal branch of the vagus nerve, except the cricothyroideus, which is innervated by the external laryngeal branch of the superior laryngeal nerve of the vagus.

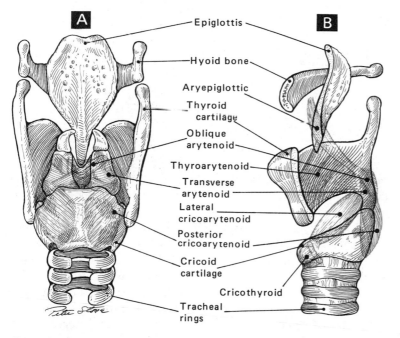

Figure 7-47. *Muscles of the larynx.* **A.** *Posterior view.* **B.** *Lateral view. The larynx is shown in sagittal section.*

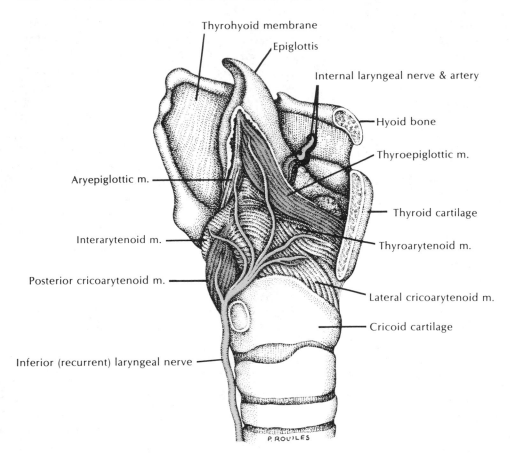

Figure 7-48. *Muscles and nerve supply of the larynx.*

Arteries and Nerves (Fig. 7-48)

The innervation of the larynx is from the **vagus nerve.** The **superior laryngeal branch,** originating high in the cervical region, passes inferiorly to divide opposite the hyoid bone into **internal** and **external laryngeal nerves.** The former pierces the thyrohyoid membrane to supply sensation to the mucous membrane above the vocal folds, while the **external laryngeal nerve** passes along the external surface of the larynx to supply the cricothyroideus muscle. The **right recurrent (inferior) laryngeal branch** of the vagus nerve loops around the right subclavian artery, and the **left recurrent (inferior) branch** loops around the arch of the aorta. Both nerves then ascend in the tracheoesophageal groove to supply all the intrinsic muscles of the larynx except the cricothyroideus, and to supply sensation to the interior of the larynx below the vocal folds (Table 7-34).

The **superior laryngeal branch of the superior thyroid artery** accompanies the internal laryngeal nerve (Table 7-35). It passes through the thyrohoid membrane to ramify and supply the internal surface of the larynx. The **inferior laryngeal branch of the inferior thyroid artery** accompanies the recurrent laryngeal nerve in the tracheoesophageal groove to supply the larynx from its inferior aspect.

Table 7-34
Innervation of Larynx

Nerve	Origin	Course	Distribution
Internal laryngeal	Superior laryngeal branch of vagus (X)	Pierces thyrohyoid membrane to reach interior of larynx	Sensory to larynx above level of vocal folds
External laryngeal	Superior laryngeal branch of vagus (X)	Follows course of superior thyroid artery	Cricothyroid
Inferior (recurrent) laryngeal	Vagus (X)	Right hooks around subclavian artery, left hooks around arch of aorta; both ascend in tracheoesophageal groove to reach larynx	Sensory to larynx below level of vocal fold; motor to all intrinsic muscles of larynx except cricothyroid

Table 7-35
Arterial Supply to the Larynx

Artery	Origin	Course	Distribution	Anastomoses
Superior laryngeal	Superior thyroid	Pierces thyrohoid membrane; runs in floor of piriform recess to mucosa of larynx	Larynx above vocal folds	Inferior laryngeal
Inferior laryngeal	Inferior thyroid	Enters larynx just posterior to the cricothyroid articulation	Larynx below vocal folds	Superior laryngeal

MAJOR ANATOMIC AND CLINICAL POINTS

Posterior Triangle

□ The phrenic nerve is "tacked down" by two arteries, the transverse cervical and suprascapular, as it crosses the superficial aspect of the anterior scalene muscle.

□ Structures traversing the scalene gap include the trunks of the brachial plexus and the subclavian artery but not the subclavian vein.

□ The spinal accessory nerve is very superficial. It lies just deep to the external investing layer of deep fascia and supplies the trapezius.

□ The cords of the brachial plexus lie deep to the clavicle and are named for their relationship to the axillary artery.

□ Four branches of the brachial plexus arise in the neck: dorsal scapular to levator scapulae and rhomboids; suprascapular to supraspinatus and infraspinatus; long thoracic to serratus anterior; and nerve to subclavius.

□ The spinal accessory nerve bisects the posterior triangle, and divides it into "carefree" and "careful" zones. A superficial wound could sever the nerve and result in a "drooped shoulder."

□ Traumatic transection of the spinal cord at C_3 level or above will necessitate some type of external means to provide respiration.

□ The scalene (thoracic outlet) syndrome, caused by a narrowing of the gap between scalenus anterior and medius, may also be the result of a cervical rib, pressure on brachial plexus, or occlusion of subclavian artery. It causes symptoms in the upper extremity.

□ Inflammation in the neck may extend along the cervicoaxillary sheath to reach the axilla.

Anterior Triangle

□ Six branches of the external carotid arise near the intermediate tendon of the digastric: three anterior (superior thyroid, lingual, facial), two posterior (posterior auricular, occipital), and one medial (ascending pharyngeal).

□ The internal carotid has no branches in the neck.

□ The cervical plexus supplies the skin of the neck and upper chest (to 2nd intercostal space). The ansa cervicalis, to strap muscles, is formed completely by cervical plexus, with no contribution by the hypoglossal nerve.

□ Postganglionic nerve cell bodies of all sympathetics to the head and face are located in the cervical sympathetic ganglia.

□ Anesthesia to the skin of the neck and upper chest can be attained by blocking cutaneous nerves as they emerge at the midpoint of the posterior border of the sternocleidomastoid muscle.

□ Interruption or trauma to the cervical sympathetic trunk or superior cervical ganglion causes Horner's syndrome (pseudoptosis of lid, enophthalmos, miosis, anhidrosis, and flushed warm skin).

□ In a thyroidectomy, high transection and retraction inferiorly of the infrahyoid muscles is performed to retain their nerve supply. The inferior (recurrent) laryngeal nerve is identified before any structures are clamped, transected, or ligated.

□ Occlusion of the subclavian artery proximal to the vertebral artery may result in "subclavian steal"; increased demand for blood in the upper extremity siphons blood away from the brain by way of the basilar/vertebral system.

Face

□ All divisions of the trigeminal provide cutaneous innervation to the face: ophthalmic (supraorbital, supratrochlear, infratrochlear),

maxillary (infraorbital, zygomaticofacial and zygomaticotemporal), mandibular (mental, auriculotemporal and buccal).

□ As the facial nerve emerges from the stylomastoid foramen it gives the posterior auricular and nerves to the stylohyoid and posterior belly of the digastric; cleaves the parotid and then terminates in five branches to the muscles of facial expression.

□ Facial muscles, inserting into the skin, should be studied by the orifices (eye, nose, mouth, ear) that they open, close, or contort.

□ Inflammation of the facial nerve in the confined space of the facial canal exerts pressure on the nerve and can cause paralysis of the facial muscles, a condition known as Bell's palsy.

□ The "danger area" of the face and scalp extends from the upper lip to the vortex of the skull. Superficial venous drainage communicates with the pterygoid plexus of veins, which in turn communicates with the cavernous sinus. Inflammation of a superficial lesion can thereby cause a meningitis.

□ A lesion to any of the divisions of the trigeminal nerve can be assessed by testing appropriate areas of skin for sensitivity: the forehead for the ophthalmic division, skin of cheek for the maxillary, and the skin over mandible for the mandibular.

□ Terminal branches of each division of the trigeminal nerve can be blocked at three foramina (supraorbital, infraorbital and mental) that lie along a vertical line about an inch lateral to the midline.

Infratemporal Fossa

□ The motor root of the trigeminal nerve passes independently through the foramen ovale to join the mandibular division in the infratemporal fossa. Therefore all motor distribution of the trigeminal nerve is by way of branches of the mandibular nerve.

□ The chorda tympani leaves the facial nerve in the facial canal, traverses middle ear cavity to reach the infratemporal fossa, where it joins and is distributed with the lingual nerve.

□ The maxillary artery supplies the infratemporal fossa. In addition to muscular branches, it gives off the middle meningeal (the major supply to the meninges), the inferior alveolar (which supplies the lower teeth), and two small branches to the external ear.

□ The angle and ramus of the mandible are "sandwiched" between the masseter on their external surface and the medial (internal) pterygoid on their internal surface.

□ Anesthesia of the lower teeth can be attained by blocking the inferior alveolar nerve, at the mandibular foramen, as it enters the mandibular canal.

□ Transection of the lingual nerve as it originates would affect only general sensation to the anterior two-thirds of the tongue. Damage to the nerve distal to the point where it is joined by the chorda tympani would, in addition, block special (taste) sensation to the anterior two-thirds of the tongue and secretion of the submandibular and sublingual · glands.

□ In dislocation of the mandible the condyle slips over the mandibular eminence. To reduce the dislocation, the mandible must be protracted and depressed, allowing the condyle to move back into the mandibular fossa.

□ The masseter and temporalis are "biting" muscles. The side-to-side "grinding" movement of the mandible is accomplished by the medial and lateral pterygoids.

Cranial Cavity

□ The middle cranial fossa is bounded anteriorly by the lesser wing of the sphenoid and posteriorly by the superior border of the petrous portion of the temporal bone.

□ Six structures traverse the cavernous sinus. A mnemonic to remember these and their sequence is "amotor":

    A—Abducens
    M—Maxillary
    O—Oculomotor
    T—Trochlear
    O—Ophthalmic
    R—Artery (internal carotid)

□ Cerebrospinal fluid is elaborated by the choroid plexuses in the roof of the lateral and third ventricles. It leaves the ventricular system to reach the subarachnoid space by way of the foramina of Luschka and Magendie in the fourth ventricle, and is resorbed into the vascular system at the arachnoid villi of dural sinuses (primarily at superior sagittal sinus).

□ The basilar (vertebrals) and internal carotid arteries contribute to the formation of the circle of Willis. The anterior communicating between anterior cerebrals, and the posterior communicating between the posterior cerebral and the middle cerebral complete the circle.

□ A head injury can rupture cranial vessels. In an epidural hemorrhage, blood is confined between the dura (periosteum) and bone; in a subdural hemorrhage, between dura and arachnoid; in a subarachnoid hemorrhage, between arachnoid and pia.

□ Anastomotic communications are present between the vertebral plexuses and the venous drainage of the cranial cavity. This provides a pathway for blood-borne metastases to pass from pelvic organs to the brain.

□ An aneurysm of the internal carotid artery in the cavernous sinus can involve eye movement and cause sensory disturbances in the dermatomes of the ophthalmic and maxillary divisions of the trigeminal nerve.

□ A tumor of the hypophysis can impinge on the optic chiasm and optic tract; differential diagnosis can be made from the visual fields that are affected.

Cranial Nerves

□ Bipolar neurons are associated with three cranial nerves: olfactory (I) with cell bodies in olfactory mucosa; optic (II) with cell bodies in the retina; vestibulocochlear (VIII) with cell bodies in the spiral and vestibular ganglia.

□ Three cranial nerves convey only sensory impulses (I, II, VIII), five carry only motor impulses (III, IV, VI, XI, XII), and four carry both (V, VII, IX, X).

□ Four cranial nerves (III, VII, IX, X) transmit preganglionic parasympathetic fibers; oculomotor (cell bodies in Edinger–Westphal nucleus) synapse in ciliary ganglion; facial (cell bodies in superior salivatory nucleus) synapse in pterygopalatine and submandibular ganglia; glossopharyngeal (cell bodies in inferior salivatory nucleus) synapse in otic ganglion; vagus (cell bodies in dorsal motor nucleus X) synapse in disseminated ganglia of the neck, thorax, and abdomen.

- The trigeminal nerve is the great sensory nerve of the head. All motor distribution of this nerve is by way of a branch of its mandibular division.
- Embryonic derivation of muscles can be deduced from their innervation: the trigeminal nerve supplies muscles derived from the first branchial arch; the facial nerve supplies muscles from second arch; the glossopharyngeal nerve, from third arch; and the vagus nerve from the fourth arch.
- Excepting the palatoglossus, all tongue muscles are supplied by the hypoglossal nerve.
- Excepting the stylopharyngeus, all pharyngeal muscles are supplied by the vagus nerve.
- Excepting the cricothyroideus, all intrinsic laryngeal muscles are supplied by the inferior (recurrent) laryngeal branch of the vagus.

Orbit

- Seven bones contribute to the skeleton of the orbit. The margin is formed equally by three bones, the frontal, maxillary, and zygomatic bones. The other four are the lacrimal, ethmoid, sphenoid, and palatine.
- Six cranial nerves supply structures in the orbit: optic—vision, ophthalmic (V_1)—sensory, facial—lacrimal gland, abducens—lateral rectus, trochlear—superior oblique, oculomotor—rest of eye muscles and levator palpebrae superioris.
- A mnemonic to remember innervation to eye muscles is $LR_6 (SO4)_3$:

    Lateral rectus by abducens (VI)

    Superior oblique by trochlear (IV)

    All other muscles by oculomotor (III)

- The optic "nerve" is a brain tract conveying axons of tertiary neurons in the visual pathway, and surrounded by meninges and cerebrospinal fluid.
- The external layer of the eyeball is comprised of the sclera and cornea; the middle layer is the vascular and nerve pathway (by way of the area cribrosa of sclera), with a ciliary body that provides attachment anteriorly for the iris and suspensory ligament of lens; the internal retinal layer extends to the pupil, only the posterior portion has visual receptors.
- The floor of the orbital cavity is formed by a thin layer of maxillary bone. In a "blow-out"

fracture the eyeball can herniate into the maxillary air sinus.
- Damage to cranial nerves III, IV, and VI can be assessed by the loss of specific eye movements.
- Edema of the optic disc ("choked disc") is indicative of increased intracranial pressure.
- Obstruction of the central artery of the retina causes instantaneous blindness.
- Anastomoses between terminal branches of the ophthalmic artery and the superficial temporal, maxillary, and facial arteries provide collateral circulation between the internal and external carotids.

Ear

- The ear is located in the temporal bone.
- The middle ear (tympanum) is an air-filled cavity with a chain of three bones (malleus, incus, and stapes) that transmit vibratory motion from the tympanic membrane (external ear) to the oval window of the internal ear.
- The tympanic cavity communicates anteriorly with the nasopharynx (pharyngotympanic tube), and posteriorly with mastoid air cells (epitympanic recess and aditus).
- The osseous labyrinth is a series of channels in the petrous portion of the temporal bone. The membranous labyrinth is a closed system of tubular and saclike structures located within the osseous labyrinth.
- Three branches of the facial nerve (major petrosal, the nerve to the stapedius and chorda tympani) arise in the facial canal that is related to the walls of the tympanic cavity.
- Otitis media (middle ear infection) blocks the pharyngotympanic tube and the tympanic cavity fills with fluid and/or pus. The inflammatory process can extend into the mastoid air cells (mastoiditis) or erode the tegmen tympani and reach the cranial cavity (meningitis).
- The tensor tympani and stapedius muscles have a protective function. They limit excursion of the ear ossicles when a loud noise causes an excessive vibratory motion of tympanic membrane.
- Ankylosis of the footplate of the stapes and sclerosis of the secondary tympanic membrane block movement at the oval and round windows, resulting in transmission deafness.

□ Inflammation of the facial nerve in the confined space of the facial canal causes pressure on the nerve and results in Bell's palsy, a paralysis of the muscles of facial expression.

Nasal Cavity/Pterygopalatine Fossa

□ The ethmoid bone contributes to the roof, lateral wall, and septum of the nasal cavity.

□ The paranasal sinuses drain onto the lateral wall of the nasal cavity; frontal and maxillary sinuses drain into the middle meatus; most of the ethmoidal air cells, into the superior meatus; the sphenoid, into the sphenoethmoid recess. The nasolacrimal duct drains into the inferior meatus.

□ A branch of the ophthalmic nerve (V_1), the continuation of the anterior ethmoidal nerve, supplies sensation to the anterosuperior aspect of the nasal cavity and the skin on the dorsum of the nose.

□ The maxillary nerve (V_2) is the nerve of the pterygopalatine fossa. From the fossa, distally it provides sensory branches (and with the pterygopalatine ganglion autonomic fibers) to the nasal cavity, palate, upper teeth, pharynx, lacrimal gland, and skin of the mid-region of face.

□ Trauma of the nose frequently dislocates (fractures) the nasal cartilages. If untreated, this can result in a deviated septum because the external nose is attached to the septal cartilage.

□ With trauma to the nose, the ease of bleeding is due to disruption of the extensive submucosal vascular plexus that functions to warm the inspired air.

□ Sinus headaches result from blocked drainage of the paranasal sinuses and the concommitant pressure of the entrapped secretions. Shifting the position of the head sometimes facilitates drainage and eases the pain.

□ Rupture of the maxillary artery in the pterygopalatine fossa may result in an intractable nose bleed. Because of the inaccessability of the vessel in this region, it may be necessary to ligate the external carotid artery.

Oral Cavity/Pharynx

□ The hypoglossal nerve supplies motor innervation to all muscles of the tongue, both intrinsic and extrinsic, except for the palatoglossus.

□ Five cranial nerves supply the tongue V, VII, IX, X, and XII.

□ The pharynx extends from the base of the skull to the level of the cricoid cartilage (C_6), subdivided respectively into the nasopharynx, oropharynx, and laryngopharynx.

□ The major muscles of the pharynx, the constrictors, telescope into one another; five additional muscles, the levator and tensor veli palatini, salpingopharyngeus, palatopharyngeus and stylopharyngeus contribute to the pharyngeal wall.

□ The motor supply to the pharyngeal plexus is by way of the vagus nerve; sensory supply, by way of the glossopharyngeal. The glossopharyngeal supplies only one muscle: the stylopharyngeus.

□ A stone or blockage of the parotid duct may be mistaken for a tumor on the inside of the cheek. It will be positioned opposite the second upper molar tooth.

□ In swallowing, a contraction of the salpingopharyngeus tenses the mucosa of the pharyngotympanic tube, opens the tube, and hence equalizes the pressure of the middle ear with the level of outside atmospheric pressure.

□ In deglutition, the tensor veli palatini initially tautens the soft palate, then the levator veli palatini elevates the soft palate to seal off the nasopharynx.

□ To fully examine the tongue and oropharynx, a tongue blade is used to depress the tongue in its horizontal portion; this enables the vertical portion, epiglottis, and posterior wall of the oropharynx to be seen.

Larynx

□ Three paired cartilages (arytenoid, cuneiform, corniculate) and three unpaired cartilages (epiglottic, thyroid, cricoid) form the skeleton of the larynx.

□ Most of the intrinsic muscles of the larynx attach to the arytenoid cartilages.

□ The inferior (recurrent) laryngeal nerve supplies all of the intrinsic muscles of the larynx, except the cricothyroideus, which is supplied by the external laryngeal branch of the superior laryngeal nerve.

- Sensation to the larynx above the level of the vocal folds is supplied by the internal laryngeal branch of the superior laryngeal nerve. Below the vocal folds it is supplied by the inferior laryngeal nerve.
- The posterior cricoarytenoideus ("safety muscle") is the only adductor of the vocal folds.
- In emergency tracheotomies, an incision that is too high might damage the vocal cords; an incision that is too low, into the vascular anastomosis of the isthmus of the thyroid, might result in asphyxiation of the patient from bleeding into the trachea.

- Denervation of the posterior cricoarytenoid muscle (transection of the inferior laryngeal nerve) would result in a flaccid vocal fold.
- In thyroidectomy, the inferior laryngeal nerve, in the tracheoesophageal groove, is carefully dissected and identified before any structures are clamped, ligated, or severed.
- Most of laryngeal muscles have a protective role. They act to tightly close the rima glottidis and prevents foreign material from entering the tracheobronchial tree.

QUESTIONS FOR REVIEW

1. Anteriorly the posterior triangle is bounded by which of the following muscles?

   A. Trapezius
   B. Anterior scalene
   C. Middle scalene
   D. Sternocleidomastoid
   E. Inferior belly of omohyoid

2. The subclavian artery crosses the superior surface of the first rib and passes

   A. Between the middle scalene and the posterior scalene
   B. Anterior to the anterior scalene
   C. Between the anterior scalene and the sternocleidomastoid
   D. Between the anterior and middle scalene
   E. Between the omohyoid and anterior scalene

3. A "drooped shoulder" would result from damage to which of the following nerves?

   A. Dorsal scapular
   B. Supraspinatus
   C. Subscapular
   D. Axillary
   E. Spinal accessory

4. Which is the correct order from superior to inferior of the palpable bones and cartilages in the neck?

   A. Hyoid, thyroid, cricoid
   B. Hyoid, cricoid, thyroid

   C. Thyroid, hyoid, cricoid
   D. Thyroid, cricoid, hyoid
   E. Thyroid and cricoid only

5. The posterior belly of the digastric forms a boundary of both the

   A. Submental and submandibular triangles
   B. Submandibular and carotid triangles
   C. Carotid and muscular triangles
   D. Muscular and submental triangles

6. Destruction of the superior cervical sympathetic ganglion would affect all of the following except

   A. Vasculature of the face
   B. Sweat glands in the skin over the anterior aspect of the neck
   C. The radial (dilator) muscle of the iris
   D. Contraction of the thoracic diaphragm

7. Usually, the first branch of the anterior aspect of the external carotid artery is the

   A. Facial
   B. Lingual
   C. Superior thyroid
   D. Inferior thyroid
   E. Maxillary

8. The strap muscles are innervated by the ansa cervicalis which arises from cord segments

   A. C_1-C_2
   B. C_2-C_3

C. C_2–C_4
D. C_1–C_4
E. C_1–C_3

9. The terminal branch of the facial artery is the

   A. Transverse facial
   B. Superior labial
   C. Lateral nasal
   D. Angular

10. Which of the following statements about the parotid gland is false?

    A. It extends from the zygomatic arch to the angle of the mandible.
    B. It has a deep portion that lies along the internal aspect of the mandible.
    C. Its duct perforates the buccinator muscle.
    D. It is supplied by branches of the otic ganglion.
    E. Its duct opens into the floor of the oral cavity

11. The zygomaticus major muscle has its main action in

    A. Crying
    B. Frowning
    C. Laughing
    D. Kissing
    E. Smoking

12. The occipitalis muscle is innervated by a branch of the

    A. Greater occipital nerve
    B. Lesser occipital nerve
    C. Third occipital nerve
    D. Facial nerve
    E. Trigeminal nerve

13. Which muscle does not receive its motor supply from V_3?

    A. Tensor tympani
    B. Masseter
    C. Buccinator
    D. Mylohyoid
    E. Temporalis

14. Preganglionic parasympathetic fibers travel in all of the following nerves except the

    A. Facial
    B. Chorda tympani
    C. Lingual
    D. Inferior alveolar

15. The middle meningeal artery enters the cranial cavity by way of the foramen

    A. Spinosum
    B. Ovale
    C. Rotundum
    D. Lacerum
    E. Stylomastoid

16. An injury in the infratemporal fossa that affected taste would have probably injured which branch of V_3?

    A. Buccal
    B. Inferior alveolar
    C. Auriculotemporal
    D. Lingual

17. Which opening is not associated with the middle cranial fossa?

    A. Supraorbital fissure
    B. Foramen rotundum
    C. Foramen ovale
    D. Foramen spinosum
    E. Internal auditory meatus

18. Cerebrospinal fluid circulates from the ventricles into the subarachnoid space through the

    A. Central canal of the spinal cord
    B. Interventricular foramen (of Monro)
    C. Cerebral aqueduct (of Sylvius)
    D. Choroid plexus
    E. Formina of Luschka and Magendie

19. The cribriform plate is a process of which of the following bones?

    A. Frontal
    B. Nasal
    C. Ethmoid
    D. Sphenoid
    E. Temporal

20. Primary or direct branches of the internal carotid artery include all the following except the

    A. Ophthalmic artery
    B. Anterior cerebral artery
    C. Middle cerebral artery
    D. Posterior cerebral artery

21. Definitive components of the wall of the eyeball include all of the following except the

    A. Lens
    B. Sclera

C. Cornea
D. Ciliary processes
E. Choroid

22. All of the following are true of the lacrimal canaliculi except that they

A. Are located near the medial canthus of the eye
B. Open at the punctum lacrimali
C. Transport lacrimal fluid from the lacrimal gland to the conjunctival surface of the eye
D. Extend to the lacrimal sac
E. Are present in both the upper and lower eyelid

23. The posterior chamber of the eye is located between the

A. Cornea and iris
B. Iris and lens
C. Cornea and lens
D. Lens and retina
E. Iris and retina

24. Which one of the following cranial nerves does not innervate any orbital structure?

A. Glossopharyngeal
B. Trigeminal
C. Facial
D. Trochlear
E. Abducens

25. Components of the osseous labyrinth include all of the following except the

A. Semicircular canals
B. Cochlea
C. Vestibule
D. Stapes
E. Scala tympani

26. The ganglion of the internal ear associated with hearing is the

A. Geniculate
B. Auditory
C. Spiral
D. Superior petrosal
E. Cochlear

27. All of the following are associated with the medial wall of the tympanic cavity except the

A. Round window
B. Promontory

C. Ridge formed by the canal of the facial nerve
D. Pyramid housing the stapedius muscle
E. Tympanic plexus

28. The incus

A. Articulates only with the malleus
B. Articulates with both the stapes and malleus
C. Articulates only with the stapes
D. Is attached to the tympanic membrane

29. Sensory innervation to the nasal cavity is by way of the

A. Trigeminal nerve
B. Glossopharyngeal nerve
C. Abducens nerve
D. Facial nerve
E. Oculomotor nerve

30. The middle nasal concha is part of which bone?

A. Palatine
B. Maxilla
C. Sphenoid
D. Ethmoid
E. None of the above, it is a separate bone.

31. All of the following are considered paranasal sinuses except the

A. Maxillary
B. Sphenoidal
C. Frontal
D. Ethmoidal
E. Mastoid

32. The nasolacrimal duct passes through the maxillary bone, and through this duct lacrimal fluid flows into the

A. Superior meatus
B. Middle meatus
C. Inferior meatus
D. Sphenoid sinus
E. Ethmoid sinus

33. The tongue has several cranial nerves associated with it. Which cranial nerve supplies taste fibers for the anterior two-thirds of the tongue?

A. Trigeminal
B. Facial
C. Glossopharyngeal
D. Vagus
E. Hypoglossal

34. Which of the following structures is located in the vestibule of the oral cavity?

    A. Tongue
    B. Opening of the parotid duct
    C. Opening of the submandibular duct
    D. Sublingual fold
    E. Uvula

35. The motor innervation for the intrinsic muscles of the tongue is supplied by which cranial nerve?

    A. Trigeminal
    B. Facial
    C. Glossopharyngeal
    D. Vagus
    E. Hypoglossal

36. The pharynx is located posteriorly or dorsal to the

    A. Nasal cavity
    B. Mouth
    C. Larynx
    D. All of the above
    E. Answers A and C

37. The rima glottidis is bounded by the

    A. Pharyngeal pouch
    B. Epiglottis
    C. Vestibular (false vocal) folds
    D. Aryepiglottic folds
    E. Ventricular (true vocal) folds

38. A patient developed hoarseness due to paralysis of the vocal cords; which of the cranial nerves was involved?

    A. Sixth
    B. Eighth
    C. Ninth
    D. Tenth
    E. Twelfth

39. All of the following are paired structures except the

    A. Corniculate cartilage
    B. Arytenoid cartilage
    C. Cuneiform cartilage
    D. Cricoid cartilage
    E. Vestibular fold

40. All of the following muscles act in closing the rima glottidis except the

    A. Oblique arytenoid
    B. Transverse arytenoid
    C. Aryepiglottic
    D. Thyroarytenoid
    E. Posterior cricoarytenoid

41. The external investing layer of deep fascia splits to enclose the

    A. Sternocleidomastoid muscle
    B. Sublingual gland
    C. Retropharyngeal space
    D. Platysma
    E. Scalene muscles

42. The body, greater wing, and lesser wing are portions of the

    A. Thyroid cartilage
    B. Palatine bone
    C. Arytenoid cartilage
    D. Sphenoid
    E. Cricoid cartilage

43. All of the following structures pass through the cavernous sinus except the

    A. Cranial nerve III
    B. Cranial nerve IV
    C. Cranial nerve VI
    D. V_1 of trigeminal
    E. V_3 of trigeminal

44. Which of the following foramina is mismatched with its location or bone

    A. Optic foramen, sphenoid bone
    B. Foramen cecum, anterior cranial fossa
    C. Internal auditory meatus, middle cranial fossa
    D. External auditory meatus, temporal bone
    E. Superior orbital fissure, sphenoid bone

45. Which muscle in the orbit has a smooth muscle portion and dual innervation?

    A. Levator palpebrae superioris
    B. Inferior rectus
    C. Inferior oblique
    D. Superior rectus
    E. Lateral rectus

46. The basilar artery terminates by dividing into which arteries?

    A. Posterior cerebral
    B. Middle cerebral
    C. Anterior inferior cerebellar
    D. Posterior communicating
    E. Pontine

47. Which one of the following pairs is incorrectly matched?

    A. Temporal bone, mandibular fossa
    B. Sphenoid bone, hiatus of the facial canal
    C. Ethmoid bone, crista galli
    D. Maxillary bone, posterior superior alveolar foramina
    E. Occipital bone, superior nuchal line

48. The muscles of mastication receive their primary blood supply from arteries arising from the _____ artery.

    A. Facial
    B. Pterygopalatine portion of the maxillary
    C. Second part of the maxillary
    D. Superficial temporal
    E. Angular portion of the facial

49. A fracture through the roof of the maxillary sinus might result in sensory loss to the

    A. Skin of the forehead
    B. Upper molar teeth
    C. Upper incisors and canine teeth
    D. Upper eyelid
    E. Lacrimal gland

50. The palatoglossal fold overlies a muscle whose motor innervation comes from the

    A. Ansa cervicalis
    B. Hypoglossal nerve
    C. Branches of the vagus nerve
    D. Branches of the carotid plexus
    E. Glossopharyngeal nerve

Answer Section

Chapter Two

1. D	11. D	21. E	31. A	41. C
2. C	12. D	22. A	32. A	42. A
3. E	13. B	23. D	33. B	43. E
4. C	14. A	24. C	34. D	44. C
5. A	15. C	25. B	35. D	45. C
6. C	16. C	26. B	36. B	46. A
7. C	17. A	27. A	37. E	47. B
8. D	18. B	28. A	38. C	48. A
9. C	19. B	29. C	39. D	49. E
10. D	20. C	30. C	40. B	50. E

Chapter Four

1. A	11. E	21. D	31. B	41. D
2. B	12. A	22. A	32. D	42. A
3. C	13. E	23. A	33. D	43. C
4. E	14. B	24. C	34. A	44. E
5. E	15. E	25. E	35. B	45. D
6. E	16. C	26. B	36. A	46. E
7. B	17. A	27. E	37. D	47. C
8. B	18. E	28. C	38. B	48. D
9. D	19. E	29. C	39. B	49. D
10. B	20. B	30. D	40. C	50. B

Chapter Three

1. E	11. C	21. A	31. C	41. B
2. C	12. B	22. C	32. C	42. A
3. E	13. B	23. D	33. C	43. B
4. B	14. C	24. D	34. E	44. C
5. E	15. B	25. A	35. D	45. D
6. C	16. B	26. A	36. C	46. D
7. C	17. C	27. B	37. C	47. D
8. D	18. A	28. D	38. B	48. C
9. B	19. D	29. E	39. D	49. C
10. D	20. D	30. B	40. C	50. E

Chapter Five

1. E	11. E	21. D	31. E	41. B
2. B	12. B	22. A	32. D	42. E
3. C	13. B	23. B	33. A	43. E
4. B	14. C	24. A	34. B	44. B
5. A	15. D	25. D	35. D	45. C
6. A	16. B	26. B	36. C	46. A
7. E	17. A	27. E	37. D	47. A
8. C	18. A	28. B	38. C	48. E
9. E	19. D	29. B	39. D	49. B
10. A	20. D	30. A	40. D	50. D

Chapter Six

1. C	11. B	21. E	31. E	41. C
2. D	12. C	22. A	32. A	42. E
3. D	13. A	23. D	33. C	43. D
4. A	14. A	24. D	34. D	44. A
5. D	15. C	25. B	35. C	45. B
6. A	16. D	26. B	36. D	46. D
7. B	17. A	27. A	37. A	47. A
8. A	18. B	28. B	38. C	48. D
9. D	19. A	29. A	39. C	49. E
10. B	20. D	30. C	40. C	50. C

Chapter Seven

1. D	11. C	21. A	31. E	41. A
2. D	12. D	22. C	32. C	42. D
3. E	13. C	23. B	33. B	43. E
4. A	14. D	24. A	34. B	44. C
5. B	15. A	25. D	35. E	45. A
6. D	16. D	26. C	36. D	46. A
7. C	17. E	27. D	37. E	47. B
8. E	18. E	28. B	38. D	48. C
9. D	19. C	29. A	39. D	49. C
10. E	20. D	30. D	40. E	50. C

Index

The letter *f* after a page number indicates a figure; *t* following a page number indicates tabular material.